It is a wid
urban for
retaining
so heavil
it on cons

Conserva
the built
activities
to explor
change, i

Examinin
conserva
systems
practice.
conserva
question

Conserva
studying
theoretic
change a

Peter J. Larkham is a Senior Lecturer in the School of Planning, University of Central England.

CONSERVATION AND THE CITY

PETER J. LARKHAM

LONDON AND NEW YORK

First published 1996
by Routledge
11 New Fetter Lane, London EC4P 4EE

Transferred to Digital Printing 2003

Simultaneously published in the USA and Canada
by Routledge
29 West 35th Street, New York, NY 10001

Typeset in Garamond by
Solidus (Bristol) Limited

British Library Cataloguing in Publication Data
A catalogue record for this book is available from the British Library

Library of Congress Cataloguing in Publication Data
A catalogue record for this book has been requested

ISBN 0-415-07947-0 (hbk)
ISBN 0-415-07948-9 (pbk)

Printed and bound by Antony Rowe Ltd, Eastbourne

CONTENTS

FIGURES

TABLES

PREFACE

On looking around in any city, town or even village, it is evident, sometimes painfully so, that our built-up areas are changing fast. Despite the supposed increase in support for conservation, many fine buildings and their settings have been lost or maimed. Conversely, it can be argued that an overwhelming respect for the old and fear of the new has stifled architectural creativity in recent years. Are we now creating buildings that will, in half a century or so, be worthy of statutory protection?

This book addresses the problems of old and new in urban areas. First, by providing a background to conservation, then with studies of how much change is occurring, of what type, and who is involved in the processes of change. Lastly there is commentary on changes, on the forces that produce them, and on the lack of any theory that would support the management of change. The arguments are based largely on micro-scale studies in UK conservation areas, but examples from further afield are used as appropriate.

This is primarily an academic book, of interest to scholars and professionals involved in urban change. But it is also intended for a wider public readership, for there is an undeniable public impact of continuing change in historical areas. Academics, professionals and the general public should all be aware of what is going on, and why. It must be more generally understood that not all urban landscapes can be preserved without change, but the practical implications of this, the decisions underlying the selection of what is kept, and the relating of new development to historical urban landscapes, must also be clearly understood. There are differences between these concepts as applied to towns and buildings, and to other fields such as works of art and machinery; these parallels are also briefly examined.

This approach to change in the historical urban landscape requires the making of value judgements – something that academic study is often reluctant to do. But all of us, every day, make value judgements in our own responses to buildings old and new, and the relationships between them. This book seeks a basis for making *informed* judgements. Although some parts contained here may be familiar from journal papers appearing over the last decade, I hope that readers will find something new in the bringing together for the first time of these disparate ideas and examples. Yet, because this is an exploration of a topic, not merely description or hypothesis-testing, the book's structure is deliberately more circular than linear. In the end, we must return to the concerns from which we began.

Peter J. Larkham
Birmingham
August 1995

ACKNOWLEDGEMENTS

This book is the result of a number of years of research. Projects have been funded by the Economic and Social Research Council, the Leverhulme Trust, the British Academy and the Royal Town Planning Institute. Their support is much appreciated. Much of the work was carried out in the School of Geography, University of Birmingham, and the School of Planning, University of Central England; to my friends and colleagues there I express my thanks for moral and practical assistance. The ideas, research and writing have benefited greatly from discussions with members of the Urban Morphology Research Group, in particular Professor J.W.R. Whitehand, Dr Terry Slater, Dr Nick Pompa, Dr Mike Freeman, Dr Phil Hubbard, Dr Karl Kropf, Heather Barrett and Professor Gordon Cherry. Collaboration with Dr Andrew Jones, at Birmingham and latterly of Chesterton Consulting, has been particularly fruitful with respect to conservation area planning practice. Newer colleagues at the School of Planning, University of Central England, have also been supportive, as have my students, on whom sections of this book have been tested. Helena Plant, a former UCE student, produced some ideas which I have quoted at length in Chapter 1. Parts of this book first appeared as a paper in *Progress in Planning* (1992), and I am grateful to Professor Derek Diamond for his encouragement to prepare that paper, and for publishing it.

Assistance in drawing and photographing the figures came from Geoff Dowling and Kevin Burkhill at the University of Birmingham, and Sally Jones and Steve Roddie at the University of Central England.

Permission to reproduce copyright material is gratefully acknowledged to the following individuals and organisations: English Heritage, the International Commission on Monuments and Sites (ICOMOS), *Private Eye*, SAVE Britain's Heritage, Brindleyplace PLC, AG 1824 (a Fortis company) of Brussels, Coventry City Council, North Norfolk District Council, Vale of White Horse District Council, Joe Holyoak, Antique Buildings Ltd, Medieval Oak Homes, Mike Freeman, David Austin and Karl Kropf.

Key quotations introducing each chapter have been largely drawn from those listed in Knevitt (1985, 1986), where their full references may be found; some, however, being only reported, do not have full bibliographic references.

Lastly, the encouragement and patience of Tristan Palmer and Matthew Smith at Routledge have helped the book's concept become reality.

Part 1

THE CONSERVATION BACKGROUND

1

INTRODUCTION: CONFLICT AND CONSERVATION

> Architectural structures which because of their specific technical features frequently survive for centuries, belong to the basic monuments of the past and the cultural development of the entire nation. At the same time they are evidence of its economic and social development. The works of architecture and town planning thus embody the history of the people that created them.
>
> (Adolf Ciborowski, 1956)

INTRODUCTION

This introductory chapter sets out the dilemma at the heart of this book. There is a widespread agreement that urban areas must change, or they will stagnate. Yet, at the same time, there are growing pressures for preservation from both the general public – or, at least, an educated and vociferous minority – and from increasing elements of the design and planning professions. For the most part, Western cities are creations of the capitalist order, where investment fuels an economy and becomes part of the cycle of wealth creation. This capitalist imperative runs counter to the set of values based on aesthetic, environmental, non-quantitative criteria. So there is a clash of values: land and property exploitation for capital gain versus consideration of art, aesthetic and historical appreciation. There is also, in aesthetic terms, an essential tension between the old and the new, the familiar and the unfamiliar (Figure 1.1). This heightens our reactions to, and colours our enjoyment of, urban landscapes as with other fields of aesthetic endeavour. There is tension, too, between the uses of heritage to legitimate socio-political positions and conflicting ideologies of dissenting groups; hence the targeting of élite architecture to represent the national identity, and the targeting of those same structures for demonstrations or destruction. It is the nature and scale of this conflict which are major problems, remaining unresolved in theory – for there is no generally accepted theory of how to manage urban landscapes for conservation – and practice – as many post-war buildings in historic areas demonstrate. But the production and management of the changing urban landscape are processes in which conflicting ideologies are deeply embedded, and the common depiction of tension as a simple

Figure 1.1 Visual contrast: tension in the urban landscape: Swansea Castle and overshadowing office block (author's photograph)

dichotomy of retain or redevelop is a gross over-simplification.

The production and maintenance of the physical fabric of the urban environment absorb a large amount of the wealth of the Western world, and have done so for centuries, giving rise to the historic compositeness of many urban landscapes. Furthermore, a strong case has been made for the social, cultural and psychological significance of the townscape. Many studies show the need, in these terms, for the preservation of historic townscapes, in outward appearance at least. Economic reasons for preservation also exist and while practically strong, they are not intellectually as compelling as these other reasons. Substantial planning problems arise as townscapes age and as the social and economic conditions under which they were created change.

Buildings become structurally, functionally and economically obsolete.

Adaptation of the townscape becomes necessary, but this is hard to achieve without some wastage of the investment of previous societies. During periods of great capitalist investment and development in property (for example the 1950s and 1960s in many Western countries) such wastage received relatively little consideration, pressures were high on historic sites, and the philosophy of 'comprehensive redevelopment' was dominant. At other times, for a variety of socio-cultural reasons, the conservationist ethic will be dominant. For example, interest in conservation in the Western world is at its highest for a century, and this can readily be demonstrated for the UK. In addition, increasingly the reality of urban development in Western societies is about the re-use of urban sites inherited from previous generations. Some two-thirds of residential developments in south-east England, for example, are within existing urban areas. This marked change of attitudes has helped to highlight the problem of accommodating the requirements of previous societies within the townscape legacy from previous generations.

There is thus a multiple problem. First, what is to be preserved? Closely allied to this is the question of who identifies the preservation-worthy buildings and areas, and whether this identification meets with the approval of the population living, working and recreating in these areas. Indeed, as a second facet of the problem, to what extent do those influencing development and those affected by it have consistent views about the area in which development is proposed? Thirdly, how is conservation/preservation to be carried out: are the buildings and areas identified in any way removed from the natural life-cycle of construction, use, obsolescence, decay and demolition? Fourthly, what is the nature and scale of changes proposed and carried out to the physical urban fabric? These are the questions addressed throughout the main body of this book.

How can this problem best be examined? Recent research in the UK has demonstrated the importance of a grouping of those involved directly and indirectly with change, the 'agents of change'. Those directly involved are principally the owner, developer and architect; those indirectly involved include those affected by and commenting upon changes (neighbours, the public, local amenity societies); those making the decisions are the professional staff and elected representatives of the planning authority. In the UK there are two tiers of decision-making: the Local Planning Authority (LPA), usually the District Council, and central government, represented in England by the Department of the Environment (DoE). The manner in which these agents of change inter-relate is crucial to the outcome on the ground. This book draws upon a growing body of research dealing with these agents of change, their characteristics, interactions, and the outcomes in a variety of historic townscapes.

The book deals deliberately with UK examples for the bulk of its argument, for it is here that much of the detailed morphological research has been carried out, and here where there is a long-established legal-administrative system to promote conservation. Many key concepts of conservation have, since the mid-1980s, been tested in court. Some cases have even reached the highest court in the land, the House of Lords; so the legal interpretations of conservation-related actions and duties are well argued. Illustrations and examples of other systems, derived from other cultures and experiences, are used wherever appropriate. The concept of conservation used here is also deliberately restricted. As will be shown in this chapter, psychological and aesthetic arguments form some of the key justifications for conservation; both professional and public attention is often directed principally at the external appearance of structures and areas. This fits well with the 'townscape' tradition in UK planning thought, although this emphasis on the visual aesthetic is rather different to some other national approaches (cf. Bandini 1992). In this way, this investigation uses micro-level examinations of activities in conserved districts to address the multiple questions posed above. Large-scale changes affecting the historic or architectural character of preserved areas or buildings are significant, but so, too, are the myriad individually small-scale changes that may have significant cumulative impact. Changes to conserved structures, to structures not specifically conserved but which are within designated areas, and the design of wholly new structures are all significant in this exploration of the visual/aesthetic impacts of conservation.

ARGUMENTS FOR CONSERVATION

The growth in public support for conservation, and its increasing enshrinement in planning legislation throughout the developed world, may owe much to a growing awareness of basic reasons for conservation, and an examination of recent literature suggests that psychological, didactic, financial, fashion and historical reasons are of greatest significance.

Psychology and conservation

In his all-encompassing examination of civilisation, Lord Clark (1969) suggested that civilisation could be defined as 'a sense of permanence', and that a civilised man 'must feel that he belongs somewhere in space and time, that he consciously looks forward and looks back'. Various studies in environmental psychology, particularly those of P.F. Smith (1974a, b, 1975, 1977), strongly suggest that this 'looking back' is psychologically very important: even, perhaps, a necessity. There is a human need for visual stimuli to provide 'orientation' – the observer's awareness of his or her own location in a given environment – and 'variety' (Lozano 1974). These needs are met, in part at least, by historical areas that have survived relatively unchanged, providing symbols of stability: 'the visual confirmation of the past provides a fixed reference point of inestimable value' (P.F. Smith 1974a: 903). Historical areas act as cushions against Toffler's concept of 'future shock' (Toffler 1970). In a wider context, Lowenthal (1985) used a detailed presentation of minutiae and ephemera to support his relating of ordinary everyday experiences to the need for some tangible 'heritage'. Not content with expressing this need, he also showed in some detail how we use the heritage and shape it to our own ends.

There is some empirical evidence to support this theoretical work. Taylor and Konrad (1980), for example, constructed a scaling system to measure the dispositions of people towards the past, and found strong sentiments in favour of conservation and heritage. Morris (1978) analysed reactions towards slides of different types and ages of buildings, finding that contemporary buildings are dismissed as discordant intrusions, while mediaeval buildings in particular, and classical styles to a lesser extent, possess considerable historical and architectural interest. Holzner (1970) gave an interesting description of conservation in practice in post-war Germany. He noted that after wartime destruction of over one million buildings, every effort was made to repair as many of the old buildings as possible and to reconstruct them in the inherited manner. Holzner concluded that this (and, by implication, any) preservation was not rational, but that a popular social need overcame rational approaches to post-war reconstruction. This appears to be a practical and large-scale demonstration of the psychological argument for conservation, which may also be applicable in other countries, such as the case of the post-war rebuilding of Polish towns.

Many of the psychological arguments are well summarised by Hubbard (1993) in a review of behavioural research and conservation attitudes. He suggests that the conservation of the familiar is of value in, for example, stabilising individual and group identities, particularly in times of stress. However, experimental psychology has shown that although familiar townscapes and buildings are highly rated, so too are some modern structures. Further, sample groups have also shown profound ambivalence towards the historicity of townscapes and buildings: in some cases, it is outward appearance only which was perceived to be important. Hubbard also warned that perceptions of place and conservation vary between groups of different socio-cultural backgrounds: the valued landscapes of one group may thus represent different things to another, and 'the increasing fragmentation of British society exposes the inadequacy of conservation of traditional national and local cultures' (Hubbard 1993: 370). The same is true of any multicultural society. This introduces the concept of 'heritage' into the conservation argument.

Didactic: teaching, learning and conservation

The notion that there is a moral duty to preserve and conserve our historic heritage, to remember and pass on the accomplishments of our ancestors (Tuan 1977: 197), is a common argument. The main reason underlying this moral duty appears to be pedagogic: the physical artifacts of history teach observers about landscapes, people, events and values of the past, giving substance to the 'cultural memory' (Lewis 1975). However, Faulkner (1978) showed that the historical heritage can be divided into the heritage of objects and that of ideas, with the implication that one may be conserved without the other. 'This inevitably leads us to the question of architectural design and whether we can ... separate the design from the building, and claim the design as a concept has a separate existence from the stones of which the building is built' (Faulkner 1978: 456). If so, then the handing down of the heritage could easily be achieved by the comprehensive recording of buildings or areas. Preservation of the actual fabric becomes no longer an absolute necessity in original or rebuilt form.

Finance and conservation

It has long been recognised that at least some aspects of conservation can be profitable. Hence an increasing number of country house owners began opening houses and grounds as tourist attractions, beginning with the Marquess of Bath at Longleat in 1949. Indeed, Lord Montagu of Beaulieu wrote a book with the significant sub-title 'How to live in a stately home and make money' (Montagu of Beaulieu 1967).

Yet because of the involvement of central and local governments in identifying structures and areas for conservation, and in imposing constraints upon owners, the economics of urban conservation become complex. Nevertheless, key issues of gains and losses can be represented quite simply (Figure 1.2).

> While being property which is owned and managed like other real estate, government claims via 'heritage tenure' of its cultural quality, which it proposes to conserve for the public, contemporary and prospective, means that the owners of the property can only manage it subject to constraints.... But the costs of conservation are by comparison localised. They fall on the owners and occupiers of the property and also on the community of the administrative area in which the property happens to be.
>
> (Lichfield 1988: 201)

The key issues of the cost implications of conservation, and the nature, extent and distribution of gains and losses, can to some extent be assessed through standard techniques of cost-benefit analysis. Research for the Royal Institute of Chartered Surveyors took a qualitative view of the financial impact of listing buildings, and from interviews gave a series of reasons why such recognition of historic value may increase or decrease building value (*Chartered Surveyor Weekly* 1991: 19) (Figure 1.3). However, political ideologies do distort considerations based purely on financial data, raising issues such as the following conflicts:

> owners of heritage property should not be denied the rights available to others, so that any loss incurred should be *compensated* by the rest of the community;

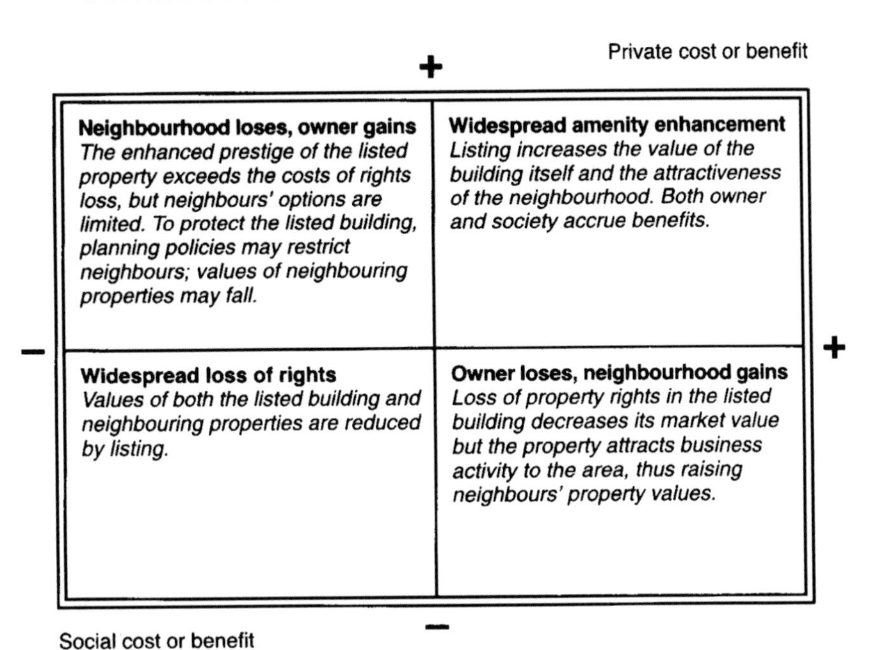

Figure 1.2 Gains and losses in a financial analysis of conservation (adapted from Scanlon *et al.* 1994)

> but where this loss is accompanied by increase in value to other owners, a betterment charge should be levied, to offset the burden on the rest of the community;
>
> since the benefit is for posterity, the community finding the compensation should be that of the future, so that the cost to the contemporary generation should be met out of long-term loans to be paid off by future generations.
>
> (Lichfield 1988: 209–10)

In practice, these dilemmas are little addressed. More practical considerations are more influential. For example, much impetus was given to conservationists by the 1973 oil crisis and subsequent concerns for energy efficiency. Local groups campaigned against numerous demolitions on the grounds that buildings were capable of economic re-use. However, at that time, mainstream economic arguments centring on the functional efficiency of maintenance, layout and adaptability dominated. Older buildings were expensive to restore and maintain, and could not provide accommodation to contemporary standards (Dobby 1978).

Attitudes have now shifted further towards the 'green' arguments. The concept of 'embodied energy' is now more widely accepted: calculating the energy cost of building an existing structure and modifying it, compared to the energy costs of its demolition and replacement, often suggests that the

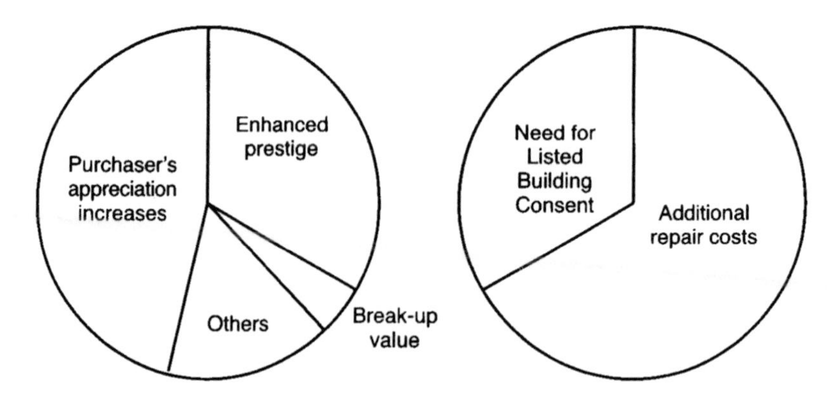

Figure 1.3 Perceptions of reasons why listing may increase or decrease building value (redrawn from *Chartered Surveyor Weekly* 1991)

former is a more energy-efficient solution.

> Historic buildings, like businesses, often demand an entrepreneurial approach. Those who see them principally as a burden are of two types. One is the speculator, the developer who wishes to be rid of them for financial gain. The other is the administrator, who grudges the time and money absorbed in looking after older buildings or who lacks the expertise, advice or imagination to see how they could be adapted in a practical and economic way. Both types are often obsessed by the idea that a new building replacing the old one will somehow be magically maintenance-free.
>
> (Binney 1981)

Certainly, low costs and freedom from maintenance are key conservation issues. They have recently been reasons widely promoted to sell uPVC glazing. This is incongruous in historic structures and areas, has led to campaigns against the legal freedom to make such alterations without prior LPA consent (English Historic Towns Forum 1992), and an information campaign from English Heritage on appropriate glazing and its costs (English Heritage 1991). Maintenance has always been a financial problem to conservation, with Knevitt (1985: 119) quoting a proverb that 'it is easier to build two chimneys than to maintain one'.

Conserved properties, despite their costs, can be financially efficient. Estate agents in the UK have commonly argued that location within a conservation area, or designation as a listed building, can improve the asking price of a building by perhaps 10 per cent, although research is required to substantiate this claim. Research on the investment performance of listed office buildings across the UK does suggest that, over the period 1980–92, the listed structures held their own against non-listed, and more modern, structures. Indeed, listed offices attracted as much occupier demand and achieved rates of rental growth as good as, or better than, those of other office categories. This research, based on a sample of 2,650 buildings, challenges the general perceptions that (a) listed offices produce lower rents or (b) if rents are high, this is solely because listed buildings tend to occupy advantageous sites (Royal Institution of Chartered Surveyors/English Heritage 1993; see also Scanlon *et al.* 1994). Thus there is considerable economic support for conservation, both in theory via the embedded energy and societal investment in existing structures, and in practice through property prices and investment performance.

Fashion and conservation

When conservation activity is high on the political agenda, as it is at present (see Chapter 3), fashion becomes a reason supporting some conservation activity. Whereas two or three decades ago there was considerable corporate prestige involved in building and occupying evidently new and up-to-date office blocks built in the Modernist styles from new materials, as Marriott (1967) suggested in his analysis of the post-war property boom, this is not now the case. Even a cursory examination of advertisements in the professional property press in the late 1980s and 1990s shows the popularity of restored buildings for prestige headquarters and office accommodation for a wide range of companies. The Tarmac construction group, for example, moved from its corporate tower block in Wolverhampton town centre to a restored moated manor house in a neighbouring village. Even the 'high-tech' industries are not immune from this desire to gain corporate prestige through occupying historic buildings.

Closely related to this phenomenon is the rise in popularity of pastiche and neo-historicist architectural styles for both offices and residential use. Quinlan Terry's celebrated neo-Georgian office development at Richmond Riverside (1984–7) hides a large office development behind a façade giving the appearance of 15 separate Georgian/Regency houses. Georgian and Tudor were favourites of speculative residential developers in the 1970s and 1980s. Even some of the 200 or so large country houses still being built in the late twentieth century are deliberately historicist: one argument being that these styles, far more than the Modern style, complement the collections of furnishings, paintings and other *objets d'art* of the families concerned (Robinson 1984).

Another fashion is the colonisation and piecemeal improvement of various residential areas by members of the professional classes, having either the leisure time or the disposable income to effect such improvement. It is argued that this displaces indigenous populations, usually of lower social class and with lower incomes, particularly when local property values rise beyond their means.

> it is the formation of new housing classes – 'gentrifiers' – in the inner city ... which has turned conservation and preservation from an aesthete's dream into a normal feature of local government.... There is nothing 'historic' in any archaeological sense about many of the districts, or enclaves, now designated as conservation areas. What marks them out is the class of people who have moved into them, and the idiom of their expression. Conservation ... turns the humblest dwelling into a period residence, façades into 'historic' fabrics. In a triangular exchange of prestige it invests property with a pedigree and newcomers, if only by proxy, with roots, while allowing council officers to notch up 'environmental' gains for what might otherwise be a deteriorating locality.
>
> (Samuel 1990: 52)

However, although much is written about this gentrification/displacement issue, and the localised improvements are not in doubt, little by way of a causal link has been found between this fashionable social/economic movement and conservation *per se* (cf. Gale 1991).

History, historicism and conservation

It could be argued that what is to be retained should be restored to its original state and use where possible, that is, 'preserved'. This view is encapsulated in Ruskin's comment that

> it is ... no question of expediency or feeling whether we shall preserve the buildings of past times or not. *We have no right whatever to touch them*. They are not ours. They belong partly to those who built them, and partly to all the generations of mankind who are to follow us. The dead still have their right in them.
>
> (Ruskin 1849; his emphasis)

Yet this is an extreme and impracticable view. Alternatively, there is the view that changes should be permitted to allow a new function to inject life into the building and its surroundings, even at the cost of some alterations. This is the dichotomy of 'conservation' versus 'preservation' and, although this has long been an issue in UK planning, other countries' systems have not viewed it as such a problem. It was well expressed in the influential study of Bath, one of four government-sponsored city studies published in 1968: 'conservation is bound to involve preservation, but it is more than preservation: it is bringing an area back to life' (Buchanan 1968). Michael Middleton, then Director of the Civic Trust, clearly expressed the dictionary-derived view which has become embedded in UK official guidance:

> to *preserve* may be taken to mean as far as possible to retain intact the total integrity of the structure, with its original finishes, decorations, its setting and so on. To *conserve* has come to have a wider meaning which

can include the sensible use, re-use, adaptation, extension and enhancement of scarce assets.

(Middleton, his emphasis, quoted in Young 1977: 68)

These attempts at definitions are found wanting whenever a severe conservation-related crisis occurs, as when fire gutted the National Trust's Uppark House. Here, the key decision whether to restore to near-original, to consolidate the ruin or to demolish was based largely on the insurance company's interpretation of the policy: it would pay for complete reconstruction but nothing else. The house is now reconstructed, but how much is a replica, how much original? Is this conservation or preservation?

In the case of the conservation and/or alteration of buildings, it is suggested that since the 'architectural morality' of both 'function equals form equals beauty' and 'the exterior should reflect the interior structure' (two of the principal tenets of Modern architecture) is unrelated to the manner in which buildings are perceived (Smith 1975: 79), there should be little opposition to either facsimile reproduction if the original is unsalvageable, or 'façadism', the rebuilding in forms suitable for modern functions behind a retained and restored façade. Pastiche new architecture would also be acceptable. Both of these suggestions seek to retain the visual appeal of a townscape – the 'aesthetic texture' (Lewis 1975), the 'aesthetic justification for preservation' (Tuan 1977), while permitting the necessary modern uses. The behavioural research which has demonstrated a degree of public indifference to historical authenticity would also support this approach. Yet even this preservation of style destroys the 'patina of age', the 'aura of history', that is also of considerable importance; something clearly demonstrated in the 'scrape vs. anti-scrape' debate over church restoration in the nineteenth century (Pevsner 1976).

The argument of conserving for posterity is not wholly convincing, except possibly in the case of undisturbed archaeological deposits. This is because modern urban functions require some change in the townscape, and the selection of what is to be preserved and/or conserved, together with differences over how such conservation should be effected, does interfere with the pure notion of preservation for posterity. More so than other arguments for conservation, the historicist arguments highlight the fact that an overall philosophical framework for conservation is lacking. Faulkner, in particular, has attempted to formulate such a framework (Faulkner 1978), but his theory has not been translated into practical terms.

HERITAGE AND CONSERVATION ISSUES

'Heritage' has become an increasingly significant term in the conservation/preservation debate since the early 1980s. In the UK, de-industrialisation has led to a growing reliance on service-sector industries, of which heritage tourism is very significant (Herbert 1995). Polemics such as Hewison (1987) argue against the rise of 'heritage', the selectivity and sanitisation of the

Figure 1.4 Processes of selection and targeting in the heritage industry (adapted from Ashworth 1994: Figure 2.2)

images of past places presented, and the dependency on a museum-based industry and culture.

'Heritage' is neither history nor place; it is a process of selection and presentation of aspects of both, for popular consumption (Figure 1.4). 'Heritage is history processed through mythology, ideology, nationalism, local pride, romantic ideas or just plain marketing, into a commodity' (Schouten 1995: 21). It is a form of commodification. Therefore, 'heritage' means something quite different to 'conserved relict historical resources', and Ashworth (1992) argues that selection is central to the process. Although Lowenthal (1985) argues that we all interpret the past in some form, its management and interpretation as heritage bring problems:

> Some of what now purports to be heritage has been antiqued, not only in appearance but, rather more sinisterly, in being presented as if it was significant historically as well as being ennobled by time.
>
> (Fowler 1989: 60)

The concepts of conservation and heritage are thus quite separate, although in recent years there has been a tendency to confuse, if not conflate, them.

Heritage has become increasingly used in the marketing of products and, especially relevant in the current context, places. Regeneration efforts in several neglected urban quarters have used heritage as a key component of the place-marketing and revitalisation strategies, and this can clearly be seen in Bradford's Little Germany, and Birmingham's Jewellery Quarter (cf. Falk 1993). Yet criticism surrounds the selectivity of the heritage, which excludes aspects of local heritage deemed 'unsaleable' to tourists or investors (Kearns and Philo 1993; Gold and Ward 1994), and its sanitisation, for example in pseudo-historicist street furniture and enhancement schemes (Booth 1993). But such is the competitive nature of contemporary place-marketing that such questioning is seen as unwelcome, even traitorous (see, for example, the furore surrounding Loftman and Nevin's 1992 study of Birmingham's flagship regeneration projects).

The selectivity inherent in the heritage concept is especially problematic in a multi-cultural and historically diverse context, and Tunbridge (1994: 123) suggests that 'the political implications of culturally selective identification and interpretation, conservation and marketing of the inherited built environment are profound and potentially deadly'. He particularly addresses the question of 'dissonant heritage': the heritage of oppression and war is difficult to present without causing offence to some socio-political groups. Yet such heritages may be present in subtle, but nevertheless readily perceived, ways: in Poland, the vast Palace of Culture in Warsaw is a reminder of Soviet cultural domination, while the architectural style chosen for the post-war rebuilding of the Old Town square in Łódź also reflects a Russian rather than Germanic culture.

After 'the heritage' is defined, those parts of it worth saving must be identified. Heritage conservation must be selective, or there would be little change in urban structures, which would then ossify: some change is a necessity. Further, if everything were to be conserved, 'then we would be faced with retaining in its original condition everything left to us by previous generations ... warts and all' (Cantell 1975: 6). Yet the very acts of selection and designation for conservation imply assumptions about meaning and significance held by those making the selection, although such assumptions are rarely made explicit.

Thus, although the heritage concept could be argued to have popularised conservation, its ideologies have clearly restricted choice and freedom. The power of heritage selection and promotion is vested in powerful élites, whether multi-national corporations or municipal authorities. Little effective consultation with local groups takes place. The commodification of heritage is a further part of the impact of the capitalist system on the built environment.

CONSERVATION AND PLANNING THEORY

Conservation as an activity does not appear to relate well to established theories of planning as an activity (Plant 1993, whose work forms the basis of this section), and it is instructive to examine this clash. Three main theoretical approaches to planning, and the place of planning in society, can be identified; each is discussed in turn.

Social theory and consensus

Society has been seen as a stable system within which there is common acceptance of values. Planning may thus operate as a technical exercise in the interest of the general 'public good'. For example, writing of procedural planning theory, Faludi commented that 'a planning agency is simply an organisational unit specialised for the formulation of programmes designed to solve problems in the most efficient way' (Faludi 1973: 84). Scientific principles would result, in the words of the Nuffield Foundation's study of planning, in 'the efficient use and distribution of available resources' (Nuffield Foundation 1986: 7).

In this tradition of thought there was considerable 'consensus about what society wanted, what direction it was going in, and how it was to be achieved' (O'Rourke 1987: 13). Part of this consensus was the protection of the built environment, and this was achieved through legal and administrative action. Areas and buildings were identified and designated by 'experts' from central or local government, and national law and local policy determined the types of development which would be acceptable in the conservation context.

However, even very early in the development of the UK conservation area concept, some felt that this restrictive expert-driven approach had been taken too far in practice (Rock 1974; Eversley 1974; Price 1981). The level of public and informed consensus thus appears to have been eroded with time. Commercial interests have been increasingly aware of the economic requirement to balance conservation with development and economic viability and this has become reflected in government policy advice (DoE 1994). It has become harder to sustain the 'public good', consensus-based argument for conservation.

Social theory and conflict

This perspective accepts that incompatible interests do exist within society. These interests compete for scarce resources, and thus conflict is endemic. The concept of planning being in the public interest is thus eroded (Bruton and Nicholson 1987: 159). This approach is linked to the critique of capitalism, where power (a result of the control of production) enables one class to exploit another.

The designation of conservation areas, for example, selectively advantages some members of society, most particularly those with interests in property within the designated area (since, as has been shown, financial benefits may accrue). Some individuals therefore have a vested interest in encouraging the designation of areas. Planners are seen to support and enable these capitalist pressures by serving the interests of the powerful élite (Blowers 1986: 17), in this case principally the middle-class amenity societies, by taking the line of least resistance when faced with such dilemmas. It is thus insufficient to argue that conservation is for the benefits of 'society' – the consensus view – or that it is in the 'public interest', as these groups are not socially or physically identifiable.

Social theory and pluralism

The pluralist views of society have been given little weight in conservation planning. For some time, government advice was that LPAs should be encouraged to make designations, only advertising their decisions later (DoE 1987a: paragraph 58). It is unlikely that the statutory publication of an area designation notice in the *London Gazette* and a local newspaper would inform all those parties with an interest in the designation.

Once the LPA has resolved to make a designation, the involvement of different interest groups may rise. Development proposals are advertised, and resulting comments must be considered in making a decision. Minor and local élites, such as local amenity groups, must be consulted. Major élites, such as English Heritage and the statutory amenity consultees may also be contacted, depending upon the LPA's view of the scale and nature of the work proposed.

Thus, in establishing a conservation area, the LPA is a key player in influencing the subsequent role and extent of the pluralist discussion. Such decisions may also be influenced by more academic concerns. For example, purists may argue that the compounded effects of piecemeal repairs and alterations to a building could be sufficient to destroy the heritage of the object (Faulkner 1978: 461–70), and thus the structure should not be listed or be included within a designated area. There is thus a key issue of whose heritage is to be preserved. Which groups in the pluralist society are included in these debates, and which excluded? The traditional view is very narrow, treating conservation as a simple, monolithic issue (Tunbridge 1984: 171). However, by inference, this dismisses society's minority groups, whose perceptions of what is worthy of conservation – and how to conserve – may differ.

The conservation–planning relationship?

Perhaps, in the early stages of the development of conservation legislation, the consensus view was dominant. In the UK, a former Chief Executive of English Heritage felt able to say that conservation was 'a conventional wisdom so unarguable that its virtue is simply not questioned' (Peter Rumble, English Heritage; quoted by Bradshaw 1989: 7). Yet, evidently, this is now untrue. It is not unarguable; its virtues are fundamentally questioned by particular groups in society who, admittedly, may be arguing from the standpoint of vested interests. Yet this is part of the pluralist perspective. 'At root, conservation like any other planning activity, is political. It cannot succeed in a socially acceptable way without political support' (Worskett 1975: 9). The implications of political support for the consensus and conflict interpretations of society are again profound. And, as have been explored here, some of the justifications for conservation centre upon 'subjective and emotional' issues (Eversley 1974: 14) and upon economic arguments. So there are, increasingly, conflicts with the redistributive aims of planning.

THE ESSENTIAL TENSION

Tensions thus abound in any discussion of heritage, conservation and urban form (Figure 1.5). The two principal trends, originality or alteration, are irreconcilable except that, in practice, most schemes take parts from each. The key question is *how much* originality, *how much* change? In the UK, there is a statutory duty imposed on LPAs to 'enhance' conservation areas, but what does this term mean? What is the 'context': the physical area made up of buildings and spaces, the socio-cultural aspects of land use and

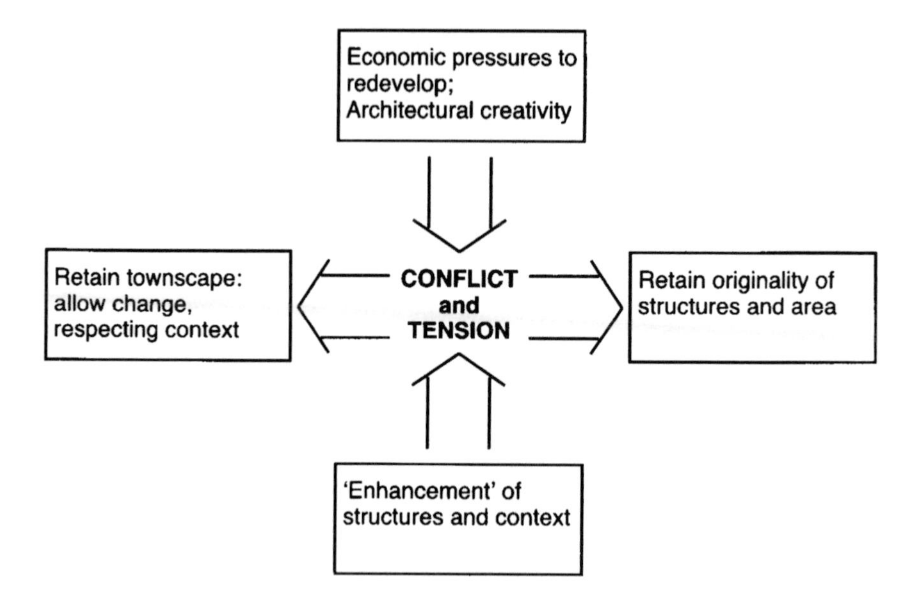

Figure 1.5 Some tensions in conserving the built environment

psychological value, or something else? And there is also the tension of design creativity versus retention of existing structures.

That there is a tension inherent in the creation of architecture (buildings) was recognised by the American architect Robert Venturi in his classic commentary on architectural form, *Complexity and Contradiction in Architecture* (Venturi 1966). There are also tensions in urban design (spaces and places). Venturi refers to the role of ambiguity in promoting richness of meaning over clarity of meaning: the tension of ambiguity leading to greater impact (Venturi 1966: 20–2, after Empson 1955). Contradiction, tension or ambiguity may lie in changing professional approaches and philosophies over time, with recent examples being the move from comprehensive clearance and redevelopment to rehabilitation and conservation; and the move in styles from Modern to post-Modern in architecture. As the time-span of any one such dominant paradigm overlaps the decline of its predecessor and the rise of its successor, there are clear conflicts of ideology in the design and production of the built environment. Since, also, urban areas are rarely created and recreated anew but are palimpsests of the achievements of successive generations, accumulations of relict, residual and modern features, with earlier features undergoing metamorphosis or partial or total replacement by later developments (Cherry 1981; Conzen 1958: 78; Martin 1968), there are clear conflicts and unusual juxtapositions in physical built forms. Old lies next to new; new adapts old; new uses old in new ways, or new ignores old.

That these tensions are, in some manner, essential is an idea derived, again, from aspects of behavioural psychology. Studies of individual reactions to both places and buildings have demonstrated that a key factor is complexity, not style or age *per se*. Thus the public tends to prefer mediaeval buildings to Georgian classical, and Georgian classical is preferred to Modern (cf. Morris 1981). Such preferences are, perhaps, uninformed and superficial. They do not arise from the type of fierce debate resting on moral and religious grounds which accompanied the 'battle of the styles' between Classical and Gothic in Victorian England (Crook, 1989; see also Chapter 2); and they ignore evident complexities in Modern architecture which are, clearly, much more subtle than in earlier styles and may be most evident only to other design professionals. Modern architecture has also suffered, in public perception at least, from its association with comprehensive clearance, failures of technology from Ronan Point onwards, and its debasement from the high ideals of its progenitors to the depths of the cheap post-war 'International style' (Esher 1981; Jencks 1985). Post-Modern styles and forms tend to be rather different in that they carry complexity, and often contradiction, within themselves; Jencks defining post-Modern architecture as being 'double-coded', carrying two meanings (Jencks 1991).

An interesting example of a post-Modern urban landscape currently emerging is that of Brindleyplace, Birmingham. A substantial redevelopment area adjoining the new International Convention Centre and adjacent to the city core, this area has been master-planned by the architect/designer Terry Farrell as a mixed-use commercial, residential and leisure district. Construction is phased over a period and, although the master-plan defined the envelope of each new building, a separate architect and style is envisaged for each. Current building designs range from a pseudo-vernacular entertainment complex to an Italianate office building by Dimitri Porphyrios and Modernist National Sea Life Centre by Sir Norman Foster (Figure 1.6). This range of new buildings has a complex design rationale. The area has to make some link with the bland Modernist towers on the other side of Broad Street, the high-tech Convention Centre, and with several listed buildings on and adjoining the site. Two have been restored and adapted to new uses; one, a church, as yet remains stubbornly vacant and derelict. Complexity, some contradiction, and tensions are evident by design throughout this 17-acre area.

It is clearly relevant that many of the ideas of tension and complexity have been adopted by urban designers, a new body emerging over the last decade (but, as yet, without professional status). A key text, now widely adopted in UK planning schools, was *Responsive Environments*, produced by the teaching team at Oxford Polytechnic's Joint Centre for Urban Design (Bentley *et al.* 1992: first edition 1985). This is an influential and well-established teaching centre, which celebrated its 21st anniversary in 1993 (Hayward and McGlynn 1993). Bentley *et al.* suggested seven key elements of good urban design. These included variety, the mixing of complementary land-uses; legibility, to improve users' understanding of places; robustness, which includes the idea of adaptability for possible future uses; richness in

Figure 1.6 Designer's impression of Brindleyplace development proposals, Birmingham (reproduced by permission of Brindleyplace plc)

surface detail, kinetic experience, smell, hearing and touch; and visual appropriateness, which includes relationship to urban context. This is urban design in a very broad sense: establishing appropriate principles but not being prescriptive about form or style.

Shortly after the publication of *Responsive Environments*, further tensions in architecture and urban design were revealed by HRH The Prince of Wales's television programme and associated book (HRH The Prince of Wales 1989). Here he explicitly laid out ten principles to guide a design code which could win public support, derived from his observations of post-war design and his assertion that 'we don't have to build towns and cities we don't want'. Considerable emphasis is given to the significance of 'place': a concept once familiar to geographers and to the British 'Townscape' school of planning (Bandini 1992), and discussed below. The Prince also gave much weight to architectural style and elevational treatment of buildings: 'a reflection perhaps of the traditional preoccupations of British townscape critics and aesthetic controllers' (Punter 1990a: 11). Immediately following the programme, Francis Tibbalds, then President of the Royal Town Planning Institute, reacted with another set of ten principles, deliberately phrased in Biblical language, but firmly based upon the design ideas of his own practice at that time (K. Murray, Director, Tibbalds Monro, pers. comm.). Tibbalds encapsulated many of the detailed concerns of the Prince in the single phrase 'visual delight', and developed other concepts of design, some familiar from *Responsive Environments* and other, more social, conceptions of design. Punter, in his critique of these and other 'Commandments', noted that

> this only emphasises that whereas the Prince of Wales remains locked into the essentially English concepts of architecture and urban design as a visual art ... Tibbalds has clearly embraced primarily American ideas of urban design as the manipulation of the public realm.
>
> (Punter 1990a: 10–11)

Table 1.1 compares the 'Commandments' of both Tibbalds and the Prince. It is salutary to conclude this review of tensions in architecture and urban design with Punter:

> all the control principles in the world will have little impact unless there are changes in the practice of development ... the biggest battles are still about the sheer scale of development, its bulk and grain, its lack of contribution to infrastructure costs (or other externality costs), its preoccupation with corporate space and the privatisation of urban space through atria and malls, [and] its desire for single use structures that epitomise lazy estate management.
>
> (Punter 1990a: 13)

Table 1.1 A comparison of the principles and commandments of HRH The Prince of Wales and Francis Tibbalds

HRH The Prince of Wales	*Francis Tibbalds*
1. The place: 'don't rape the landscape'.	1. Thou shalt consider places before buildings.
2. Hierarchy: 'if a building can't express itself, how can we understand it?'	2. Thou shalt have the humility to learn from the past and respect thy context.
3. Scale: 'less might be more: too much is not enough'.	3. Thou shalt encourage the mixing of uses in towns and cities.
4. Harmony: 'sing with the choir and not against it'.	4. Thou shalt design on a human scale.
5. Enclosure: 'give us somewhere safe to play and let the wind play somewhere else'.	5. Thou shalt encourage freedom to walk about.
6. Materials: 'let where it is be what it's made of'.	6. Thou shalt cater for all sections of the community and consult with them.
7. Decoration: 'a bare outline won't do; give us the details'.	7. Thou shalt build legible environments.
8. Art: 'Michelangelo accepted very few commissions for a free-standing abstract sculpture in the forecourt'.	8. Thou shalt build to last and adapt.
9. Signs and lights: 'don't make rude signs in public places'.	9. Thou shalt avoid change on too great a scale at the same time.
10. Community: 'let the people who will have to live with what you build help guide your hand'.	10. Thou shalt, with all the means available, promote intricacy, joy and visual delight in the built environment.

Sources: verbatim from HRH The Prince of Wales (1989: 78–97) and Tibbalds (1988: 1). The order is as in the original and is not strictly comparable, nor, necessarily, hierarchical

THE *GENIUS LOCI*

Many writers on urban form have discussed the issue of 'character': some implicitly, others explicitly, using terms such as 'spirit of place' or '*genius loci*' (Cullen 1961; Sharp 1969; Worskett 1969; Conzen 1966, 1975). These treatments owe much to the tradition of the Picturesque in English art, architecture and landscape which, Bandini (1992) suggests, led to the Townscape polemic and visual analysis in UK planning from the 1940s. For Cullen, who transformed the polemic of the *Architectural Review* into an analytical and design tool, Townscape was 'the art of relationship'; it was important

> to take all the elements that go to create the environment: buildings, trees, nature, water, traffic, advertisements and so on, and to weave them together in such a way that drama is released. For a city is a dramatic event in the environment.
>
> (Cullen 1961: 9)

However, although most reviews suggest that Cullen's work is central to a new emerging professional and academic discipline, urban design, his work and the visual tradition represented have been criticised:

> A painter by talent and romantic by instinct, Cullen's investigations of the desirable qualities of good urban environments differed considerably from the academic analyses of Lynch [1960]. Ironically, the personal vision and graphic fluency which Cullen brought to the explanation of his ideas was to some extent a handicap, arousing suspicion in the minds of those for whom a more 'objective' explanation of the urban designer's purpose was necessary.... Nevertheless, Cullen's method introduced a rather systematic framework for those sometimes elusive qualities.
>
> (Gosling and Maitland 1984: 48–9)

These qualities affect the emotional experience of, and reaction to, places. This concept can be applied to 'the past' and to conservation. Lowenthal (1979) has suggested that 'the past' exists as both individual and collective construct, with shared values and experiences being important within cultural groups. Group identity is thus closely linked with the form and history of place, creating a sense of place or *genius loci* (but see the contrasting uses of this term by Norberg-Schulz (1980) and Hunt and Willis (1988));

> in the course of time the landscape, whether that of a large region like a country or of a small locality like a market town, acquires its specific *genius loci*, its culture- and history-conditioned character which commonly reflects not only the work and aspirations of the society at present in occupancy but also that of its precursors in the area.
>
> (Conzen 1966)

Jakle (1987), however, emphasises the individual, subjective nature of place in his discussion of *genius loci*. In particular he emphasises the importance of the visual for, although we perceive places with other senses, there is an innate conflict between verbal and visual thinking. To Jakle, the ideal person to experience and express the *genius loci* is not the resident but the tourist, for tourism 'involves the deliberate searching out of place experience' (Jakle 1987: 8). This conflicts with the views of many others, who see the experience and perception of *genius loci* as a facet of long-term familiarity with place. In contrast, Walter (1988) implicitly uses the concept of *genius loci* in a study of the 'expressive intelligibility' of places: a quality which can only be perceived holistically through the senses, memory, intellect and imagination. In studying 'place' in classical thought, he contrasts Plato's subtle and complex views with the Aristotelian view of place as simply an empty container: this view, he suggests, informs much current planning practice.

In the increasingly heated debate over planning in UK conservation areas, the issue of the definition of character and its practical repercussions is of intense significance (Suddards and Morton 1991; Vallis 1994). UK conservation guidance referred to 'the familiar and cherished local scene' (DoE 1987a), a clear, although imprecise, reference to concepts of local place-identity much deeper than the more usual official phrase of 'character or appearance' (see Chapter 5). However phrased, all contributors to this debate deal with the importance and uniqueness of 'place'. The features which suggest that any place is of special interest and worthy of conservation should be examined; yet it is suggested that perhaps only 10 per cent of English conservation areas have character appraisals (Morton 1991). Attempts to define quality and character, to identify constituent parts which may be measured or valued, are academically unsatisfactory, however successful they may be in court or at appeal (and these are the justifications put forward by Vallis 1994).

A more fundamental approach to examining the *genius loci* is the suggestion that the key variables are unity and diversity (Smith 1981). All urban landscapes contain elements of both variables. Some are more highly uniform, such as Georgian classical layouts and terraces, where diversity may be confined to minor details and embellishments. Other areas are highly diverse, for example in architectural styles and materials, but may yet retain some uniformity in, for example, plot widths and storey heights. Tugnutt and Robertson (1987), discussing a similar concept, provide useful illustrations of the functionally similar small Midland market towns of Ledbury and Tewkesbury. One is visibly unified by style, materials, plot

Figure 1.7 Unity in the conserved townscape: market square, Bamberg, Germany (author's photograph)

Figure 1.8 Diversity in the conserved townscape: conflicting design cues for infill sites, Łódź, Poland (author's photograph)

widths, window and building heights; the other has much greater diversity. These concepts of similarity and diversity, and the confusions over which should be preserved, are visible in many urban landscapes in most countries. Sometimes the decisions may seem obvious (Figure 1.7), but in other cases the variety may give conflicting messages for decision-making in preservation and new design (Figure 1.8). To a considerable extent, these visual cues of similarities or differences assist in definition of the *genius loci* unique to each place. Tugnutt and Robertson begin to develop the explicit notion of how this may be used in conservation policy. If the *genius loci*, the characteristics which make a place conservation-worthy, are to be conserved (or, in English usage, 'preserved or enhanced'), do we aim for uniformity or diversity in enhancement schemes, new buildings and alterations to existing structures? Many planning policies explicitly press for unity, to the point of pastiche and replication of existing vernacular details (which are often spelled out in detail in local design guidance: see Figure 1.9). Yet this is precisely the origin of the 'conservation-area-architecture' despised by the design professions, with some justification (Rock 1974). This is also the philosophy underlying many more academic approaches to the urban landscape, such as M.R.G. Conzen's morphogenetic approach to townscape management (Larkham 1990a; Whitfield 1995), although here the basis for this philosophical position is not argued explicitly (Ashworth 1993).

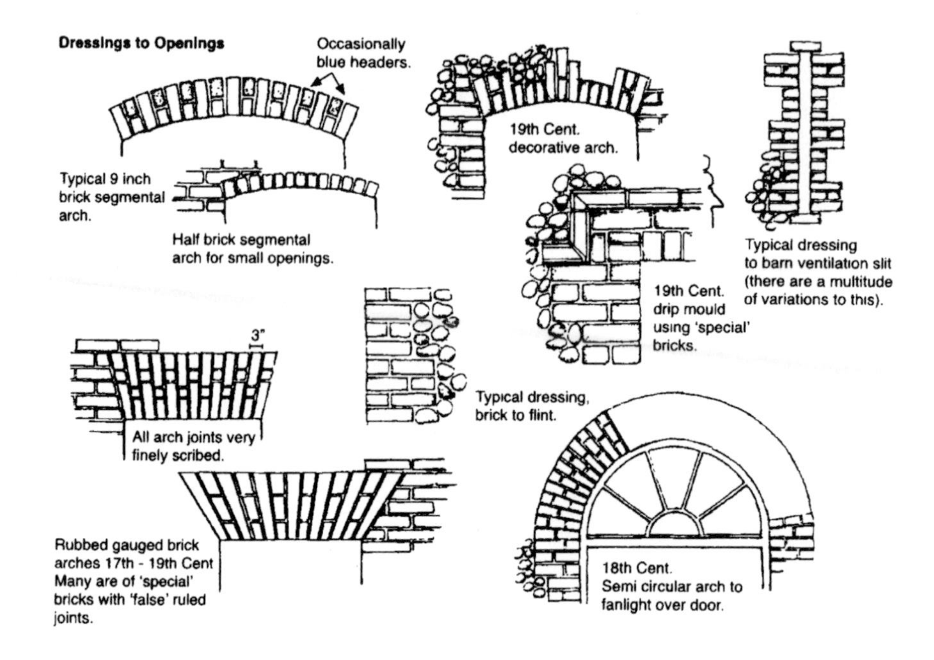

Figure 1.9 Example from LPA design guidance advocating use of vernacular idiom (reproduced with permission from North Norfolk District Council 1989)
Note: North Norfolk District Council (1989) produced an unusually thorough design guide with a particular emphasis on designing in context as 'sadly, however, some designers still seem to care little or nothing for the beauty of this comparatively unspoilt corner of England' (p. 5). The guidance on vernacular detailing, of which this figure is a sample, occupies 10 pages, or 16 per cent of the whole publication, and provides similar detailed pictorial guidance on gables, eaves, chimneys, porches, dormers, doors, windows, roof structures and fireplaces.

This volume throughout takes the Conzenian position that, on the whole, new buildings introduced into areas protected for their architectural or historical significance should respect the character of the existing built form. It is the nature and extent of this respect, or 'contextual compatibility' (Groat 1988), which is at issue. The detailed studies of town centres and residential areas explore case studies of change and their compatibility (or lack of it) in morphological and/or architectural features, and discuss the processes of decision-making which result in these built forms. It is accepted that, at times, developments which may appear grossly incompatible may, nevertheless, be constructed for a variety of reasons: politics and practicalities may, for example, outweigh aesthetics. The Pompidou Centre in Paris is one such example, although Richard Rogers, one of its architects, has argued for its contextual compatibility! As perceptions and values change over time, such incongruous buildings may become valued and even protected, as was the case with the non-contextual Willis Faber Dumas office building in Ipswich (by Foster Associates, completed in 1975), designated a Grade I listed building by English Heritage in response to threatened interior alterations.

URBAN MORPHOLOGY

The study of urban form, otherwise known as 'urban morphology', has a long tradition within historical geography. Indeed, 'it belongs as much to historical geography as to urban geography; a fact that reflects the longevity of the urban landscape that is the urban morphologist's object of study' (Whitehand 1987b: 250). In particular, its roots are in the German-speaking morphogenetic research tradition of central Europe, dating back to the work of Schlüter. He postulated a morphology of the cultural landscape (*Kulturlandschaft*) as the counterpart in human geography of geomorphology within physical geography (Schlüter 1899). In urban terms this made the *Stadtlandschaft* (urban landscape) a major research topic. In the German-speaking countries, the topic of urban form remains close to the mainstream of historical and urban geography, and there is evidently less distinction between the study of present-day towns and their historical aspects than is the case elsewhere (Whitehand 1987b: 250). Recently, the study of urban form has developed in several directions, among which the historical element remains particularly strong. A notable addition is the work on 'contextual' architecture and the planning, or management, of urban landscapes, much of which, as in this volume, gives considerable weight to the significance of existing urban forms.

With the exception of the German tradition, the 'schools' of morphological research in Britain and North America were still small-scale in the numbers of practitioners and publications in the early 1960s. In the later 1960s and early 1970s, therefore, research on urban form was vulnerable to the then current fashion in geographical analysis, the 'quantitative revolution'. Analyses were largely morphographic, describing features rather than analysing their origins and development; various quantitative methods were developed. Such studies were largely ahistorical, even when they considered the survival and distribution of historical buildings (for example Davies 1968; Johnston 1969). Contemporary with this trend was the development in the USA, and its widespread diffusion, of concepts based on economics and land-use patterns. Developed from the Chicago School of human ecology, the perspective of these urban socio-economic geographers was 'morphological only in its concern with land-use patterns: town plan and building form were generally treated only as land-use containers, if considered at all' (Whitehand 1987b: 255). A positive facet of this development was a greater reluctance to draw wide inferences from the awkward types of data often encountered in the study of urban form. The number of active researchers and publications having any historical perspective on urban form diminished, and recruitment to their ranks was minimal. By 1970 urban morphology was characterised as a 'barren outpost of urban geography' (Carter 1970). Fourteen years later the position had apparently not changed greatly, as the subject 'has been largely unaffected by those changing or shifting paradigms which supposedly have dominated geographical methodology' (Carter 1984: 145).

By 1978, however, M.P. Conzen was able to detect a resurgence of research activity in urban morphology after this period of quiescence (Conzen 1978: 135). Publications in this topic throughout the 1980s, while hardly numerous (only 12 per cent of papers on the internal structure of cities in the middle of the decade: Whitehand 1986), became more evident. In Britain, the centre of academic activity became the Urban Morphology Research Group of the School of Geography in the University of Birmingham (Larkham and Pompa 1988). Urban designers and contextual architects, led in particular by the late Francis Tibbalds (Past President, Royal Town Planning Institute), and pushed by the interventions of HRH The Prince of Wales (1989; see also Jencks 1988), have become increasingly aware of the significance of urban history and urban form in designing future urban landscapes. This resurgence in urban morphology-related studies and ideas parallels what has been seen as the increasing importance of 'place' in geography (e.g. Johnston 1984).

The strongest research tradition in the UK has been that introduced by M.R.G. Conzen (Whitehand 1987b, 1987c) and developed by members of the Urban Morphology Research Group. This volume is grounded in that tradition, which merits brief explanation. Conzen combined his research interest in urban morphogenesis with his practical planning experience in his participation in the *Survey of Whitby*, which was aimed at producing the basis for an integrated plan for the town (Conzen 1958). Notable throughout his contribution are numerous references to period buildings still surviving in the townscape, the importance of their continued preservation, and the importance of townscape as a composite historical monument (Larkham 1990a: 352). Conzen returned to conservation as a theme in his paper on historical townscapes (Conzen 1966), using as illustrations some of the smaller towns surveyed in detail earlier. The concept of 'management' was introduced, and the key attribute of a townscape that required management was identified as its 'historicity'. Conzen sees three principal factors as making up a townscape's historicity: the town plan, building fabric and land-use. In a second paper on this subject (Conzen 1975), the concern changed from a delimitation of aspects of historicity to a concern for how this historicity is shaped. It is noted that the three aspects of historicity possess differing degrees of persistence in the townscape, with town plan and building form being most persistent. These two elements in particular form the 'morphological frame', constraining future development to some degree. This concept is particularly useful in smaller towns before 1939 (the circumstances under which it was devised). The rate and scale of late twentieth-century development particularly in larger centres, and occasions where large-scale change is brought about by some catastrophe, are circumstances where the morphological frame tends to exert less influence (Larkham 1995b).

Conzen's ideas on conservation and historical townscapes are further discussed in Larkham (1990a).

The changing urban landscape

This volume develops from the Conzenian approach to urban morphology as outlined above and, more comprehensively, in Whitehand (1987a, 1992a). An understanding of the processes of morphological change, including cycles in the economy, building industry, but also in thought, legislation, architectural style and taste, is vital. But so, too, is knowledge of the identity and actions of the agents of change: organisations, institutions, individuals; particularly those directly active in the decision-making process, but also those more indirectly active in, or only passively affected by, change should be considered. Whilst the former may be examined using large-scale surveys of trends, the latter lends itself only to small-scale case studies, owing to the labour-intensive nature of data recovery and analysis at the local level of operation of the development process. The types of areas and cases studied are held to be typical, although their representativeness cannot be proved statistically since insufficient of these studies have been carried out across the country. Nevertheless, they provide useful insights into the processes and agents involved in the making of the modern urban landscape. The urban landscape itself, representing a cultural asset and significant past investment, is not merely of intrinsic importance in this perspective but is also the key to formulating theories and processes of urban landscape management. It is to this end, and the particularly problematic aspect of conservation and planning, that this volume is aimed.

THE STRUCTURE OF THIS VOLUME

This volume explores ideas of continuity and change in the urban landscape through close examination of development trends and case studies. Although largely set in the UK, it also draws on experiences in Europe and elsewhere. In particular, it aims to use academic concepts and studies in urban morphology to demonstrate the scale, nature and direction of urban landscape change, focusing particularly on aspects of conservation, with particular relevance to urban geography and planning.

The volume itself is divided into three parts. The first is introductory, setting the context of conservation, its history, actions and implications. Chapter 2 discusses the history of conservation, the way in which changes in architectural tastes and development forms can be related to conservation, and the evolution of statutory approaches to conservation over the past two centuries or so. Chapter 3 examines the spread of conservationism, some of the influential actors, and the implications of processes such as legislative development and property obsolescence cycles. Chapter 4 deals with the English system in greater detail, both because a number of other countries operate basically similar quasi-judicial procedures (Malta introduced an explicitly English-based planning system in the early 1990s), and also because the very quasi-judicial nature of this system has meant that here, perhaps more than elsewhere, some of the fundamental ideas and approaches have been argued through at appeal and in court and, owing to the system of precedent, have thereafter acted as rigid guides to practice.

Part 2, 'The changing conserved town', moves to examine some of the urban landscapes produced by the involvement of processes of conservation. Chapter 5 begins with a review of area-based conservation approaches, detailing the advantages and shortcomings of the highly developed UK system. Chapter 6 introduces the Conzenian morphological approach to studying built fabric changes and the decision-making processes shaping them, which will be developed in the three following chapters. Chapter 7 first examines the role of the LPA as a 'direct' agent of change through consideration of local authority-led 'enhancement' schemes in the UK and their recent criticisms, together with a study of how small-scale development control decisions reflect stated conservation policy. Based largely on detailed examination of English historic towns through local authority development control records, Chapters 8 and 9 are able to study processes and types of change in both historic town cores and residential areas. Chapter 10 draws some wider implications for the forms in which changes are currently occurring. Of importance throughout this part is the continuing nature and profound impact of incremental change to the character and appearance of the conserved urban landscape, together with the way in which the planning processes shaping built fabric change relate to explicit conservation policy.

Part 3 contains one concluding chapter examining the ethical, theoretical and practical implications of conservation for the continued evolution of the built fabric. Conservation remains an empirical activity, rather than one firmly grounded in ethics or theory. Even lessons from the conservation of other artifacts, such as machinery and art, cannot readily be transferred to the built environment. Unfortunately, therefore, the management of UK conservation areas appears more dominated by mismanagement than by widespread and acceptable conservation.

2

THE HISTORY OF URBAN CONSERVATION

Conservation is a comparatively new idea.

(Michael Manser, Past President,
Royal Institute of British Architects, 1984)

INTRODUCTION

Before plunging into detailed investigations of conservation-related agents and types of change, it is important to set the scene with an examination of the entire background to urban conservation. Fashions in appreciation of art, architecture and landscapes change, and the relationship of these changes with conservation action is important. This chapter discusses the idea of the conservation of buildings and entire urban areas through the medium of changes in social attitudes and official action, particularly in the form of legislation. This is achieved through a general survey of the development of ideas and legislation in the past two centuries in a variety of countries with differing socio-cultural contexts, and a detailed comparison of recent legislation in Britain and the USA. Both the overview and detailed comparison allow reference to be made to one of the key questions posed in Chapter 1, principally in identifying what mechanisms have developed to allow the identification and protection of the built heritage. This forms a necessary background to studies of current attitudes and conservation actions, which will address the other key questions posed.

In the case of conservation, the social attitudes depicted were, until relatively recently, those of a small, educated, wealthy and influential élite, which both set and reacted to changing fashions. This is an example of the 'dominant ideology thesis' first formulated by Marx and Engels (1864): 'the ideas of the ruling class are in every epoch the ruling ideas, i.e., the class which is the ruling material force of society, is at the same time its ruling intellectual force'. Yet in this example, through time, the dominant group became weakened through the increasing activities of the middle class in conservation activity. The working class is represented relatively poorly in conservation, although mainstream activities now encompass industrial heritage and archaeology. But this process of élite domination has operated further down the social hierarchy, and a good example of fashion-following lower in the socio-economic hierarchy is given in Tyack (1982).

As the élite came to recognise and appreciate older buildings, monuments and landscapes, many of the examples of its preservation efforts were monumental and well known; these include preserved churches, castles, and the creation of new landscape parks. Landscape parks in particular were often created in close association with smaller towns, having considerable impacts on the form of those towns and on the rural agricultural and settlement landscape (Slater 1977). With the exception of some of the landscape parks, these examples of élite activity were often isolated and free-standing, and their impact upon urban landscapes, now a leading focus of conservation effort, was relatively small. Indeed, in contrast to the changing views of the élite, the lower social classes, including many civic corporations (then consisting mostly of self-elected businessmen), were more interested in the opportunities for spacious urban redevelopment afforded by the demolition of the picturesque but aged building stock. This is well shown by the actions of Birmingham Corporation, in particular the Mayor, Joseph Chamberlain, in demolishing large areas of artisan housing under the Artisan's Dwellings Act 1875 and replacing it, not with new housing, but with a spectacular civic showpiece, the retail area of Corporation Street (Briggs 1952: 18–22). This tendency is perhaps most clearly seen in the case of fortified towns. In York, the proposals to demolish the walls and redevelop the land made available failed (Curr 1984), but on the European mainland these urban defences were frequently redeveloped, and leave a legacy in the current urban landscape in the form of 'fixation lines' which, to some extent, constrain future development (Conzen 1960). The Vienna Ringstrasse, developed between 1857 and 1914, is probably the most spectacular, and best-known, example (Olsen 1986: 58–81).

Using Britain as the principal example, with illustrations from other countries, the attitudes among this small, but influential, proportion of the population can be seen to have changed from the end of the eighteenth century to the present, with significant consequences for legislation and the conserved landscape. In particular, there have been significant moves from rural landscape protection to urban areas, most recently residential suburbs; and from the eighteenth-century Picturesque tradition of landscape vision has descended the Townscape view, which in the UK has led to popular and professional concentration on the outward visual appearance of conserved urban areas. Although innovators are significant, perhaps more important are the publicists who give a wider impetus to a movement. The nineteenth century is portrayed as a period of gathering momentum, leading to a first phase of conservation legislation in a number of countries towards the end of the century.

In the present century, conservation, as with planning as a whole, underwent what could be interpreted as a cyclic process of change (Sutcliffe 1981a). Moreover, the rise of a widespread consciousness of history or, perhaps more correctly, heritage, has led to a rapid broadening of conservation-related legislation and associated activities in the past two decades or so. The public, particularly as represented by pressure groups, has been more vociferous in expressing views on conservation, and in pressing for particular buildings and areas to be protected. But, as has been suggested, this broad interest and concern remains fixated on superficial external appearance rather than on questions of authenticity: heritage townscape rather than history. Furthermore, although conservation is now at a high point in the socio-political agenda, it may readily become displaced by other wide-ranging concerns, for example that for sustainability in planning and development (Elkin *et al.* 1991).

THE HISTORICAL BACKGROUND TO CONSERVATION

Conservation-related actions have a long history. It has been suggested that the earliest actions were spurred almost entirely by concern, respect, even piety, for the past and its people. Thus the Greeks preserved the Hellenic monuments with honour, Roman emperors such as Hadrian also respected these '*exemplaria Graeca*', and even their successors, Teutonic chieftains such as Theodoric of Rome, acted to preserve the monuments of that ancient city (Brown 1905: 13). Nevertheless, economic influences were significant at an early date, with a Roman inscription predating AD 63 recording that 'if any person for the sake of traffic [i.e. profit] should have purchased any building, in the hopes of gaining more by pulling it down than the sum for which he bought it, that he shall be obliged to pay into the exchequer double the sum for which he purchased it' (recorded in Russell 1750), while the Emperor Majoram (AD 457–61) took action to halt the plundering of Rome's monuments, which were seen as convenient sources of ready-quarried building stone and marble (Gibbon 1782–8). In 1450 alone, Pope Nicholas V is supposed to have removed two thousand cartloads of marble from the Colosseum in a single year, while in about 1500, Pope Alexander VI leased the ruins for exploitation as a commercial quarry (Kennet 1972: 11–12). Popes Pius II and Leo X took action to preserve Rome's monuments in 1462 and 1515 respectively, but these early preservationist pronouncements were very much the exception and, as Alexander VI's actions show, were largely ineffectual. It was not until the mid- to late eighteenth century that the attitudes of the social élite towards the monuments and inheritance of the past began to change significantly. During that period, Pope Benedict XIV declared the Colosseum sacred to the Passion of Christ, owing to the martyrdom of the early Christians there, and forbade anyone to carry away any further stone from it (Kennet 1972: 12).

THE NINETEENTH CENTURY: A KEY PERIOD

Changes in ideas concerning building and landscape styles are evident in the late eighteenth and early nineteenth centuries. The leading architectural style immediately prior to the Victorian period was Classical, with its variants of Palladianism, neo-Classical and Georgian. Classical forms had been re-introduced into Britain in the late sixteenth and early seventeenth centuries, when prevailing styles were condemned as being unsuitable and ugly. The Classical was a particularly ordered and symmetrical approach to architecture, in which all details should obey set laws of proportion to achieve aesthetically satisfying results. This attitude to current architecture also coloured the perception of historical, and merely old, buildings. For example, the twelfth-century interior of the parish church at Shobdon, Herefordshire, was set up on a hill overlooking the church following the eighteenth-century remodelling of the church itself into a more acceptable style. At the time, this was seen as praiseworthy (Thompson 1981: 17).

By the end of the century, however, the prevailing aesthetic mood had undergone an almost complete reversal. This was spurred on by the activities of the more innovative of the social élite, who had begun to introduce the aesthetic of the 'beautiful', 'sublime' and 'picturesque' into late eighteenth- and early nineteenth-century philosophy and psychology, and these concepts also became evident in architecture (Price 1794; Hipple 1957; Pevsner 1968). This was first apparent in small developments by large landowners even in the eighteenth century, as the fashion-conscious followed the lead of the trend-setters: Lord Bathurst built a Gothic folly, 'Alfred's Hall', in about 1740 (B. Jones 1974), and Marie Antoinette had a picturesque cottage built at Versailles in the 1780s (Fitch 1982: 19). Darley (1975) deals with the assimilation of ideas generally by landlords, who as a group were particularly responsive to changing architectural fashions. Their interpretations of such ideas are well shown in 'model' rural, industrial and religious settlement forms. The asymmetry and seeming disorder of the Picturesque was, in some ways, a reaction against the strict order of Classicism. These different aesthetic ideals were then popularised by a vociferous group of architects and architectural writers, who had a profound influence on the manner in which architectural styles were perceived and on how restoration and conservation should be carried out. These changing attitudes led, towards the end of the century, to the first general phase of conservation legislation.

The writers themselves were a mixed group. Not only were Victorian architects in general developing a notable propensity for writing in journals and in books to explain their approaches to architecture or to particular buildings (see the examples discussed in Pevsner 1972), but various groups such as the Ecclesiological (formerly the Cambridge Camden) Society exerted some considerable influence (Muthesius 1972). Many urged the adoption of the Gothic style, and this important transition is discussed in some detail by Kain (1986). Four individuals are identifiable as being of particular significance.

Of singular importance in the change of ideals from Classical to Picturesque and Gothic is the architect and writer Augustus W.N. Pugin (Stanton 1971; Muthesius 1972; Watkin 1977: 17–23). A convert to Catholicism, Pugin fulminated against Classical architecture: it was adopted in Britain at the same time as the Reformation, it was therefore a consequence of the Reformation and, since the Reformation was morally reprehensible (at least to a Catholic), so must its architecture be. In its place Pugin, an influential writer, widely publicised the virtues of Gothic, 'pointed', architecture. Indeed, he wrote about architecture as if it were part of a religion, with Gothic styles being almost a Divine revelation (Pugin 1836, 1841; Watkin 1977: 17–19). The significance of his proselytising is that Gothic styles were far more amenable to Picturesque thinking than Classicism, and its popularity in ecclesiastical and polite architecture was paralleled by its adoption into model village architecture, as is seen at Holly Village, Highgate, of the mid-1860s. This supplanted the vernacular Picturesque of, for example, Blaise Hamlet, developed earlier in the century (Darley 1975: Chapter 4). Gothic styles were also particularly suitable for large civic buildings, as is shown by its use in civic buildings such as town halls and new university buildings. The influence of 'medievalism in Manchester', to take one notable example, is particularly well discussed in Dellheim (1982: Chapter 4): the architect Alfred Waterhouse built the Town Hall (Archer 1982), Assize Courts and Owen's College (later Manchester University) in the Gothic style.

Of similar significance is John Ruskin, and in particular his book *The Seven Lamps of Architecture* (Ruskin 1849). Of the seven sections, or 'lamps', three are particularly relevant to this discussion (Steegman 1970: 87ff). The section on 'life' defines this as vitality, imagination and sensibility, but only as these were to be found in the architecture of the past. Modern architecture had 'a sickly look to me. I cannot tell whether it be indeed a springing of seed or a shaking among bones' (Ruskin 1849: 230). The following section, 'memory', stresses the desired longevity of buildings: it was an evil sign when houses were built to last for only one generation. The last section, 'obedience', shows Ruskin apparently as an avowed revivalist, without hope or desire of evolving any new, or contemporary, architectural style.

> A day never passes without our hearing our English architects called upon to be original, and to invent a new style. . . . We want no new style of architecture. . . .There seems to me to be a wonderful misunderstanding . . . as to the very nature and meaning of Originality, and of all wherein it consists . . . our architecture will languish, and that in the very dust, until . . . a universal system of form and workmanship be everywhere adopted and enforced.
>
> (Ruskin 1849: 309–10, 314)

Ruskin was thus insistent upon revivalism, of Gothic and not the pagan Grecian-Classical styles. This one book had a great public impact: it 'taught for the first time the spiritual and historic importance to a nation of its architecture' (Steegman 1970: 89–90).

Another significant architectural theorist was Eugène Emmanuel Viollet-le-Duc. In several works, he attempted to construct a rational theory of architecture, in harmony with prevailing contemporary thought (Viollet-le-Duc 1854–68, 1863, 1872; see also Watkin 1977: 23–31). As with Pugin and the Ecclesiologists, Viollet-le-Duc espoused Gothic architecture. His was, however, a rational and mechanistic approach, rather than the religious interpretation of the Ecclesiologists. His theory was that the architecture of the present must be derived from that of the past, but should not be mere revivalism. Past masterpieces should be analysed, and their lessons applied to current problems. This is far from the Gothic beliefs of his contemporaries, including Pugin and the architect Gilbert Scott (Summerson 1980; and see also the exchange of letters between Viollet-le-Duc and L.-A. Boileau in *Architectural Design* 1980a). This difference in approach, leaving behind the mysteries of religion, was not Viollet-le-Duc's only contribution to the history of architectural thought. Of greatest value to conservation, under the heading '*Restauration*' (vol. VII of his *Dictionnaire Raisonné*), he suggested that a new art had been born – that of the scientific restoration of monuments. A detailed knowledge and survey of the remains as presently existing would enable an informed reconstruction to be carried out. This involved the rebuilding of the monument to a conjectured complete state, rather than the simple consolidation of the existing ruins (Dupont 1966). His approach is typified by his work on the Château de Pierrefonds from 1857 to his death in 1879 (*Architectural Design* 1980b). A British example of this type of 'restoration' is the reconstruction of Castell Coch near Cardiff by William Burges for the third Marquis of Bute (Thompson 1981: 20).

Thus both Ruskin and Viollet-le-Duc are proponents of High Gothic styles, but for very different reasons. Their whole conception of Gothic differed radically. Ruskin admired the Gothic building as being alive with the life given to it by the carver who, loving his work, endowed it with beauty. Viollet-le-Duc admired the designer for his grasp of the logic of construction (Pevsner 1980: 49).

The influence of William Morris was also of considerable importance in changing attitudes in the latter part of the nineteenth century (Thompson 1967; Pevsner 1972: 269–89). The architect Philip Webb designed a solid brick house with Gothic touches for him (the Red House, Bexleyheath, 1859–60), which is now identified as a key building in the study of changing architectural styles. Morris was a fervent Socialist, had a 'hatred of modern civilization' and described himself as 'fairly steeped in medievalism' (Morris 1910–15; W. Morris, letter to a student quoted by MacKail 1899, vol. 1: 203). His ideals were inspired by Gothic architecture, which he saw as an architecture based on good craftsmanship provided within the framework of the mediaeval guild, social organisation and good craftsmanship. Given this background, it is hardly surprising that in 1877 he founded the Society for the Protection of Ancient Buildings (SPAB), a landmark in the popularising of concern for conservation and historic buildings. The impetus for this, as his original well-known letter of 5 March 1877 to the *Athenaeum* shows, was the contemporary treatment of genuine Gothic architecture: 'Sir, My eye just now caught the word "restoration" in the morning paper, and, on looking closer, I saw that this time it is nothing less than the Minster of Tewkesbury that is to be destroyed by Sir Gilbert Scott.' Morris's own philosophies of architectural aesthetics, based on Gothicism and Romanticism, influenced the Society to a very considerable extent. Buildings should not be entirely renovated, as this 'takes away the appearance of antiquity ... from such old parts of the fabric as are left'. Owing to changed social conditions and methods of production, it is not possible to restore faithfully. Even were this possible, it would be undesirable and deceptive.

> The conditions and surroundings of every period are different, so that the motives which act on men of one age cannot govern the production of genuine work conceived in the spirit and embodied in the forms of another.... Even were it possible to reproduce lost work, it may be said that ... artistic honesty is the best policy, just as much as in other affairs of life. The restorer is in reality committing a forgery.
>
> (Morris 1877; SPAB 1903: 12–13)

The SPAB has remained a particularly vociferous influence in British thinking on conservation since Morris's death. Its views, closely following his dicta, have been adopted by the State's conservation bodies to such an extent that, in the Report of the Inspector of Ancient Monuments for 1913, restoration was a most heinous offence making a foreman liable to instant dismissal (Thompson 1981: 20). This shows a considerable change from the earlier ideas of Viollet-le-Duc. The SPAB has advised central government bodies and local planning authorities on conservation matters ever since, and indeed since 1968 they have been one of a small number of expert bodies to be consulted by statute for their opinions on such matters.

EARLY LEGISLATION

After this brief examination of the various influences active in changing philosophical and architectural attitudes towards the past in general, and to monuments in particular, an examination of the official movements to which they had given rise by the end of the nineteenth century is appropriate. The late nineteenth century can generally be termed the first period of conservation legislation, despite earlier pioneering but isolated enactments. The workings of this initial phase are best illustrated through a comparison of the experiences of several countries.

Changes in legislation are notorious for usually following some way behind changes in attitudes and in society, and the case of conservation is no exception. There was no action in Britain until the 1870s. In 1875 the Austro-Hungarian Ambassador wrote to the then Foreign Secretary to inquire what was being done to protect ancient monuments. The reply, given in a letter from the Society of Antiquaries, 'deeply regretted that no legislative machinery was available' for monument protection. From 1873, Sir John Lubbock MP had introduced various Bills into the House of Commons, but all were opposed since it was considered that they involved unjust and unwarranted interference with private property. It was not until 1882 that one such Bill became law in the form of the Ancient Monuments Protection Act 1882. This was largely a passive piece of legislation. Although some monuments could be maintained at public expense, property owners were under no compulsion to maintain monuments owned by them, and the types of monuments qualifying for protection were few and mainly prehistoric (Kennett 1972: 21–30).

The Netherlands had a similar story of neglect. There was a legislative relic of the Napoleonic occupation, dated 1814, forbidding destruction of historic monuments. Nevertheless, by the middle of the nineteenth century, Prime Minister Thorbecke was able to state that 'art is not the business of government' while, during the great period of urban growth in the late nineteenth century, many civic defence works were demolished in the cause of widespread rebuilding. In 1875 a new department, the *Rijkscommissiei*, was charged with overseeing the preservation of ancient monuments. Again, as in Britain, this department was small and without financial resources. It spent much of its time compiling a list of monuments (Ashworth 1984: 606–7).

In France, the first reports of the Inspector-General of Ancient Monuments, appointed in 1831, drew attention to the decayed state of the country's mediaeval buildings, and the Chambers voted money for their preservation. Neither of the first two Inspectors-General had any technical knowledge of building – indeed the second was the author Prosper Mérimée – but their enthusiasm did lead to some necessary conservation work. In 1837 a more authoritative body, the *Commission des Monumentes Historiques*, was established. This spent much of the next 76 years in listing buildings of significance, which did at least give some measure of protection (Kain 1982).

At this time, Germany was divided into several States, each having different approaches. Bavaria and Württemberg began official inventories of ancient monuments in the sixteenth and seventeenth centuries respectively. Württemberg passed the first pioneering statute governing monument protection in 1790. Friedrich Wilhelm III of Prussia passed a law in 1855 allowing authorities to intervene where public buildings of historic significance were threatened. The first conservation officials employed on a full-time basis seem to have been in Bavaria (1835) and Prussia (1843). The Provinces were made responsible for monument conservation and the appointment of conservation officials in 1875 (von der Dollen 1983; Haines 1974). During this period, public appreciation of landscapes and environmental issues grew with the increasing popularity of walking and mountaineering societies. The completion of Cologne Cathedral (1840s–1880s) was a significant landmark in concern for historic buildings (Muthesius 1981: 38; Cologne Cathedral is discussed in Crewe 1986).

From even these few examples, it can be seen that most action came in the mid- to late nineteenth century. The German states were exceptional in their early action. In many cases, action was spurred by the pressure of a key individual, notable examples being Sir John Lubbock in Britain, Friedrich Schinkel, Head of the Building Authority in Prussia, whose Memorandum inspired Wilhelm III's action, and J. de Stuers in the Netherlands, whose publications and reputation led, in the mid-1870s, to his appointment as head of a Department of Arts and Sciences, with responsibilities for historical monuments. However, much of the action at this time concerned ancient monuments rather than buildings. These were mostly individual features, and almost exclusively pre-mediaeval in date. There was no notion of the conservation of groups of buildings, or of entire townscapes. Urban conservation *per se* was absent, despite the general expansion of towns in the nineteenth century and the consequent threats to the older buildings. York provides a rare example of the deliberate retention of older features, in this case the City Walls (Curr 1984; Palliser 1974). The Rows of Chester are an equally rare townscape feature that attracted antiquarian interest from an early date, leading to their pastiche reconstruction (Chester Archaeological Society 1984). Early legislation was also particularly toothless, and its principal achievement in many countries was the commencement of detailed inventories of monuments.

THE TWENTIETH CENTURY: LEGISLATION, ACTION AND EXPLANATIONS FOR CONSERVATION

This state of affairs has changed drastically in the present century. In many countries, the nineteenth-century legislation was updated and strengthened in the early years of this century. In France, the *Société pour la Protection des Paysages de France* was created in 1901. Its first President was Charles Beauquier, Parliamentary representative of the Doubs. The Society promoted a Bill providing for the classification of beautiful and historic landscapes in the same way that monuments were listed under the 1837 system. Beauquier, an experienced parliamentarian, was fortuitously chairman of the committee examining this Bill, which became law in 1906. This new law was a key piece of legislation.

> Like much private members' legislation, the new law proved weak and difficult to enforce, but it represented an important extension of the principle that restrictions could be imposed in the public interest on the free use of private land. As such, it was to be much cited in later years by advocates of general town-planning controls.
>
> (Sutcliffe 1981b: 142–3)

This period of modified legislation, typified by the Danish Preservation of Buildings Act of 1918 and the British series of Ancient Monuments Acts (1900, 1913, 1931), was continued throughout the middle of the century, although attention was, of course, diverted by two World Wars. After the Second World War, there was another major period of legislation, based on an accumulation of experience with existing laws, and reinforced by an awakening interest in historic preservation on the part of the public at large. This is reflected in Britain, for example, in the rapid growth in the numbers of local environmental and amenity groups formed to put pressure on local authorities, particularly following the formation of the Civic Trust in 1957

(Barker 1976; Lowe and Goyder 1983). The aims of these societies were much more active and interventionist than were those of the local historical societies that had become popular during the Victorian period (Dellheim 1982).

This growth is but one facet of a continued interest in the past. Fitch (1980, 1982) argues that the West's seemingly inexhaustible fascination with its own and with other peoples' pasts began to wane in the present century. This could be supported by local incidents, such as the campaign to preserve the Old Deanery House, Wolverhampton, in the early twentieth century. A short history of the house was produced in support of the campaign. This provoked public comments of the nature of 'I didn't know it had so little to recommend it: knock it down'. A local historian, G.P. Mander, suggested that 'it would appear that for the safety of old buildings the darkness of superstition were preferable to the light of truth' (Mander 1933: 3). However, there is a wealth of evidence to contradict Fitch's assertion, much of it cited by Lowenthal (1985) in his exhaustive discussion of why we both need and change the past. For example, the majority of the country houses built in England since 1900 have been in variations of historicist styles, with Tudor styles being popular early in the century and Georgian/Classical styles most popular in recent years (Aslet 1982; Robinson 1984). In 1924, the Federated Home-Grown Timber Merchants Association built a neo-Tudor house as an exhibit at the Wembley Exhibition, and Liberty's of London, usually in the forefront of popular taste, rebuilt their store in Tudor style using old ships' timbers (Aslet 1982). This style was also popular in speculative house-building through the 1930s, and is currently undergoing a revival (Horsey 1985; Larkham 1988a, 1988b). A considerable amount of housing built after the First World War was of historicist, particularly neo-Georgian, design; this style again had a resurgence, this time peaking during the 1970s (Pepper and Swenarton 1980; Larkham 1988a, 1988b). Modern-styled building, on the other hand, made little impact upon contemporary experience in the inter-war and early post-war periods. It was used only sparingly by speculative builders; many of the houses in this style being bespoke, purpose-designed for wealthy clients, and there were relatively few of these (Gould 1977). Only during the post-war building boom, from the mid-1950s, did the Modern style become familiar in the commercial centres of most towns, and even here it was a debased form of Modern that was most commonly used by large developers and retail chain stores (Larkham and Freeman 1988). Thus the argument for a decline in interest in the past in terms of building style – and thus by inference, in attitudes towards conservation and preservation of the genuine heritage – is not proven. Indeed, this continued fascination with the past, its evocation and recreation, has been described as a fatal obsession, and a symptom of national decline and loss of confidence in the future (Hewison 1987; MacCannell 1976).

Instead of a decline in interest in the past, there appears to be an upsurge in the number of explanations for conservation. Whilst these explanations would essentially be valid throughout the period being considered, they are most apt in the twentieth century. This is a period of particularly rapid social, economic and physical change, with correspondingly high threats to the historical urban fabric in particular. In Britain, this threat was quantified early in the current period of conservation concern by Heighway (1972) in a report for the influential archaeological body, the Council for British Archaeology. It was particularly intense during the immediate post-Second World War phase of demolition and comprehensive redevelopment in town centres and areas of sub-standard housing (1945–70) (Esher 1981). These views towards the past and the preservation of its heritage have been explained, in part at least, through a variety of studies in environmental psychology, which strongly suggest that this 'looking back' is a psychological necessity (see also Chapter 1). This need can be met by historical areas that have survived relatively unchanged and which provide symbols of stability against 'future shock'. Even areas undistinguished in age or merit, whether architectural or historical, are psychologically useful in providing 'the anonymous familiar' because of the experiences associated with them by groups or individuals (Smith 1974a: 903). It is significant that one of the earliest scholarly works on the history of conservation identifies this psychological desire underlying the construction and later preservation of monuments, and indeed attributes the ubiquitous Dr Johnson with its identification (Brown 1905: 12).

In a similar manner, the resistance to change evident in much of the rebuilding of Germany after the Second World War has been explained in terms of an anti-progressive, irrational, traditional arrogance, and the historic and humanistic responsibility for the preservation of the cultural heritage (Holzner 1970). These are powerful German intellectual movements, reflected in Nietzsche's philosophy of the German 'hinterland'. It is suggested that this preservation, and in some cases recreation, of the historic past is not rational, that post-war planning in Germany could have been better, but that a popular social requirement for conservation overcame rationality. This appears to be a practical and large-scale demonstration of the psychological argument for conservation, and seems essentially similar to the reasons of national identity advanced for the rebuilding in historical forms of Polish cities devastated during the Second World War (Lorentz 1966: other treatments of post-war rebuilding in Europe are given in Diefendorf 1990).

An alternative explanation of the rise of conservationist ideals has been given in terms of symbolism, whereby certain attributes of features of past landscapes are imbued with a symbolic significance for the present landscape and its inhabitants (Rowntree and Conkey 1980). It is postulated that this symbolisation is a response to increasing stress. In a case study of Salzburg, the stress is shown to be an increasing pressure for change and redevelopment following the arrival of the railway in 1860. This stress resulted in the attaching of symbolic significance to the walled *Altstadt*, and its preservation to a considerable extent.

This widespread interest in the past has been evident virtually throughout the Western world and with the benefit of the experience gained by many countries in the operation of basic conservation legislation, another phase of more sophisticated legislation was developed. The 1961 Monuments and Historic Buildings Act of the Netherlands, for example, dealt with both architecture and town planning, so that both valuable individual buildings and planned units of townscape could be preserved. A system of licences was developed, and no alterations could be made to monuments without the Minister's permission. Preservation of whole areas was possible, but this concept was adopted only slowly. Although this measure was a considerable advance on past protection, it was itself seen as inadequate within two decades (van Voorden 1981).

Denmark's Preservation of Buildings Act (1966, as amended in 1969) provides for the scheduling of buildings, monuments and fortifications for preservation. Emphasis is placed on 'great' or 'especially great' architectural or cultural/historical value in their selection. The legislation provides financial assistance for restoration, but this is inadequate. In its basic concepts, this legislation appears not to have developed far from the original 1918 Act (Skovgaard 1978).

The principal piece of conservation legislation in Britain is the Civic Amenities Act 1967. This was not sponsored by the government, or even by a major political party. It was originated by one man, Duncan Sandys MP, who won first place in the annual ballot for parliamentary Private Members' Bills in 1966. As founder and President of the Civic Trust, it is perhaps hardly surprising that his Bill reflected their concerns. This Bill passed through Parliament without a vote being recorded against it, and became law in 1967. This was the innovative Act that extended consideration from individual buildings to the conservation of entire areas, the beginnings of effective urban conservation *per se* in Britain. This was a most significant personal political initiative, going beyond the framework of the planning system as it then existed. Duncan Sandys has been identified as a significant figure in Britain's planning history, not least for this Act (Cherry 1982); the Act itself is described by the Civic Trust (1980a).

The French appear more innovative in conservation legislation. A law of 1913 consolidated the older procedures for listing monuments, and provided for the establishment of 'protected perimeters' around them. Groups of buildings are eligible for protection under a law of 1930, but it was not until 1958 that finance was specifically allocated for the restoration of buildings noted principally for their value as part of a group. In 1962 the 'Malraux Act' introduced the widely emulated system of *secteurs sauvegardés*. This was designed to protect older urban centres from the onslaught of modern development pressures, and such measures continued into the 1970s. However, this is essentially negative protection, ensuring only the survival, not necessarily the restoration, of the historic urban fabric. By 1976 the once-innovative system was being criticised on the grounds of cost, slow progress and the injustice of focusing attention on very small areas. Since then, attention has turned away from the spectacular but limited schemes to more general *opérations programmées d'amélioration de l'habitat*: social criteria are beginning to match those of architectural and historical values in French rehabilitation (Stungo 1972; Kain 1981).

This overview has been useful in showing how the dominant socio-political ideology has adopted conservation in a variety of different cultural contexts. In the majority of cases original legislation, grudgingly passed, was merely permissive; doing little more than identifying a small number of monuments. Later refinements, and in many cases completely new legislative systems, greatly broadened the scope of legally constituted conservation – wider definitions, for example of historical and architectural value, broader areas, introduction of grant aid, and so on. However, a more detailed study, particularly of the more recent past, suggests that the dominant ideology thesis is now, at least in part, superseded by other considerations. Among these are the uses of conservation in other governmental areas such as housing or economic regeneration, and significant economic changes. This is in line with other critiques of the dominant ideology thesis as being overtaken by 'the dull compulsion of economic relations' (Abercrombie *et al.* 1980; the interpretation of Abercrombie in conservation and legislative development is the author's).

EXAMPLES OF LEGISLATIVE ACTION: GREAT BRITAIN AND THE USA

Conservation-related legislative actions in the USA and Great Britain have been examined in order to show the links between legislative and social processes in greater detail. Cullingworth's brief survey suggests that US approaches to historic preservation have changed dramatically during the twentieth century. Originally, the emphasis was on key landmarks of historical and cultural value. However, the government intervened little until the mid-1960s, when public concern over the destructive effects of urban renewal and highway construction became evident. 'A veritable orgy of legislation followed' (Cullingworth 1992: 65–72).

Newcomb (1983) has shown fluctuations in Federal conservation-related enactments for the period 1966–82 (Table 2.1). Three distinct legislative periods are identified, quite different in intent and in resulting impact. The first is by far the longer period, from the National Historic Preservation Act 1966 to the Archaeological Resource Protection Act 1979. This was a period of comprehensive pro-conservation activity, lasting through four Presidential administrations of quite different styles. This would appear to indicate a wide support for the concept of conservation. This idea is strongly supported by discussions of the American attitudes to history and its preservation and display, most particularly in the period leading up to the Bicentennial celebrations (Lowenthal 1966, 1977). It does not, however, show that opinion was undivided in the national or State legislatures.

The second 'action period' overlaps the first, extending from 1976 to 1980. An emerging preoccupation with financial rewards is clear, as efforts were made to produce an effective legislative programme with tax incentives, and

Table 2.1 Legislative periods in US conservation 1966–82

Date	*Legislation*	*Effects on conservation*
1966	National Historic Preservation Act	Cornerstone for modern conservation in the USA. Expands National Register and sets up Advisory Council on Historic Preservation.
1969	National Environmental Policy Act	Sets out requirements for environmental impact studies and declares the protection of cultural resources to be a national policy.
1971	Executive Order 11593	Protection of the cultural environment made a Federal priority undertaking and instructs Federal agencies appropriately.
1974	Archaeological Resource Protection Act	To preserve threatened sites, especially those threatened by floods resulting from governmental activities.
1976	Tax Reform Act	Establishes tax incentives for preservation of certified historic buildings and their rehabilitation.
1978	Revenue Act	Amends 1976 Act.
1979	Archaeological Resource Protection Act	Re-enacts 1974 Act.
1980	National Historic Preservation Act	Addresses conservation of intangible cultural resources. Also revises procedures for nominations to the National Register, allowing property owners more opportunity to make their preferences known.
1981	Economic Recovery Tax Act	Increases tax credits for qualified rehabilitation expenditures and makes such expenditure more attractive under Federal tax laws.
1982	Department of Interior and Related Agencies Appropriations Bill	Prolongs Federal support of conservation through Fiscal Year 1983 (a victory for conservationists during times of political conservatism).

Source: Adapted from Newcomb 1983

there is also a preference for the use of private finance in place of Federal money. Other studies have supported the assessment of the effects of United States fiscal legislation, and in particular of the Tax Reform Act 1976, during this period (Andrews 1980; Collins 1980; Holmes 1982). A conservative attitude towards private property is firmly expressed in the 1980 Amendments to the National Historic Preservation Act, whereby an individual's rights to do as he or she wishes with their property were strengthened where an owner does not wish a property to be placed on the National Register, the equivalent of the British statutory list of buildings of architectural and/or historical interest.

Newcomb identifies a third phase from 1980 onwards, where financial restrictions for conservation work arise as a result of national recession. The importance of the 1981 legislation is that, to an extent, it reworks the 1976 Act, then due for renewal, so that changes in economic circumstances since 1976 may be incorporated. Although conservationists may be effective lobbyists, the 1982 Appropriations Bill shows that 'the expansionist sentiments of the 1960s have given way to a fiscal bunker-mentality ... and conservationists seek to hold on to whatever can be salvaged out of the wreckage of recessionary times' (Newcomb 1983: 8).

More recently, the concept of the donated development right (or 'conservation easement') has become widely used, particularly for protecting natural environments. Wright (1994) suggests that US planners should be more widely aware of this tool, which could be more positive than regulatory land-use control. Yeomans (1994) also details the tax credit system and the increasingly important role in the 1980s of the National Park Service in implementing the Secretary of the Interior's *Standards for Historic Preservation* – a set of rules which determines eligibility for tax credits.

A number of criticisms may be made of the approach of identifying such 'action periods'. First, the time period is so short, and so close to the present, that it would appear to be very difficult to differentiate clearly between periods. Secondly, the periods identified by Newcomb are short and have considerable overlaps, again raising doubts over the distinctiveness of the periods. Thirdly, the periods are drawn from an analysis of only the specifically conservation-related pieces of enacted legislation. It is highly probable that other pieces of legislation, not given in Table 2.1, would have some effects upon conservation activity, and that non-legislative publications would more clearly show the attitudes of governmental departments towards conservation issues. Legislation alone rarely reflects the entire view of an elected legislature or its professional staff. A good example of this latter point in Britain was the reaction of the Permanent Secretary of the Ministry of Housing and Local Government both before and after the Civic Amenities Act 1967. 'This kind of work [conservation/preservation] was utterly despised by Dame Evelyn. She regarded it as pure sentimentalism and called it "preservationism", a term of abuse' (Crossman 1975: 623). Lastly, in the American system, Federal pronouncements are apt to be overtaken by the activities of individual States. The States would not all react in the same way to the enactments that are cited by Newcomb: he realises this but does not deal with the point: 'the precise degrees of enthusiasm and efficiency with which the Federal mandates were met varied widely from State to State with some leading the way with innovative programs while others were content to assume a position of minimal compliance' (Newcomb 1983: 7).

Nevertheless, there is truth in the general point that attitudes, as reflected by Federal legislation, do change over time in relation to the general economic health of the nation. So it is useful to compare the American changes as identified by Newcomb with the progression of British legislation, and to investigate whether any similar phases of activity are identifiable.

Table 2.2 UK conservation legislation post-1967

Date	*Legislation*	*Circulars etc.*	*Principal contents*
1967	CIVIC AMENITIES ACT	53/67 CIVIC AMENITIES ACT 1967 – PARTS I and II	Conservation Areas introduced. Grants and loans for repair and maintenance of listed and unlisted historic buildings: grant aid to local preservation societies.
1968	TOWN AND COUNTRY PLANNING ACT	61/68 TOWN AND COUNTRY PLANNING ACT 1968 – PART V HISTORIC BUILDINGS AND CONSERVATION AREAS	Listed Building Consent applications introduced. Conservation Area Advisory Committees should be introduced.
	Housing (Financial Provisions) (Scotland) Act		Extends to Scotland the provisions of the Housing (Financial Provisions) Act 1958, covering Local Authority grants for the improvement of older houses.
1969	Housing Act		General Improvement Areas introduced: grants for environmental improvements in GIAs.
	Development of Tourism Act		Section 4 provides for grant aid for conversions of old buildings to new uses: administered by English Tourist Board.
		Development Control Policy Note 1969/7: PRESERVATION OF HISTORIC BUILDINGS AND AREAS	General advice note.
1971	TOWN AND COUNTRY PLANNING ACT		Section 277 is slight revision of 1967 Act. Section 58 introduces Building Preservation Notices. Section 114 provides for the acquisition of derelict listed buildings.
		56/71 HISTORIC TOWNS AND ROADS	New road development should respect historic towns.

1972	TOWN AND COUNTRY PLANNING (AMENDMENT) ACT	86/72 TOWN AND COUNTRY PLANNING (AMENDMENT) 1972 – CONSERVATION	Section 8: demolition control over unlisted buildings may be extended to cover entire Conservation Areas. Section 10 permits grants to be made for preservation or enhancement in 'outstanding' Conservation Areas.
	Local Government Act		Care for the conservation of existing communities: professional advice for conservation: European Architectural Heritage Year Grants.
1973		46/73 CONSERVATION AND PRESERVATION: LOCAL GOVERNMENT ACT 1972	
		Building Societies Association: agreement with Government	Mortgage application by purchasers of new buildings given priority second only to first-time buyers.
1974		(LOCAL GOVERNMENT REORGANISATION)	
		102/74 TOWN AND COUNTRY PLANNING ACT 1971 – HISTORIC BUILDINGS AND CONSERVATION	Criteria for listing buildings: warning against excessive restoration; overlap between listing of Historic Buildings and scheduling of Ancient Monuments.
	TOWN AND COUNTRY AMENITIES ACT	147/74 TOWN AND COUNTRY AMENITIES ACT 1974	Blanket control of demolition in all Conservation Areas; duty of Local Authorities to publish proposals for preservation and enhancement in Conservation Areas; publicity for applications affecting listed buildings and their settings; failure of Local Authorities to set up CAACs noted. Pressure to designate further Conservation Areas.
	Housing Act		Housing Action Areas introduced, with 75% grants (90% in hardship cases). Local Authorities have new compulsory purchase powers. Grants for GIAs introduced at 65% (75%). Improvement, Intermediate, Special and Repair Grants at 50% (65%) introduced.

Table 2.2 Continued

Date	*Legislation*	*Circulars etc.*	*Principal contents*
1975		14/75 Housing Act 1974: Parts IV, V, VI: Housing Action Areas, Priority Neighbourhoods and General Improvement Areas	
		17/75 TOWN AND COUNTRY AMENITIES ACT 1974	
1976			ARCHITECTURAL HERITAGE FUND established: £1,000,000 revolving fund, administered by Civic Trust.
1977		7/77 TOWN AND COUNTRY PLANNING ACT 1971: DEVELOPMENTS BY GOVERNMENT DEPARTMENTS	Crown buildings and listing.
		23/77 HISTORIC BUILDINGS AND CONSERVATION AREAS – POLICY AND PROCEDURE	Consolidation of all earlier circulars. Various factual appendices.
		26/77 Compulsory Purchase Orders – Procedures	Limits work required under section 115 of the 1971 Act (Repairs Notices).
		1977 General Development Order	Permitted Development defined. Article 4 deals with cases where this may be controlled in Conservation Areas if a Direction made.
		HM Customs and Excise Note	VAT at current rate (15%) levied on all repairs and maintenance.
1978	Inner Urban Areas Act		Sections 5 and 6: loans/grants in specific areas for improvement of amenities.

		TOWN AND COUNTRY PLANNING (LISTED BUILDINGS IN CONSERVATION AREAS) REGULATIONS	Listed Building Consent to be sought from Local Authority. For their own buildings, Local Authorities must apply to Secretary of State. Appeals may be made to the Secretary of State re demolition control of unlisted buildings in Conservation Areas.
1979	ANCIENT MONUMENTS AND ARCHAEOLOGICAL AREAS ACT		Section 48 inserts into the 1972 Act provision for recovering grant monies if property sold within a given period, or if terms of grant not complied with.
1980	NATIONAL HERITAGE ACT		Section 3: the National Heritage Fund may make grants or loans for the acquisition, maintenance or preservation of buildings, objects, etc. of particular interest.
	Housing Act		Improvement Grants (cf. 1974 Act). Higher eligible expense limits within HAAs.
	Local Government, Planning and Land Act		Adds Section 54A to 1971 Act: application may be made to the Secretary of State to say that a building will not be listed for five years. All consents limited to five years or other specified period. Listing criteria extended. 'Outstanding' Conservation Area designation previously necessary for Section 10 Grants withdrawn. Areas of Special Advertising control.
		21/80 The improvement of older housing	House Renovation Grant system.
		22/80 Development control: Policy and Practice	Aesthetic control of development: alternative uses for historic buildings.
1981		12/81 HISTORIC BUILDINGS AND CONSERVATION AREAS	Explanation of 1980 Act.
		TOWN AND COUNTRY PLANNING (NATIONAL PARKS, AREAS OF OUTSTANDING NATURAL BEAUTY AND CONSERVATION AREAS) SPECIAL DEVELOPMENT ORDER	Relaxation of Permitted Development outside Conservation Areas, and in those areas designated after April 1981: within pre-April 1981 areas, control remains strict.

Table 2.2 Continued

Date	*Legislation*	*Circulars etc.*	*Principal contents*
	Local Government, Planning and Land (Amendment) Act		Enforcement procedure for unauthorised works. Removal of some (dormant) powers of Secretary of State regarding Conservation Areas.
1983	NATIONAL HERITAGE ACT		Set up Historic Buildings and Monuments Commission for England (English Heritage): which may make grants for the preservation of historic buildings, may acquire them, or may give grants to Local Authorities or National Trust for acquisition.
1984	FINANCE ACT		Levied 15% VAT on new construction, passed in Commons 19/3/84: Listed buildings and churches exempted by Amendment, May 1984.
1986	Housing and Planning Act		Section 40 and Schedule 9 Part 1 amend legislation on listed buildings and Conservation Areas.
1987		8/87 HISTORIC BUILDINGS AND CONSERVATION AREAS – POLICY AND PROCEDURES	Explains provisions of 1986 Act and consolidates all extant advice, replacing Circular 23/77.
		TOWN AND COUNTRY PLANNING (LISTED BUILDINGS AND BUILDINGS IN CONSERVATION AREAS) REGULATIONS	Updating of regulations. Minor changes.
		TOWN AND COUNTRY PLANNING ACT 1971: CURRENT WORDING OF PROVISIONS RELATING TO HISTORIC BUILDINGS AND CONSERVATION AREAS	HMSO publication consolidating various amendments of 1971 Act.
1990	Town and Country Planning Act		Supersedes most of the 1971 Act.

	PLANNING (LISTED BUILDINGS AND CONSERVATION AREAS) ACT		Supersedes previous legislation. Incorporates rest of 1971 Act, and parts of other Acts including 1974 Town and Country Amenities Act.
1992		PLANNING POLICY GUIDANCE NOTE 1 (revised)	Annex A now states that design is a material consideration in planning decisions.
1993		PLANNING POLICY GUIDANCE NOTE 15 (draft)	Draft PPG largely replacing 8/87, incorporating notes of implications of recent case law.
1994		PLANNING POLICY GUIDANCE NOTE 15	Final publication; structure much amended.
1995		GENERAL PERMITTED DEVELOPMENT ORDER	New rules for Article 4 Directions: LPAs may make Directions in Conservation Areas without reference to Secretary of State in certain circumstances.

Table 2.2 presents major British legislation (in upper case), and also includes some relevant non-legislative publications such as Circulars of the Department of the Environment, together with legislation and publications that have some indirect impact upon conservation (in lower case). The table covers a period similar to that covered by Newcomb, beginning with the Civic Amenities Act 1967, which introduced the concept of area conservation into British planning legislation, even though this was foreshadowed by Circular 51/63 of the Ministry of Housing and Local Government in 1963.

With the inclusion of some of the more peripheral Acts, the meaning of 'conservation' is, in a sense, being broadened. Little work exists on the relationship between conservation and employment or industry, but there is a considerable body of work relating conservation to housing provision (Maguire 1980; Thomas 1983). This is a reflection of the noticeable change in emphasis over the last two decades or so from comprehensive redevelopment towards rehabilitation. This policy of rehabilitation is also carried over into more general legislation such as the British Inner Urban Areas Act 1978.

A consideration of recent rehabilitation policies does show that more finance is available for conservation and related work than is provided by the purely conservation-related legislation. This legislation itself tends to be largely permissive, allowing local authorities to designate conservation areas, but is not useful in the provision of finance for conservation work. Conservation grants are available, for example under the Local Authorities (Historic Buildings) Act 1962, and under Section 10 of the Town and Country Planning (Amendment) Act 1972. Yet the total finance thus provided is minimal. Grants under the 1962 Act are discretionary and not mandatory, and the policies of individual authorities vary widely (Yeomans 1974; Wools 1978). It was estimated in 1974–5 that the total amount spent on building conservation was between £3 million and £3.5 million, while at present, although government allocations rise roughly in line with inflation, the total available per listed building is under £50 (*English Heritage Monitor* annual). However, the inclusion of rehabilitation funds allocated under the Housing Acts for improvements in designated Housing Action Areas and General Improvement Areas would make the total effectively spent upon conservation rather greater. Moreover, some local authorities are adept at using these related sources of finance to fund conservation, while others still rely on specific conservation allocations.

Table 2.2 does show some general trends in the legislation. Superimposed on the period 1967–79, which in its general legislative support of conservation may be akin to Newcomb's first enactment phase, there are two non-legislative measures that have had considerable implications for conservation and finance. The first is the informal agreement between government and the building societies in 1973 to give preferential treatment to potential purchasers of new property. However, not all potential purchasers require new property, and 'older houses are worth retaining, for they are usually soundly built, offering better space and flexibility. Many have intrinsic qualities of character and townscape value, and some are of real architectural merit'

(Freeman 1982). Furthermore, much of Britain's housing stock consists of older property: 31 per cent being built before 1919, and 22 per cent between 1919 and 1944. Nevertheless, owing to building society lending restrictions, 75 per cent of first-time buyers are being forced to purchase newer houses than they desire, as the building societies are biased in favour of new houses and against older building, particularly listed buildings, buildings requiring repair, housing in inner cities and converted buildings, especially flats over shops (Alliance Building Society 1978; Freeman 1982). These lending restrictions, which may take the form of 'red-lining' entire districts, are significant since the building societies are the largest source of housing finance in Britain (Freeman 1982; Williams 1978). Thus, although social support may favour conservation and rehabilitation during this period, the economic climate was unfavourable, and private finance for older and/or historic buildings was difficult to obtain. This was an important consideration when, as has been shown, financial support from both local and national government was minimal. The second significant non-legislative measure was the interpretation of Value Added Tax (VAT) regulations by HM Customs and Excise in 1977. Alone among European Community countries, Britain penalised restoration and encouraged alterations to buildings by levying VAT on maintenance but not on new building: a potentially unfavourable trend when finance was scarce (Figure 2.1).

As the general economic recession took hold in the late 1970s, the trend is perhaps best shown by the 1979 Ancient Monuments and Archaeological Areas Act, whereby grant monies may be recovered if the property is sold within a given period, or if the grant terms are contravened. There are also similar provisions for the recovery of housing rehabilitation grants. The 1980 Act and its Amendment Act of 1981 continued this trend to some degree, in that they are not as wholly pro-conservation as were earlier Acts. The 1980 Act gives the building owner some security against listing, if an application is made to the Secretary of State for the Environment for a Certificate of Immunity from listing: a declaration that the building will not be placed on the Statutory List for a five-year period. There is thus some similarity with the US National Historic Preservation Act Amendments 1980, but this would represent a dubious parallel, since there is still no right of appeal in Britain from the Secretary of State's decision to list a building (and, if a Certificate of Immunity is not granted, the building in question is automatically listed!). The 1981 Act was not anti-conservationist, but in fulfilment of Conservative election pledges to cut central government controls and to speed the planning process, Schedule II of the Local Government Planning and Land (No. 2) Bill proposed the removal of dormant powers of the Secretary of State relating to conservation areas. Critics of the Bill protested that this might lead to local authorities feeling that conservation areas were unimportant. Later proposals were to reduce planning controls outside conservation areas, rather than to see these designated areas as having more planning controls than is usual. This may seem a pedantic point, but 'it would be unfortunate if areas not blessed with designation came to be thought of

Figure 2.1 The implications of levying VAT on restoration but not on alteration: VAT *is* levied on repair/restoration (A) but *not* on conversions and extensions (B) (reproduced with permission from SAVE Britain's Heritage 1980)

as expendable' (Cantell 1981). This trend in legislation, whereby conservation is not as favoured as was previously the case, does not continue to the present, as the National Heritage Act 1983 makes some provision for grant aid for the restoration and purchase of historic buildings. Further, the 1984 Budget added VAT to general building work (previously zero-rated), although an amendment to the Finance Act in May 1984 exempted listed buildings (although not unlisted buildings in conservation areas) from this increase. This tax regime is thus a disincentive to building owners and conservation areas (Nelson 1991).

In the case of both Great Britain and the USA, therefore, detailed examination shows a progression of legislation, although hardly a 'cycle', in which the details of legislation are considerably influenced by wider pressures. Originally these were more social, for example when Duncan Sandys MP originated the Civic Amenities Act in 1967. More recently, however, economic considerations appear to have had a greater effect upon legislative action affecting conservation. Governments appear to have attempted to minimise the direct use of state finance and to encourage private efforts, and there was considerable pressure to minimise expensive delays in acquiring the appropriate permissions by streamlining the planning system.

RECENT TRENDS IN GREAT BRITAIN AND THE USA

Most recently, the dictates of current US planning practice, and important court decisions, have come to be of greater significance than national legislation. The strict implementation of special district zoning regulations, Historic Preservation Controls, Transfers of Development Rights from historic buildings to other developments, and Environmental Impact Statements have become important (although they still treat entire urban landscapes in terms of fragments of historic significance) (see Boyer 1987; Knox 1991). The move towards mixed- and multi-use land zoning has been seen by planning agencies as attractive, as it may initiate urban revitalisation and thus enhance a city's tax base (Lassar 1989).

> Increasingly, mixed-use downtown zoning is becoming combined with incentive systems in which developers are awarded additional height or density allowances in exchange for specified building features (Seattle, for example, offers density bonuses for ornate building cornices and rooflines).
>
> (Knox 1991: 188)

These techniques may assist new buildings to blend with historic landscapes, and play a part in the adoption of post-Modern architectural styles. But many of the favourable tax incentives developed during the 1970s vanished during the Reagan era and have not reappeared.

In Britain, social attitudes towards architectural design and conservation are changing rapidly. The direct intervention of HRH Prince Charles in this debate, by a series of provocative speeches, a television programme and book, and the response of the architectural and planning professions to this, have raised the public awareness of these issues in a manner that hardly seems possible in the USA. Jencks (1988) discusses the Prince's architectural speeches, while his own views are given in HRH Prince Charles (1989). Hutchinson, then President of the Royal Institute of British Architects, counters this attractive and lavishly illustrated volume with a small paperback, densely and confusingly written, and containing not one illustration (Hutchinson 1989). To show but one example of increasing public awareness of these issues, a series of efforts by the development company London and Edinburgh Trust (LET) to redevelop the Bull Ring market area in central Birmingham has led to public meetings and protests, the formation of a pressure group, Birmingham for People, which has prepared alternative plans for the site (Figure 2.2), and the distribution of a publicity brochure describing the then-current proposals to some 44,000 homes in the city by the local planning authority. LET's first proposals were for a single monumental building, and these have been amended in line with public opinion to form a series of smaller buildings and a re-creation of part of the mediaeval street plan, obliterated by 1960s redevelopment. By 1995 the proposals had again been amended, with extensive public participation, although the retail recession had postponed actual development.

CONCLUSIONS

This chapter has examined the linkages between wider social, philosophical and aesthetic ideals and contemporary legislative and planning ideals in the history of urban conservation as a theoretical and practical concept. Drawing on a wide variety of examples from several countries, a distinct trend in conservation-related thought and action is discernible. It was not until the late eighteenth century that there was an increasing awareness of landscape, and particularly of its picturesque and romantic aspects, and even then this was shown only by a small social élite. Classical architectural styles, and their Georgian derivatives, were being derided in several countries; mediaeval styles, particularly Gothic-inspired, were finding increased favour. Generally, by the mid-1800s there was a concern for the identification, listing and preservation of 'ancient monuments', which lasted to the end of the nineteenth century, and even later in some cases. This listing formed the basis of much of the first phase of conservation legislation.

The main concern for historicist styles has continued through to the present with various revivals, notably of neo-Tudor and neo-Georgian, to the historicist aspects of current post-Modern styles (Jencks and Chaitkin 1982). Interest in these revival styles and in conservation generally is now being shown by a large proportion of the population. Modern-styled architecture, the antithesis of historicism, had a wide spatial impact, especially in the post-Second World War period, as this clean, up-to-date, easy-to-construct style was adopted by many leading property developers (Bentley 1983, 1984), and spread throughout the historical city centres of Europe and beyond. Its

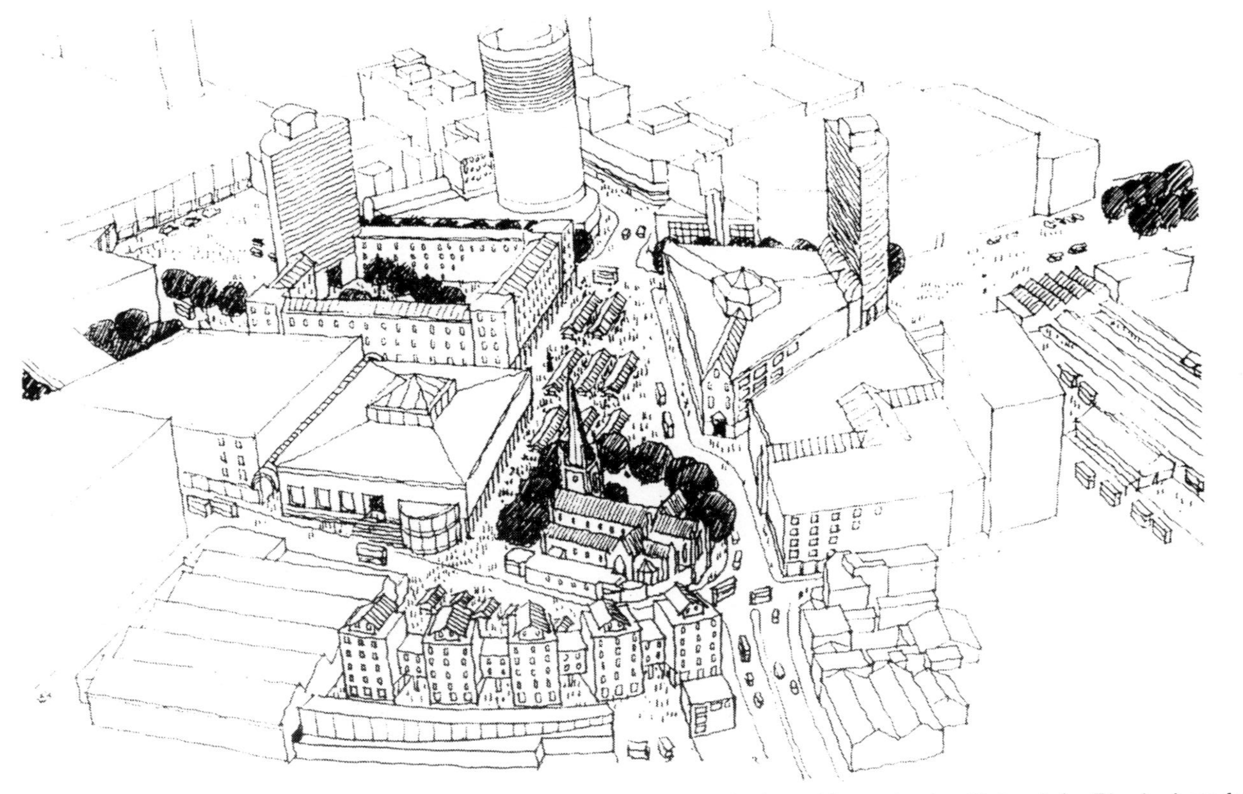

Figure 2.2 An alternative view of the redevelopment of the Bull Ring area, Birmingham (drawn by Joe Holyoak for Birmingham for People, reproduced with his permission)

limited temporal distribution and rapid displacement by post-Modern in the 1980s suggests that the Modern style did not have a profound cultural significance. Meanwhile, the accumulation of experience with the first phase of simple legislation led to various modifications early in the twentieth century.

Immediately following the Second World War, there was a period of demolition and rebuilding in many countries. The rebuilding took many different forms, both conservationist in seeking to reconstruct vanished townscapes, and Modernist, in seizing the opportunity to build new urban infrastructures. Even as this process was in full swing, a variety of countries considerably updated their systems of conservation legislation. Public interest in conservation and preservation has continued to grow, fostered by events such as the founding of the Civic Trust in Britain (1957), European Architectural Heritage Year (1975), the Bicentennial celebrations of the USA (1976) and even by archaeological events such as the televised raising of the *Mary Rose* (1982) and the uncovering of the Rose Theatre remains (1991). There have been few significant modifications to the planning legislation of the 1950s and 1960s, but a continual series of smaller changes has developed through to the present. A detailed examination of these recent changes in the USA and Great Britain reveals their increasing relationship with large-scale economic, rather than social, trends. The social atmosphere of concern for conservation has hardly changed direction over the past two or three decades. If anything, it has increased in strength. Nevertheless, some legislation of the early 1980s would seem to indicate a reduction in official support for conservation, at least as expressed in the availability of financial assistance from governmental sources. Some dissatisfaction with the second phase of legislation is also becoming evident, for example in France and the Netherlands. The continuing social expectations may bring about a third phase of legislation, more strongly coercive than the previous largely persuasive phases but, owing to unfavourable economic circumstances, relying little upon State finance. The economics of conservation are now more the economics of development and project management than State financial assistance (Lichfield 1988).

Thus some form of cyclic process may be seen in operation, in addition to an absolute overall growth in attitudes in favour of conservation. It has been suggested that planning in general follows the Kondratiev cycle of approximately 25-year booms and 25-year slumps, and that the present concern for conservation, historicism and even post-Modern architecture can be seen to echo the conservationist concerns of the late nineteenth century (Sutcliffe 1981a). This does appear to be true on a large scale, but at smaller spatial and temporal scales the time-lags in the diffusion and widespread adoption of innovative ideas through social strata and between countries distort this pattern. The ideas of Camillo Sitte in the 1880s thus took some time to spread, and similarly the questioning of the entire Modernist attitude towards historic city planning by Italian architects such as Muratori and Rossi is still little-known outside Italy, since translations of even major works may take up to 16 years (Collins and Collins 1965; Samuels 1985). The study of legislative development in Britain and the USA suggests that some quite small-scale, but nevertheless significant, legislative fluctuations may be apparent, making the generalised depiction of large-scale trends appear somewhat simplistic. However, this is not to deny the existence or conceptual utility of trends apparent at larger scales.

Owing to the complexity of the processes shown at work, there remains wide scope for further detailed studies of the factors influencing the timing and nature of planning legislation. In the case of conservation, the historical significance of changing social trends has been suggested, with economic factors becoming significant during the recession of more recent years. Also of interest is the importance of certain key personalities in the instigation and development of legislation. These key figures appear, in many cases, to act as catalysts in the reaction between changing social expectations and legislative action, and the differences in the timing of action between different countries may owe as much to the presence or absence of such a catalytic personality in a position of appropriate influence as to the differing social histories and the undoubted lags in communications between countries.

3

THE SPREAD OF CONSERVATIONISM: THE BRITISH EXPERIENCE

> The conservation movement is an expression of the concern that material progress should not be self-defeating.
>
> (the Burton St Leonards Society, Sussex, 1972)

INTRODUCTION

This chapter develops directly from Chapter 2 in its consideration of the rise of conservationism as a movement. Beginning in élite social concerns, during the twentieth century there has been a significant rise in the numbers and membership of local amenity societies, particularly following the foundation of Britain's Civic Trust in 1957. This closely parallels the rise of local historical societies in the nineteenth century. Interest in revival architectural styles and in conservation in general is now being shown actively by a large proportion of the population: this is shown, for example, by the numbers of housebuilders specialising in 'Heritage Cottages' and 'Medieval Oak Homes', and advertisements for these relatively new specialisms are common in the fashionable coffee-table magazines catering for old houses and interior design (Figure 3.1). Although some of these magazines ceased publication in the recession of the late 1980s, in the mid-1990s they are beginning to be replaced with, for example, *Perspectives on Architecture*, a populist built environment magazine linked to the new Prince of Wales's Institute of Architecture. Heritage issues generated their own BBC series, 'One Foot in the Past', in 1993.

The commercial interest in the past – its 'commodification' – is well shown by the Past Times company. This highly successful company markets what its catalogue describes as 'fine and unusual gifts inspired by the past'. Some, such as some of the jewellery, are accurate replicas; others, such as the cardigan based on a seventeenth-century Mughal wall-hanging, are more distanced from their historic original. Originally primarily a mail-order business, the company now has 53 retail shops located in key historic cities (except, perhaps, Preston) and at Heathrow and Gatwick airports. In this instance, commodification is clearly popular.

The rise of heritage tourism, heritage theme parks and similar manifestations

Antique Buildings Limited

DUNSFOLD SURREY – TEL: 048 649 477

We Specialise Exclusively in Ancient Timber Framed Buildings

We are architectural consultants who specialise exclusively in the restoration, extension, conversion, dismantling and re-erection of ancient oak framed buildings. We have received six conservation and historic building, design awards for our work.

Our immense stock of ancient oak beams (over 15,000 cubic feet stored under cover) handmade bricks, peg tiles and walling stone is second to none.

We have over thirty barn, cartshed granary, hovel and house frames available for re-erection.

Each building has been painstakingly drawn, sketched, photographed and numbered before being carefully dismantled.

Whether we supply a beam or a barn we are always pleased to offer professional advice.

POUGHLEY BARN - Circa 1650 Is just one example of the splendid oak framed buildings which we have available for re-erection.

New Oak Homes Built Traditionally by Craftsmen

Authentically styled half-timbered homes at prices comparable to modern housing. Built to your personal specifications and using traditional construction methods, your new home could be a unique and magnificent example of this architectural heritage and as individual as you.

For further information contact

Medieval Oak Homes

Judd House, Ripple Road, Barking,
Essex IG11 0TU.
Telephone: 0277 651196

A PART OF THE SILCOCK EXPRESS GROUP

Figure 3.1 Advertisements for old and new timber-framed buildings, from *Traditional Homes* magazine, 1990 (reproduced with the permission of Antique Buildings Ltd and Medieval Oak Homes) also shows the level of public interest in the past: it has been estimated that a new museum opens every week in Britain. But we are also criticised as being a backward-looking country of museums (e.g. Hewison 1987; Lumley 1988). This great rise in heritage-related interest leads to a concern for understanding who is involved in the process of conserving the townscape, and how and why they become involved. Studies of local amenity societies, admitted as representatives of the general public within the British planning system, show a vociferous, well-educated, minority: another conservation élite. Whose heritage is being conserved, and for whom?

The British official conservation systems reinforce these concerns. Listed buildings are designated according to national, not local, criteria, and designation is by a small number of well-educated people: it is very much a 'high Art' approach, frequently unresponsive to local wishes. The designation of conservation areas is similarly erratic; sometimes by detailed survey, sometimes *ad hoc*; rarely are the public consulted. The take-up rates and local problems caused by these two official systems will be examined.

THE CONSERVER SOCIETY?

The concept of conservation is now an accepted part of urban planning in most developed countries. This is a reflection of the widespread interest in aspects of the past, and how it is viewed, used and changed (Lowenthal 1985). There now exists a 'conserver society' that creates its own landscapes and which is particularly manifest in the rapid growth of the conservation movement (Larkham 1988a; Relph 1982). This implies that conservation is having a significant effect upon urban form, in urban landscapes ranging from central areas to high-class residential suburbs and industrial areas. The current concern with conservation appears to be allied strongly with the post-Modern reaction in architecture against bland Modernism, which has been seen by many critics as eroding the unique attributes of places by the imposition of uniform building types and house styles in new materials alien to the locale. In Britain at least, interest in conservation seems also to be a reaction against the trend in post-Second World War town planning towards comprehensive clearance and redevelopment (Esher 1981), whereby large areas of often historical town cores were redeveloped at a scale rarely seen before, except after natural or man-made disaster. The effects of this comprehensive redevelopment have, in fact, been compared directly, and unfavourably, to the devastation of war (HRH The Prince of Wales 1987).

Conservation is one reaction, particularly common in the late twentieth century, to the problem of ageing urban landscapes. The production and maintenance of the physical fabric of the urban environment absorb a large amount of the wealth of the developed world. Substantial problems arise as townscapes age and as the social and economic conditions under which they were created change. Adaptation or renewal of the townscape becomes necessary, but this is hard to achieve without some wastage of the investment of previous societies. The profession of town planning has developed during the twentieth century and has a leading role to play in addressing these issues. However, the dominant paradigm in the planning profession and the planning system – their ideological and sociological structures (Simmie 1981; Low 1991) – has often favoured radical urban reshaping rather than gradualist adaptation. During some periods, for example the 1950s in many Western countries, this wastage received relatively little consideration, as the philosophy of comprehensive clearance was dominant. During the 1980s, in contrast, it seems that interest in conservation in Britain was at its highest for a century (Sutcliffe 1981a). This change of attitudes has helped to highlight the problem of accommodating the requirements of present societies within the townscape legacy from previous generations.

This interest in the past has, indeed, reached such a point that some critics are protesting strongly against the concept of conservation in principle, for example in terms of the way in which the forms and uses of the built environment should develop (Price 1981). Britain's 'museum-based culture' in its various manifestations is roundly criticised as being backward-looking, portraying an unauthentic, interpreted, sanitised version of the past (Hewison 1987; Lumley 1988). The new architectural forms and styles which have come to dominate in conservation areas were critically termed 'conservation-area-architecture' as early as 1974 (Rock 1974). Glancey (1989: 27) agreed, stating that 'a rather self-conscious hackneyed architecture emerged in the early 1970s, designed primarily to slip through local authority procedures with the least friction'. For the *Sunday Times*, 'too many new buildings ... are just formulaic, designed for committee decisions, and therefore either routinely mediocre or, on a good day, merely uncontroversial' (*Sunday Times* 1993: 24).

Since, as Whitehand (1987a) demonstrates, a number of factors are known to affect the urban landscape in a cyclical manner, it is important to assess the background to the current interest in urban conservation in order to suggest the long-term nature and impact of this trend on the urban landscape, which will be developed through detailed micro-scale case studies in Part 2 of this volume.

WHAT LIES BEHIND THE CONSERVATIONIST TREND?

Public attitudes

In general terms, the British public appears to be inherently conservative, strongly resistant to change especially on a large scale. For the urban environment, this is reflected in a number of books published from the late 1960s onwards; anti-development polemics whose titles contain emotive, value-laden words: *The Rape of Britain* (Amery and Cruickshank 1975), *The Sack of Bath* (Fergusson 1973) and *The Erosion of Oxford* (Curl 1977) are good examples of this literature. All are copiously illustrated chronicles of the destruction of old and historic buildings, and their post-war replacement generally by large, Modern-styled buildings; all are produced by local or

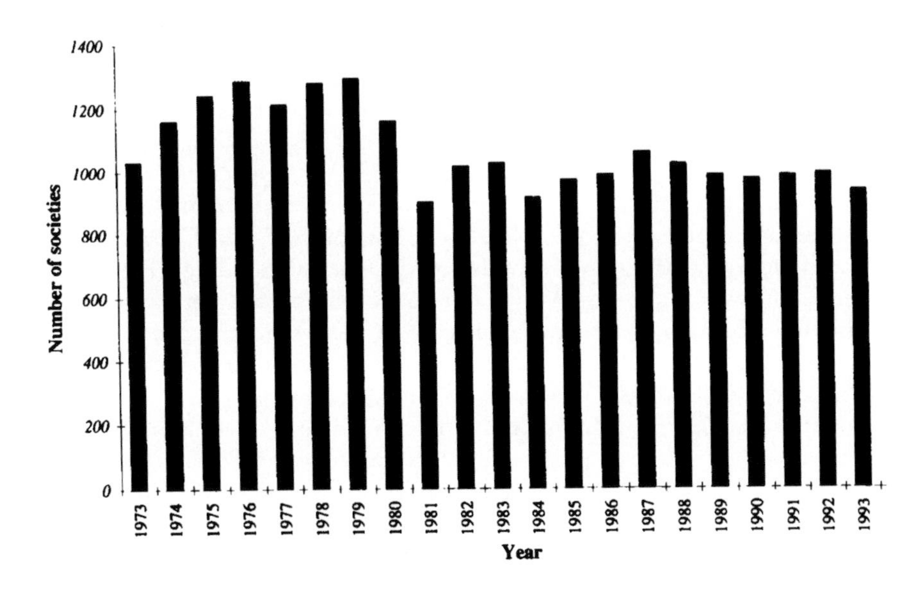

Figure 3.2 Local amenity societies affiliated to the Civic Trust (drawn from data provided by the Civic Trust)

minor publishers, but all have been influential in shaping the attitudes of the conservationist lobby. In conservationist writing, these are 'citation classics'.

Particularly since the late 1960s, the British planning system allows public involvement within the planning process, particularly through inspection of planning applications and participation in the preparation of Structure and Local Plans. In the majority of cases, public opinion is represented by local voluntary amenity groups, particularly Civic Societies. These societies have grown rapidly in number since 1957, when the national Civic Trust was formed (Lowe and Goyder 1983: their figure is updated by Figure 3.2). The number of these local societies registered with the Civic Trust reached a peak in the mid- to late 1970s, and declined slightly into the 1980s. It has seemed to some commentators that interest in 'the environment' in general, and as a political issue in particular, emerged quite suddenly during the late 1960s (Allaby 1971). However, a content analysis of *The Times*, chosen because of its reputation as a 'newspaper of record' and for its serious treatment of items, indicates that there has been a fairly constant level of interest in environmental matters throughout the 1950s and 1960s. Despite a marked rise in coverage *circa* 1969, this view of the 'explosive' growth of environmental interest must be qualified (Brookes *et al.* 1976).

It is useful to examine the spheres of interest of some of these societies. The first national amenity bodies in Britain were a product of concern in the mid-nineteenth century over the industrial and residential encroachment into common lands and historic places (Barker 1976: 17), and this is reflected in the first national body, the Commons Preservation Society of 1865. The circumstances of the foundation of the second body, the Society for the Protection of Ancient Buildings, by William Morris, have been mentioned in Chapter 2. There is a considerable gap between the SPAB in 1877 and the formation of the next national body concerned largely with architectural, as opposed to rural, preservation. The Ancient Monuments Society, founded in 1924, had interests far wider than architectural and urban conservation, including the conservation of 'fine old craftsmanship'. The Georgian Group was founded in 1937 to preserve the many Georgian buildings threatened by urban growth and, by the very nature of the Classical style of architecture and town planning termed for convenience 'Georgian', this group has often to concern itself with the preservation of entire townscape units – squares, terraces and crescents – as well as individual buildings. As time passed, architecture once thought modern and unworthy of preservation became the subject of conservation interest. In 1958 the Victorian Society was formed, similar in nature and aims to the Georgian Group, and owing much to the efforts of John Betjeman in publicising Victorian architecture. More recent still is the Thirties Society, whose sphere of activity grew to such an extent that, in 1992, it was renamed the Twentieth Century Society.

These societies claim to represent public opinion, but it is clear that they are directly representative, in terms of numbers of members, of only a small proportion of the population. This is particularly true of the local voluntary amenity groups. This proportion is greatest in smaller towns and villages, and minuscule in larger industrial towns (Table 3.1). Indeed, many of the criticisms of these societies are that they represent the views of a minority:

> the ever-present ancient establishment, the landed aristocracy, the products of Oxford and Cambridge, the landowners, the officer-class, and, behind them, their hangers-on: the trendy academics with less pretensions to gentility who prove their club-worthiness by espousing these élitist views.
>
> (*New Statesman* 1973: 146)

In other words, these groups are accused of representing the well-educated, vociferous élite, rather than the public at large (Crosland 1971; Eversley 1974). It is certainly true that surveys of membership have shown predominantly middle-class occupations and values (Barker 1976: 25–6; Barker and Farmer 1974). But it is characteristic that, in general, voluntary organisations are largely formed and supported by the middle class (Stacey 1960; Goldthorpe *et al.* 1969). The important question is whether the 'attentive public', in this case those members of the general public interested in conservation, is also predominantly middle class. Opinion poll research in the USA suggests that the attentive public is more socially representative than environmental groups themselves (Lowe and Goyder 1983: 12), and thus benefits from the activities of the 'fortunate minority'. It should, perhaps, not be surprising that conservation is caricatured as an élitist activity: it appears to be following a general trend of middle-class activity.

Table 3.1 Membership of amenity societies in the English Midlands

Society	*Population of settlement*	*Membership of society*	*% of population*
Hampton in Arden	1,500	518	34.5
Knowle	12,000	1,208	10.0
Moseley	10,000	776	8.0
Ludlow	9,000	586	6.5
Upton on Severn	3,800	221	6.0
East Lindsay Villages	4,000	108	2.4
Louth	13,000	241	1.8
Arkwright	14,000	215	1.5
Drayton	9,500	100	1.0
Burton on Trent	60,000	392	0.71
Uttoxeter	10,000	70	0.70
Vale of Evesham	15,000	90	0.65
Dronfield	26,000	156	0.64
Shrewsbury	65,000	375	0.62
Bromsgrove	90,000	397	0.45
Stafford	60,000	210	0.35
Nottingham	278,000	943	0.30
Chesterfield	90,000	107	0.12
Leek	36,000	45	0.12
Rugby	85,000	97	0.10
Penkridge	12,000	11	0.10
Wolverhampton	255,000	50	0.02

Source: Unpublished survey carried out by the Midlands Amenity Societies Association in 1990. Figures have been rounded

> In the second half of the twentieth century, what seems to be happening is that the middle class is adding to its traditional concern for others a lively concern for its *own* welfare. . . . It is developing organisations that are designed both to provide some kind of service and to appreciate critically what the ordinary consumer gets.
>
> (Broady 1968: 54)

Many of these local societies regularly comment on planning proposals, while representations made by individual members of the public are less common. Many critics, not least developers, complain that these amenity societies are wholly negative, anti-development, in their views. 'They have one common denominator . . . the lowest: that no change is always better than change, that their taste is always better than that of any architect or planner in the public service' (Eversley 1974: 15). Another critic notes that an 'élitist delight in the archaic' makes the destruction of an old building exceptionally difficult, and attributes this to a varied 'range of prejudice, conceit, ignorance, sloth and feeble thinking' (Price 1981: 39). Indeed, studies of the comments made by amenity societies are few, and the impact of these comments is almost impossible to measure. Nevertheless, an

Table 3.2 Comparison of Wolverhampton Civic Society's responses to planning consultations with Planning Committee decisions

Civic Society response	*Planning Committee decision*	*Number*	*Comment*
Refuse	Refuse	24	No conflict with LPA; but 2 refusals overturned at Appeal, so conflict with DoE
Refuse	Grant	13	Conflict
No comment/ no objection	Refuse	9	No major conflict (1 refusal overturned at Appeal: still no conflict)
No comment/ no objection	Grant	58	No conflict
Accept	Refuse	5	Conflict with LPA (1 refusal overturned at Appeal: no conflict with DoE)
Accept	Grant	60	No conflict
Accept principle but refuse on details	Refuse	7	No conflict
Accept principle but refuse on details	Grant	38	Conflict if proposal not amended

Source: Civic Society comments from minute books and planning correspondence in possession of the Hon. Secretary; Committee decisions from Planning Register

examination of the comments made by one society, Wolverhampton Civic Society, show that the Society advised against planning proposals in only 15.4 per cent of cases between January 1974 and June 1985; in only 18 of 214 cases was the Civic Society's view in significant conflict with the eventual decision of the LPA (Table 3.2; Larkham 1985).

Similarly, there are accusations from aggrieved applicants and others that this form of public consultation induces undue delay into the planning system. A further examination of the same society's responses to consultation shows that, between 1983 and 1989, the number of working days taken to respond varied between one and 35, with the average for the whole period being 17.3 working days. Although this is slightly longer than the period allowed by the LPA, it should be remembered that Wolverhampton Civic Society is a small group, entirely dependent upon voluntary assistance in making its comments. The delay caused by this consultation does not appear to be significantly large (Larkham 1990b). It is certainly below the average for consultation delays found by Simms (1978) in the only other significant study of this subject.

But the criticism of the NIMBY (Not In My Back Yard) view is hard to dispel. Nevertheless, some commentators are sympathetic:

> the emergence of NIMBYism is an understandable phenomenon after decades of development of which much was inappropriate, unsympathetic or environmentally unfriendly. The NIMBY armies have firmly established themselves at the district level of local government, in alliance with groups that do not oppose all development, but seek less of it, and of better quality. Such views impact directly on local authority planning decisions; they are politically important, and local political expediency demands that they are heard.
>
> (Wallis 1991: 106)

Published cases where public views, particularly of the societies themselves and the Civic Trust, have materially assisted in the delay or withdrawal of development proposals are more common than cases where negotiation has resulted in acceptable amended plans. For example, one developer noted of a site in Ludlow that 'unfortunately, it would appear that due to pressure from the Ludlow Civic Society the opinion of the Planning Officers and certainly that of the councillors was changed and our planning application was refused' (L. Greenall, Treasures Estates, pers. comm.: see Larkham 1992). It is also evident that much of public opinion is selfish, reacting only to immediate threats of development, and rarely taking a longer term or wider spatial view. This is confirmed by detailed examination of individual comments on planning proposals. One example in Wolverhampton for the extension and alteration of a large Victorian villa produced a storm of protest from neighbours worried about being overlooked, the possible devaluation of their properties, and increased traffic in narrow lanes (Figure 3.3). An examination of the proposals suggests that much of the new construction would be hidden by the dense tree screen surrounding the site, and would not be visible from the main road (A41 'The Rock'), which runs in a cutting at this point. It is unlikely that any of the residents in the north-eastern part of Clifton Road would suffer from any extra traffic. Some objectors cited the area's status as a designated conservation area as a reason for refusing development, without recognising that the area's original characteristic of large villas in extensive grounds had been eroded by demolition and subdivision throughout the previous decade. In fact, many of the objectors themselves live in 1970s houses built speculatively on such subdivided plots. It is evident here, and in many other examples, that newcomers living in new houses are most likely to protest about continued new developments. This, and many other similar examples, reveal a very self-centred sense of place. None of the residents appear to have considered that adjoining development may indeed enhance their own properties or property values (Larkham 1990a: 359–62). These cases have strong parallels in rural development, when vociferous newcomers attempt to ensure that theirs is the last new house to be built in a village, thus preserving its character and halting the influx of outsiders (Cloke and Park 1985).

It should be noted that, in the conservation-related field of ecology, an

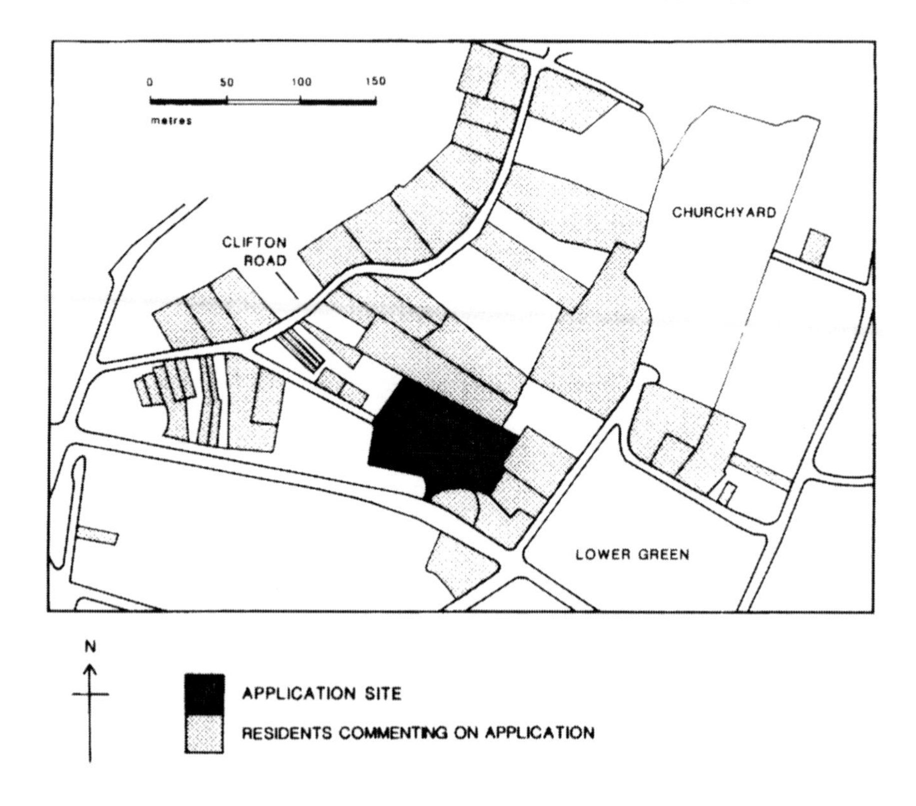

Figure 3.3 Origin of residents commenting on a planning application, Tettenhall, Wolverhampton

'issue-attention cycle' has been identified (Downs 1972). This suggests that a gradual build-up of public support for an issue may reverse quite suddenly as soon as other issues become prominent. Public interest in urban conservation and preservation between 1957 and 1975 seems to have mirrored the early stages of Downs's hypothesis (Figure 3.2). Whether this level of interest can be sustained, contrary to Downs's theory, is uncertain; indeed Figure 3.2 does show a decline in the number of societies registered with the Civic Trust in the early 1980s. What is certain, however, is that the removal of public interest in, and support of, conservation would rapidly lead to significant townscape change. Others have also warned of the dangers of conservationists over-reacting, or lapsing into an 'inglorious fetishism' (Smith 1975; Dix 1985).

Again heritage, as distinct from conservation, has a place in this discussion. The rise of heritage facilities chronicled in Hewison (1987) and of heritage tourism (Prentice 1993; Herbert 1995) suggests that many conserved sites, structures and areas can and must be viewed as heritage attractions. One of their major functions is to draw visitors (and, often, their money). Some are consistently successful in doing so, whilst others rise and fall in the listing of

Table 3.3 Visits to heritage attractions: the top 10 in England 1977–91 (paying visitors only)

1977	*1981*	*1986*	*1991*
Tower of London	Tower of London	Tower of London	Tower of London
3,089,000	2,088,000	2,019,000	1,923,520
St George's, Windsor	State Apartments, Windsor*	Roman Baths, Bath	St Paul's*
989,000	727,000	828,492	1,500,000
Stonehenge	Roman Baths, Bath	State Apartments, Windsor	Roman Baths, Bath
815,000	657,000	616,000	827,214
Roman Baths, Bath	Stonehenge	Warwick Castle	Warwick Castle
727,000	546,000	580,255	682,621
Hampton Court	Hampton Court	Beaulieu	State Apartments, Windsor
666,000	524,000	500,551	627,213
Shakespeare's Birthplace	St George's, Windsor	Shakespeare's Birthplace	Stonehenge
661,000	500,000	496,331	615,377
Beaulieu	Beaulieu	Stonehenge	Shakespeare's Birthplace
582,000	477,000	496,138	516,623
Ann Hathaway's Cottage	Shakespeare's Birthplace	Hampton Court	Blenheim
522,000	460,000	482,000	503,528
Warwick Castle	Warwick Castle	Leeds Castle*	Hampton Court
485,000	421,000	433,559	502,377
Brighton Pavilion	Salisbury Cathedral*	Tower Bridge*	Leeds Castle
428,000	358,000	419,003	497,528

Source: English Tourist Board (annual) *English Heritage Monitor*: note that not all tourist destinations responded to this ETB annual survey
Note: *Newly open, or newly making admission charges

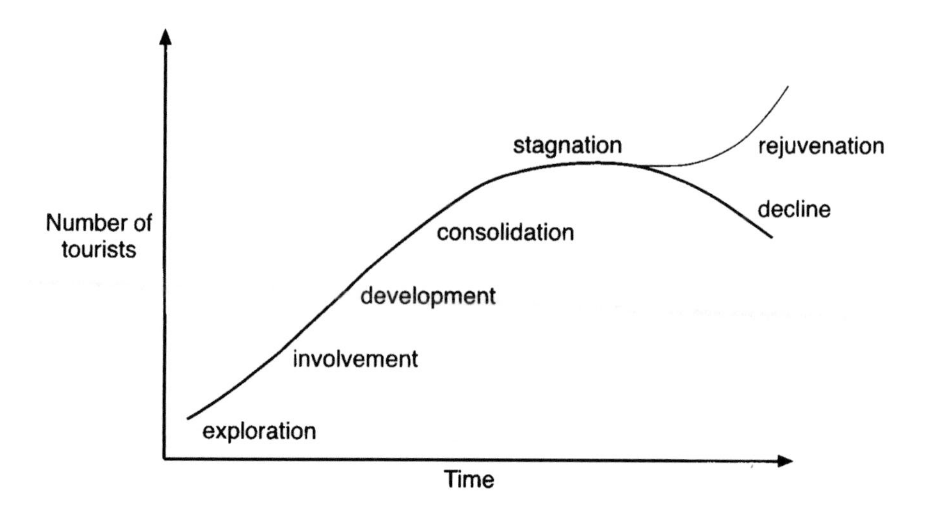

Figure 3.4 A cycle of the development of a heritage tourist attraction (redrawn from Butler 1980)

the top 10 heritage attractions (Table 3.3). But, therefore, they must also be subject to the same cyclical development process, of waxing and waning interest, as any other tourist attraction. Figure 3.4 suggests that decline can be halted by some form of revitalisation. But how far can attractions be revitalised or reinterpreted, and what further consequences will there be for the authenticity debate?

Legislation

Chapter 2 discussed the early history of conservation-related legislation. These early Acts had a major limitation: they were permissive rather than compulsive. The limitations were gradually dealt with throughout the twentieth century, most notably with the requirement of local authorities to maintain on behalf of the DoE/DNH a list of buildings of architectural and/or historical interest, and by the Civic Amenities Act 1967 which, for the first time in Britain, extended consideration from individual monuments and buildings to their wider urban settings. Action was spurred by a series of reports published in the 1960s, with the Ministry of Housing and Local Government (MoHLG) report *Preservation and Change* (MoHLG 1967b) noting the high and accelerating rate of urban redevelopment, and suggesting that conservation policies should be developed for sensitive areas. By 1970 the Preservation Policy Group reported that 'we do not think it would be an exaggeration to say there has been a revolution over the past five years in the way old buildings are regarded, and in the importance now attached by public opinion to preservation and conservation' (Preservation Policy Group 1970).

Even now, however, the provisions of the relevant legislation rarely compel action, they merely permit it. Legislation still lags behind public concern: it is notable that the major British legislation of both the nineteenth and twentieth centuries was put forward by key individuals rather than by governments or political parties. Moreover, 'policy' increasingly appears to be determined by key decisions in the courts or on appeal to the Secretary of State for the Environment (examples of such key cases are given in Chapter 4). Little finance is made available for local authorities to take the actions that are permitted to them, action involving significant expense, such as restoring buildings or enhancing areas, is thus relatively rarely undertaken by the authorities themselves (Chapter 6).

Actions permitted by legislation

The formal systems of conservation action in Britain permitted by legislation are the declaration of conservation areas and the listing of buildings. Conservation areas were introduced by the 1967 Act, defined as 'areas of special architectural or historic interest, the character or appearance of which it is desirable to preserve or enhance' (DoE 1987a). Areas must, therefore, be of *special* interest, and this interest must be specifically *architectural* or *historical*. The problems of some of these significant, but legally undefined, terms is discussed in Chapter 4. This new system was an explicit recognition that

> the public view is extending beyond buildings of architectural and historic interest to include others which, on the face of it, embody substantial resources of building materials already assembled to provide floor-space and which offer a type of accommodation which might not be provided on redevelopment.
>
> (Chapman 1975: 365)

Conservation areas are identified, delimited and designated by the LPA. It was originally felt that there would perhaps be some 1,250 designated areas. The rate of designation was initially quite rapid – quickly surpassing that estimate – with some slight diminution into the 1980s (Figure 3.5). It was noted in 1975 that analysis of the significance and distribution of conservation areas 'needs to await completion of designations' (Chapman 1975: 369). Yet Figure 3.5 shows that designations are still proceeding, with the total number in England in 1994 being over 8,300, and a further 1,100 or so in Scotland, Wales and Northern Ireland. The historic or architectural significance of these areas must now be questioned, particularly for the more recent designations. Although ideas on 'conservation-worthiness' change over time, can it be said that the areas only now being identified and designated are as significant as the historic urban cores, designated very soon after the system was introduced?

> It cannot be contested that both the original purpose of many conservation areas and the precise character and appearance they were

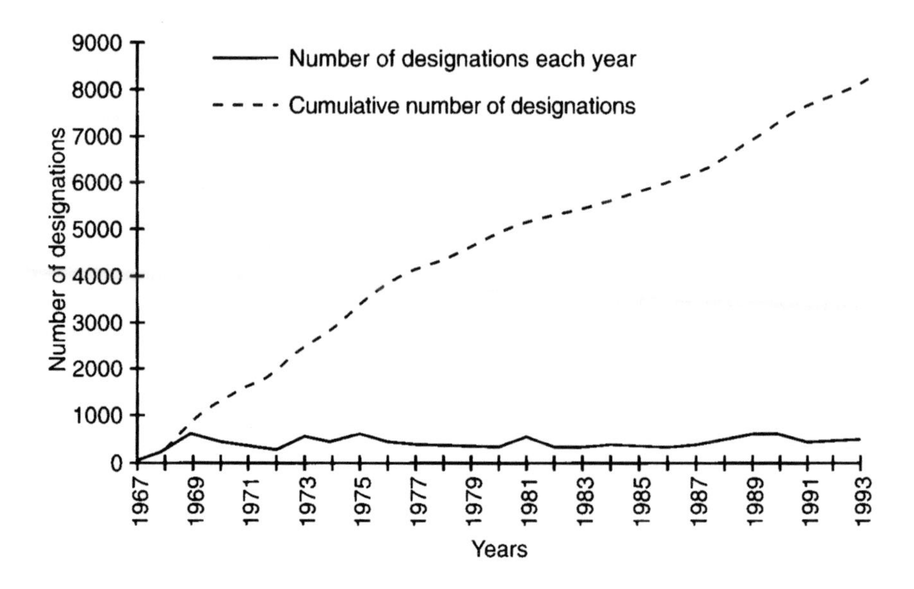

Figure 3.5 Designation of conservation areas in England (drawn from data provided by English Heritage)

> designed to conserve and enhance have been watered down by successive *ad hoc* designations in the name of additional control or local pressure. It is unfortunately the case that local planning authorities find it easier and safer to say 'no': little wonder that conservation areas are embraced for their additional controls.
>
> (Huntingford 1991: 2)

Systems whereby areas are identified vary greatly between LPAs, ranging from *ad hoc* delineations to very systematic surveys (Gamston 1975; West Midlands planning officers, pers. comms, 1980s). In cases where the public are consulted and offered alternative areas, they appear to choose the largest possible area for conservation, regardless of its intrinsic merit (West Midlands and Staffordshire planning officers, pers. comms, 1980s). Only in a very few known instances has public consultation resulted in reactions against designation.

Once designated, LPAs can exercise more control over elements of planning proposals such as materials, colours and architectural styles than they are permitted to do outside such environmentally sensitive areas (Punter 1986a). There is also a legal duty placed on LPAs to make proposals for the enhancement of these designated areas: a rather more active role than merely responding to the proposals submitted piecemeal by prospective developers.

Conservation areas are unevenly distributed throughout the country, with ten counties accounting for 40 per cent of all areas. It has also recently been revealed that rural areas accounted for 60 per cent of all conservation areas as at December 1986, whilst having only 23 per cent of the population. However, urban conservation areas contain far more listed buildings, encompass more areas nationally recognised as being of great value, and this was reflected in their preponderance in the former list of Outstanding Conservation Areas (*English Heritage Monitor*, annual; Pearce *et al.* 1990). Conservation areas are dealt with in greater detail in the consideration of area-based conservation in Chapter 5.

The listing of individual buildings is also determined by their special historical or architectural interest. In this case it is the Secretary of State for National Heritage who draws up the national list, after taking professional advice, rather than the LPA. National criteria are applied in the selection of buildings, and the list should thus include:

(a) all buildings built before 1700 which survive in anything like their original form;
(b) most buildings of 1700–1840, although selection is necessary;
(c) from 1840–1914, only buildings of definite quality and character;
(d) between 1914 and 1939, selected buildings of high quality only; and
(e) after 1939, a few outstanding buildings.

(DoE 1987a; Ross 1991: Appendix A)

Listed buildings are classified in three grades: Grade I, of outstanding or exceptional interest, of national importance, and being about 2 per cent of the total; Grade II*, particularly important buildings of more than special interest but not outstanding (4 per cent) and Grade II, the remainder, of special interest but not counted among the élite (Ross 1991). In total, by December 1992 there were 440,675 listed buildings, of which 6,068 were Grade I and approximately 23,000 Grade II* (*English Heritage Monitor* 1993: 8).

It must be remembered that this is a count of list items which may range from the Royal Crescent at Bath to a single milestone. If a count is taken of separate addresses, i.e. each house in Royal Crescent for example, then the total would be in excess of 500,000, and this is the meaning of the 'more than half a million listed buildings' which has been a widely published statement (Robertson 1993: 92).

A major problem with the listing of buildings is the application of a national scale of values to local contexts. What may be a significant building in a local area, whether as a rare local survivor or for its contribution to an urban landscape, may not meet the national criteria and may thus be denied the protection of listing. An example is the case of 64–70 Tettenhall Road, Wolverhampton. This was a terrace of grand early nineteenth-century town houses (Figure 3.6), which were not listed. Their owner had allowed them to decay for a considerable number of years, and a visual survey by consulting engineers revealed considerable dry rot infestation, to the point that 'we cannot advise at this stage that renovation be contemplated as this, in our opinion, is not a practicable proposition' (Engineers' report of survey,

Figure 3.6 Original elevation of 64–70 Tettenhall Road, Wolverhampton (redrawn from planning files)

planning files, Wolverhampton MBC). The buildings were purchased by a local developer with a good reputation for conserving buildings and after some campaigning by the developer, the LPA and the Civic Society, the buildings were listed in April 1990. The buildings are of considerable local merit, and their omission from the first survey of listed buildings has evidently contributed to their neglect and near demolition. Owing to the neglect and decay, little more than the façade was retained in a subsequent office development scheme (Figure 3.7), even with a willing developer and advice from English Heritage. Another facet of this problem is the suggestion that certain building types, unfashionable in academic architectural appreciation, may become under-represented in the statutory listings, as occurred with certain vernacular building types in Scotland (Horne 1993).

A further problem is that the criteria for listing buildings do change through time. It is only relatively recently, for example, that *any* inter- or post-war buildings have qualified (Bar-Hillel 1988a; see also Ross 1991: section 3.6). Lastly, there is the fact that the actual selection of buildings that may qualify for listing is carried out by an educated élite. It is a process more akin to the appreciation of other *objets d'art* than to any other aspect of town planning, although in the relisting survey and current practice, inspectors are trained to evaluate buildings according to a set list of criteria (Table 3.4). There is some evidence that the public do not value buildings in the same manner as those responsible for the selection of buildings for listing and, although a didactic role may remain, this reinforces the view of conservation as an élite activity rather than one for the benefit of the general public (for parallels see Carlson 1978; Bishop 1982). The actual decision to list is taken by the Secretary of State for the National Heritage, an elected politician rarely trained in architectural or historical appreciation, and thus stressing the élite nature of the process.

Figure 3.7 Redevelopment of 64–70 Tettenhall Road: retention of façade, change of use to offices and extension; completed 1993 (author's photograph)

Listing does add an administrative layer of protection to a building, associated fixtures and its curtilage, but nevertheless consent is granted each year for a number of demolitions. However, the rate of demolition of listed buildings has fallen in recent years to a 1992 total of 68 buildings: 40 per cent of the comparable figure in 1988, seemingly a product of the general recession, particularly in the development industry. Indeed, the rate of demolitions by 1992 had fallen to 20 per cent of the 1979 figure, despite the considerable increase in the number of listed buildings (*English Heritage Monitor* 1993: 8).

Building life-cycles

The amount of protection that conservation area designation or the listing of a building offers must be questioned. This attitude to conservation, the identification of areas and buildings, is typical of the first phase of legislation from the late 1800s onwards. One approach to the problem is to consider the life-cycle of buildings.

Bourne (1967) highlighted the idea that it is the age of the building stock, coupled with changes in function and economic influence through time, that lead to change in the urban landscape:

Table 3.4 Description and evaluation criteria for listed buildings

1. There is the need to convey a general, but not necessarily detailed impression of the appearance and character of the building.
2. There is the need to provide some indication of its worth as a 'building of special architectural or historic interest'.

Towards the former the notes should provide a concise and systematic description of the main facts of the building – of its history and appearance.

Towards the latter the notes should stress or emphasise those aspects of the building's history or appearance which are of more particular interest. The two purposes are best served by arranging the notes in a systematic and orderly way and by making them precise, extremely concise and objective.

B **Building type** – The original purpose for which the building was constructed (if known) followed by the present use (if different), e.g. Stableblock, now flats.

D **Date/s** – The different dates of construction as accurately as possible with the necessary explanation, e.g. Early C18, west wing 1850; or C13 restored 1875.

A **Architect/Craftsman/Patron** – The name or names will be taken to refer to architects if no profession is specified. The name should be written as fully as necessary. It should be noted if a person is connected with a part of the building only.

M **Materials** – These should be written in the order: structure, cladding, decorative treatment, roof, e.g. timber-framed with brick front, stone quoins, tiled roof.

P **Plan/Style** – Descriptive terms for both of these should be limited to those in common use.

F **Façades** – The building should be described from the ground up, main frontage first in the order storeys, bays/windows, door, roof shape.

I **Interior** – This should be limited to the briefest note of significant features which the listing would seek to preserve. . . .

S **Subsidiary features** – These are gates, railings, walls, urns, garden features, etc. Any of sufficient importance to merit listing in their own right should be itemised separately.

H **History** – This may be the history of the building, its association with well-known figures, or other relevant historical matter.

E **Extra information** – This might be any aspect of the building in the land or townscape. This should only be completed if the information is really relevant. If the building is primarily listed for group value this will be noted here.

S **Sources** – When relevant, sources should be given as briefly as possible and may refer to primary sources, contemporary secondary sources or reliable modern accounts.

Source: First appeared in the DoE (now DNH) listing *Handbook*, 1979; the page reproduced as an illustration in Robertson (1993)

The stock of buildings in a city represents an ageing and declining asset. Thus, not only is the present structure increasingly unsuited for the demands placed upon it by the market, it is becoming physically less

> suited through age and abuse, as reflected in declining values and rates of investment return.
>
> (Bourne 1967: 45–6)

Most building fabric change is brought about by some form of obsolescence, an indirect function of the ageing of the property.

Obsolescence is not a simple condition, and many factors act to cause it. Five types of obsolescence that may affect buildings have been identified. These include structural, functional and economic, the most important categories, together with rental and community obsolescence (Lewis 1965; Sim 1982). Considering structural obsolescence, it is important to recognise that modern buildings are usually designed for short life-spans, houses for some 60 years, with shops and offices having slightly shorter lives. This life expectancy is rather shorter than the ground leases of many developments, and it is a period more closely related to financial amortisement (US Department of Treasury 1947, 1962). For example, in planning the redevelopment of the Bull Ring site in Birmingham, the London and Edinburgh Trust secured a 125-year lease, and were prepared to demolish and rebuild four or five times during that period. All buildings require constant investment to ensure adequate maintenance of a sound structure and to enable the building structure to perform within acceptable limits. As buildings age, more expensive repairs become necessary as materials age, weather or decay. If repairs are not carried out this deterioration will bring the structure below the performance levels which are found acceptable by even marginal users, and the structure will continue to decline until it is abandoned and demolished. Another alternative is that when performance falls below optimum, the property will be demolished and the site immediately redeveloped (Figure 3.8).

Functional obsolescence simply indicates that, for whatever reasons, the occupier finds the building no longer suitable. A building may become functionally obsolete in the absence of structural obsolescence. It may also be depicted as a process of decline (Figure 3.9), each of the steps in the diagram indicating the introduction of a procedure or technique that suddenly lowers the functional effectiveness of the building (Cowan 1963).

Figures 3.8 and 3.9 show that for both types of obsolescence, periodic maintenance or adaptation allow the structure to continue to perform acceptably; major rebuilding considerably extends the life-span of the building, while the absence of either process leads to neglect, decay, and demolition. This idea, which was originally developed with reference to individual buildings by Cowan (1963), may be generalised to the entire historic urban fabric (see Lichfield 1988: Figure 1.3).

However, the rise of the conservation ethic and, in particular, the legal implications of the designation of listed buildings or conservation areas suggest that these buildings and areas are, to a considerable extent, isolated from the more usual cycle. The increasing structural obsolescence and decay leading towards demolition should be halted and, indeed, in general terms

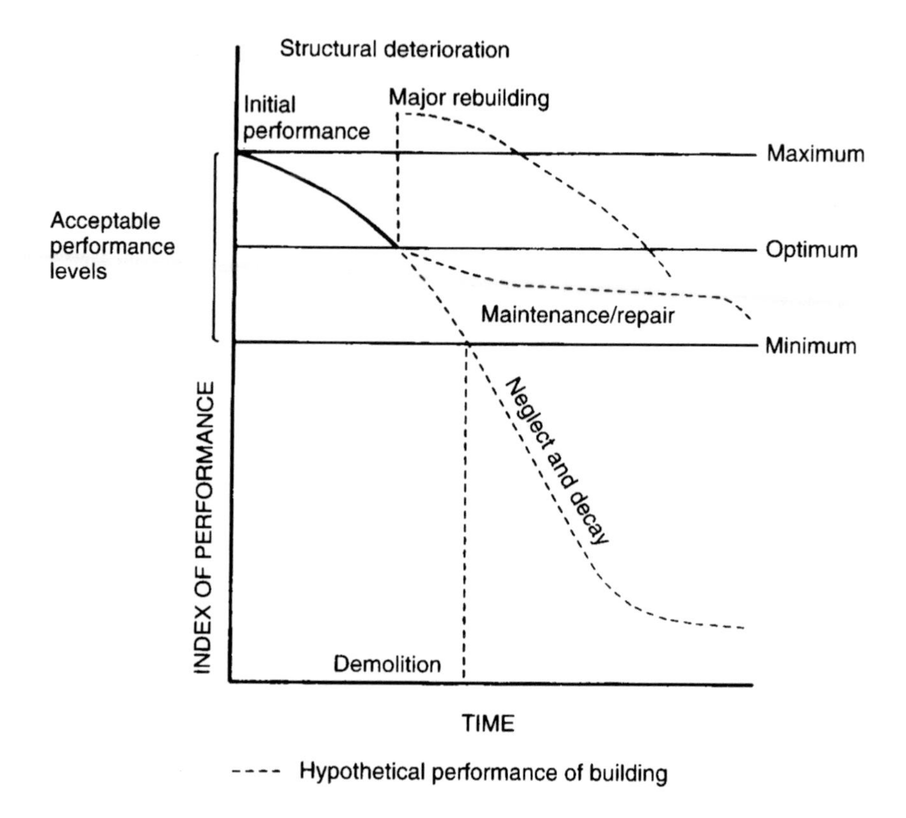

Figure 3.8 Structural obsolescence (after Cowan 1963)

there is a strong presumption against demolition. Regular maintenance, or rebuilding in the form of thorough restoration, should lead to the preservation of the building or area, or indeed its enhancement. Functional obsolescence is overcome by seeking appropriate alternative uses for older buildings: such uses may well entail changes, for example the conversion of large dwellings into apartments or offices; but the main part of the fabric, in particular the façade, should remain preserved.

In this manner, conservation prolongs the life-spans of buildings (the area in Figures 3.8 and 3.9 between maximum and minimum acceptable performance levels) beyond what would otherwise be expected. It is this prolongation that leads to periodic substantial expenditure, which is often, for public buildings at least, accompanied by appeals to raise money, as materials reach the end of their useful life. Lead roofs, for example, require re-casting every 200 years; soft sandstone decays over a slightly longer period (although a church in Wolverhampton, built in 1750, has over the past 20 years had all exterior stonework replaced owing to decay caused by pollution-induced chemical weathering). A good example of the nature and extent of repair to

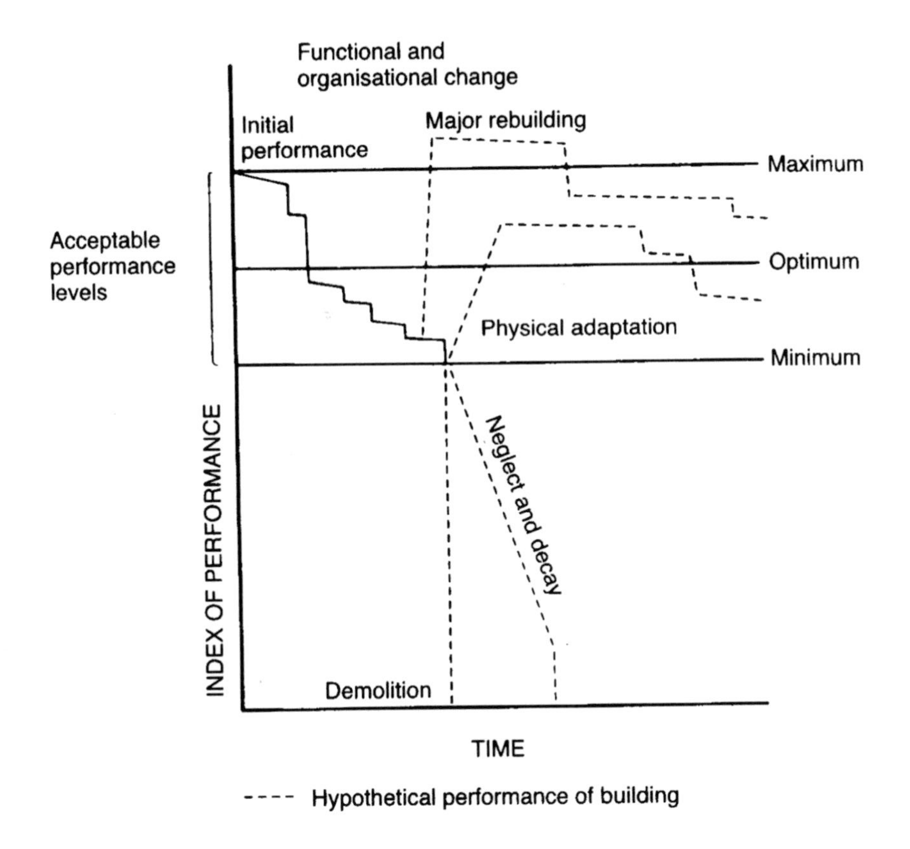

Figure 3.9 Functional obsolescence (after Cowan 1963)

traditionally constructed historic buildings is given in Oakeshott (1975), detailing the work necessary in Oxford after neglect during the years of wartime and post-war shortages of building materials. As the requirements of society change through time, this prolongation of building life-spans has serious consequences. However they may be adapted, some older buildings simply cannot accommodate many newer, often extensive, uses. If the character of areas is to be preserved by the retention of all, or a significant proportion, of the building stock, then the opportunities for new development to accommodate these uses in these areas diminish sharply. There is thus a strong tendency for such uses to migrate from the preserved historic city centre. Central business and retail functions, for example, may move from the preserved urban core; if other functions, with equal powers of investment in the urban fabric, do not replace these, then decay will take place despite conservation requirements. Buswell (1984a, 1984b) has begun a consideration of this problem of changing functional areas and the requirements of conservation; it also receives some mention in a recent examination of urban tourism, history and management (Ashworth and Tunbridge 1990). This process lies behind the creation of large new retail parks in edge- and out-of-town locations during the last decade, and the adverse effect felt in historic urban cores with the loss of investment. The high demand by developers for extensive edge- and out-of-town retail sites is shown, for the case of the West Midlands, by Larkham and Pompa (1989). The adverse effects of such developments on a particular smaller market town, Shrewsbury, are discussed by Martin and Buckler (1988). They show the economic consequences for buildings in the historic centre of new large edge-of-town stores which attract both custom and investment; the new town centre shopping centres attract some custom back, but at some cost to the historical urban fabric and to archaeological remains (Baker 1988), and there has as yet been little or no re-investment in many of the decaying historic buildings. On a larger scale, the scale and nature of the impact on numerous local traditional town centres of the Merry Hill Centre near Dudley is graphically demonstrated by Roger Tym and Partners (1993); some have lost 15–25 per cent of their trade over a decade.

Changing fashions in architecture

That architectural fashions change over time is well known, but little systematic attention has been paid to the styles preferred by the leaders in architectural taste during the last century, parallel with the rise in popular attitudes in favour of conservation. It is clear that their taste was predominantly historicist. This was briefly discussed in Chapter 2, with some examples, as a reflection of changing attitudes towards history and conservation. Recent studies of middle-class suburbia, particularly in the inter-war period, have shown how such originally élite tastes filter down the social hierarchy and are catered for by speculative builders (for example, see Barrett and Phillips 1987). During the same period, other speculative and bespoke developers used historical styles, and this is particularly noticeable in the new public houses built in the rapidly expanding Birmingham suburbs. There is also a striking example in Wolverhampton town centre of a pub rebuilt as part of a road-widening scheme in the late 1920s: its historicist style appears quite anachronistic amidst the contemporary 1920s classicist buildings surrounding it (Figure 3.10).

For commercial developments in urban centres, the setters of architectural fashion have been increasingly non-local, especially London-based, large firms that are driving out the local smaller developers and architects that shaped urban areas in the inter-war and earlier periods. To a considerable extent, they introduce novel architectural styles, especially Art Deco in the inter-war period and Modern in the post-war period. These styles are then adopted and further diffused by local firms only after some delay. An exception to this principle are the occasional owner-occupiers developing for themselves, who may seek to create a new impression with an eye-catching novel style (Whitehand 1984; Larkham and Freeman 1988). In the course of

Figure 3.10 Inter-war historicism: Giffard Arms, Victoria Street, Wolverhampton (1928–9) (author's photograph)

the last two decades, the ubiquitous Modern style has become displaced, particularly in commercial architecture, by 'post-Modern', characterised by the use of historicist styles, local idiom and materials, which in general relates far better to existing British urban contexts than did the Modern style.

These changes in preferred styles have had significant consequences for the urban cores that are now being conserved. Even the novel styles of the inter-war period were used in developments not, in general, of dissimilar size to the mediaeval plot sizes still common in many British towns. Post-Modern architecture appears to show a greater concern for context, in some respects. Extensive developments are given façades that attempt to give the appearance of comprising more than one building of traditional plot widths; while façadism, the retention of a front wall while developing anew behind it, is becoming common in conserved areas. Such techniques retain the visual appearance of historic areas although they may obliterate much of the historical and architectural significance of individual buildings. Although the utility of streets composed of historic front walls has been questioned, it is notable that several surveys have suggested that residents perceive urban landscapes on a superficial scale, with historical factors having little influence on judgements of context and style. It is also noted that the use of façadism is supported by the manner in which buildings are perceived: the public are rarely aware of, or concerned by, the architectural concept that the exterior of a building should relate in style, form and materials to the interior.

Changing fashions in town planning

Large-scale cycles have been observed in town planning in Britain (Sutcliffe 1981a). There have been up-swings in 1880–1914 and 1945–70, with down-swings between 1918 and 1939, and 1970 to the present. Such cycles are closely related to, and difficult to distinguish from, cycles in the free-market uncontrolled economy (Whitehand 1987a).

> Our present concern with conservation, rehabilitation, neo-classicism and even the pastiche, all of which the Modern Movement pretended to have seen off for all time, can now be seen to echo, at a distance of two full Kondratiev cycles, the conservationist and historical concerns of the 1880s.
>
> (Sutcliffe 1981a: 66)

Currently much planning, whether in terms of the guidance Circulars published by the DoE, or LPAs' responses to planning proposals, rejects the immediate past of the 1945–70 period, and constructively incorporates historicist elements into new planning and architecture.

Sutcliffe notes that planning professionals multiply in the up-swings when there is 'so much planning to be *done*' and, by contrast, down-swings are dominated by gifted amateurs and the like. Yet at present there is a tremendous amount of planning to be done, not least in the redevelopment of the errors of the last up-swing with its extensive clearance and system-built replacement; while the supposedly conserved urban centres are wilting through loss of investment to edge- and out-of-town sites. Certainly the amateur, in the form of the public and local amenity groups, has had some noticeable impact on planning since about 1970. How far this can continue, and to what extent public participation in planning is a myth, are hotly debated (Heap 1986).

Sutcliffe also notes the historical dominance of planning debates by 'ideal types' such as the Garden City or *Ville Radieuse*. Yet current planning practice is, to a considerable extent, dominated by another ideal type: the conserved city. The pervasiveness of this ideal is shown in practice by the spread of identical features throughout many diverse towns: identical design solutions, architectural details, cast-iron street furniture from the same specialist supplier, and so on. In the name of conservation, places run the risk of becoming placeless, of losing their uniqueness (see, for example, Ashworth and Tunbridge 1990: 199–200).

CONCLUSION

This chapter has examined a number of the major societal trends shaping the conserved urban landscape. This develops the idea expounded in Chapter 2 of the developmental process of conservation-related legislation and government guidance. Here, in contrast, are shown some of the trends underlying legislative change – broad movements in the tastes and activities in society, to which legislation eventually responds. That the development of UK conservation legislation has largely been driven by key individuals – with the 1882, 1967 and 1974 Acts all being sponsored by private members of Parliament – is one way in which legislation may respond more quickly to changing public tastes. The differing roles of the shapers of taste, legislation and policy, and those more passively affected by such changes, are important. So, too, is the perception that conservation is an activity undertaken by, and for the benefit of, a tiny but influential élite within society. Conservation thus poses considerable problems in terms of equity and social justice. A further important point is that conservation of buildings and areas does remove them from what has been shown to be a 'normal' cycle of decay and replacement, with attendant financial and social implications. But whatever its implications, whatever the costs and benefits for those directly and indirectly involved, the process of conservation is currently exerting a considerable direct influence on the urban landscape, most particularly upon its immediately visible aesthetic aspects. The next chapter will examine one facet of this, namely the inter-relationship between changing law and landscape.

4

CONSERVATION AND CHANGING LEGISLATION: IMPLICATIONS FOR THE LANDSCAPE

> Planning is the means of conservation; it is also the means of total destruction.
>
> (Pershore Civic Society, Worcestershire, 1972)

INTRODUCTION

In every country which attempts to conserve parts of its built form through legislative control, the detailed form and wording of that legislation, and the manner of its change over time, are crucial. Chapter 2 showed some generalities in the historical development of legislation, but this was illustrative of the general rising trend towards conservation as a field of interest amongst the public and relevant professions. Chapter 3 introduced some of the confusing legislative detail together with discussion of wider societal trends affecting legislative development. This chapter discusses some of the important minutiae. The precise and detailed wording of legislation, once passed, may seem trivial to the public, those whose involvement is generally limited to heritage tourism and building appreciation. However, to those deeply involved in the practicalities of conservation, words do matter.

The UK case is examined in some detail, since some of these relevant minutiae have recently come under intense scrutiny. The semantics of words, their dictionary definitions, and the meanings imputed to them by legislators, decision-makers and others have been discussed at length through the workings of the planning system by planning inspectors of the Department of the Environment when applications are taken to appeal. The UK planning system is a quasi-judicial process, and the legal processes of the High Court, the Court of Appeal and, on occasion, the House of Lords have also been involved in seeking resolutions to certain complex issues. Although the precise nature of the English legal system detailed here is not replicated everywhere, this section has a wider relevance in that the English system has acted as the model for many others, and the nature of these debates – on preservation and enhancing; the character of conserved areas; the value of listed and unlisted buildings and their potential replacements – is of much wider significance. It is also instructive to see how an established system deals with such complex issues, all of which are of relevance to the key questions posed in Chapter 1.

This quasi-judicial UK system requires brief introduction. The right to develop was effectively nationalised by the Town and Country Planning Act 1947. Planning permission was required from the LPA, although there is a right of appeal to the Secretary of State against a refusal of permission or unduly onerous conditions. Further appeal to the courts may be made only on a point of law. Some specified, relatively minor, changes are allowed without specific permission: these are 'permitted development' but, under some circumstances, even these may be brought under control. The planning system focuses most attention on external aesthetic issues in conserved areas. There are wider conservation-related factors which are sometimes at issue. For example, internal changes to historic buildings are sometimes controversial, while trees and planting are rarely directly mentioned, even in mature residential areas where mature landscaping is an important characteristic, although Tree Preservation Orders can be imposed. Generally, however, socio-economic issues such as changing property values and maintenance costs consequent upon conservation designation, and gentrification, are not seen as planning issues.

THE CIVIC AMENITIES ACT 1967 AND ITS PRECURSORS

A major legislative innovation occurred in 1967 with the consideration of areas of special architectural or historical interest, thus giving wider scope for the protection of the settings of groups of buildings or even individual buildings. This was not a completely novel idea, as it had been foreshadowed by the Town and Country Planning Act 1932, which provided for 'preservation schemes for buildings and groups of buildings', and even the provisions of the Housing Act 1923 for town planning schemes, which could prescribe the set-back, height and character of new buildings and which were used by a number of towns, principally the acknowledged key historic centres of Oxford, Winchester, Canterbury, Exeter and Stratford (Cherry 1974: 87; Smith 1974: 37). A number of local Acts, such as that for Bath in 1925, also gave specific local authorities powers similar to those of the schemes implemented under the 1923 Act. (Punter 1986b deals thoroughly with these early Acts, and aesthetic control in general in this period.) The instructions given to investigators listing historic buildings under the Town and Country Planning Act 1944 are also of relevance, since they suggested that one criterion for listing was membership in an 'accidental or pictorial architectural group' (see Harvey 1993). It was also stated that the listing might eventually result in the identification of 'towns and villages and areas of special amenity from the architectural point of view' (quoted in Smith 1974: 103). Various official publications of the early 1960s also took note of the importance of whole areas, particularly of town centres (MoHLG 1962: paragraphs 8 and 24; 1963: section 5). A Memorandum published in 1966 'began to discuss the techniques of analysis which were required to evaluate urban areas and prepared the ground for what were to become conservation areas' (Tarn 1985: 256). Interestingly, considering its later involvement, the Civic Trust introduced the ideas of areas of special development control in 1962, which would be defined on Town Maps in order to make clear to local property owners and potential developers that particularly stringent controls would be exercised (*Civic Trust Bulletin* 1962), although in this case the role of Duncan Sandys, the Civic Trust President, is unclear.

In 1957 Sandys, then Minister of Housing and Local Government, had set up the Civic Trust, against the advice of his civil servants and Ministerial colleagues. He said that at the time

> my civil servants nearly had kittens when they heard I was founding an independent society which would quite likely bring pressure to bear on the department. They went behind my back to the Prime Minister, Anthony Eden, who admitted it was a great idea but didn't think I could do it as a minister.
>
> (Sandys, quoted in Civic Trust 1988)

It is generally accepted that the genesis of the conservation area concept lies in the case of *The Earl of Iveagh* v. *Minister of Housing and Local Government* (Court of Appeal [1964] 1 QB 395). The case involved two adjoining terraced houses in St James's Square, London, owned by the Earl. Building preservation notices had been served on these houses on the grounds that their alteration or demolition would be detrimental to the square. This was challenged by the Earl: the contentious point in this case being whether a building should be listed for its intrinsic architectural or historical interest, or whether it might possess such interest merely because it was a part of a group. The Court of Appeal decided for the Minister, holding that 'a building might be of special architectural or historic interest by reason of its setting as one of a group', but the decision was not unanimous, and it was apparent that a more general power was required in the case of 'group value' (Suddards 1988: 45–6; Graves and Ross 1991: 108; Ross 1991: 30). This case led to Duncan Sandys drawing up the Civic Amenities Bill on winning first place in the 1966 parliamentary ballot for Private Members' Bills.

His Bill received support from Richard Crossman, then Minister of Housing and Local Government, and the Bill passed through Parliament with all-party support, although in slightly modified form (Smith 1974). The Act is fully described in Brown (1967), Smith (1969a, 1969b) and Telling (1967). Yet there was opposition, principally from the Permanent Secretary of the Ministry, Dame Evelyn Sharp. 'This kind of work [conservation and preservation] was utterly despised by Dame Evelyn. She regarded it as pure sentimentalism, and called it "preservation", a term of abuse' (Crossman 1975: 623). Incidentally, her strongly held personal views on architecture also led to clashes between the Ministry and professional bodies (Hillman 1990: 4). Sandys and the Civic Trust would have liked to include some mechanisms for control of the demolition of buildings in conservation areas, but the Ministry felt that this would be unacceptable and the idea was dropped (Civic Trust 1980a). The importance of Sandys personally in this legislative success was considerable (Cherry 1982: 70–1).

The new Act was thus largely permissive, and was essentially a 'declaration of interest'. One of the reasons for its easy and widespread acceptance was probably just that: it provided no new regulatory powers or onerous duties for local or central government. Local authorities were merely required, from time to time,

> to determine which parts of their area ... are areas of special architectural or historical interest, the character or appearance of which it is desirable to preserve or enhance, and shall designate such areas (hereafter referred to as conservation areas).
>
> (1967 Act, Section 1)

The word 'special' in this context, applied to areas, is vague, unlike its use in the case of listed buildings, where detailed guidelines do exist. This concept may ultimately have to be tested in the courts (Suddards 1988: 48). Likewise, there was no guidance as to what types of character or appearance it is desirable to preserve or enhance.

Nevertheless, the concept was accepted fairly rapidly by local authorities and the public, with the help of considerable publicity from the Civic Trust, to the extent that by 1980 only seven local authorities were still without any conservation areas (Civic Trust 1980a). However, it is perhaps significant that an authoritative and nearly contemporary review of conservation powers for planners (Layfield 1971) made no reference whatsoever to urban conservation save for Tree Preservation Orders.

The Ministry issued a Development Control Policy Note on the preservation of historic buildings and areas (MoHLG 1969). This brief Note contained the policy that

> the first consideration on any proposal for development in a conservation area must ... be its effect on the character of the area as a whole, and whether or not it would serve to 'preserve or enhance' its character. This would normally preclude large scale or comprehensive development schemes and the emphasis will usually be on the selective renewal of individual buildings.
>
> (MoHLG 1969: paragraph 6)

The Note also contained some guidance as to what a conservation area might contain; although this, the second-longest paragraph in the Note, consists mainly of rather bland statements.

THE POST-1967 SYSTEM

The 1968 Act

The Town and Country Planning Act 1968 (MoHLG 1968) contained the strong suggestion that local authorities should set up Conservation Area Advisory Committees. These have not generally proved popular (Ray 1978).

The relevant powers of Section 1 of the 1967 Act were re-enacted in successive planning Acts, principally as Sections 28, 29(4) and 277 of the Town and Country Planning Act 1971 (Mynors 1984) and, most recently, in the Planning (Listed Buildings and Conservation Areas) Act 1990 (Ross 1991). This developing legislation will be discussed chronologically.

The 1971 Act

The 1971 Act and the principal guiding Circular (DoE 1977a, but see also DoE 1972a) suggest that a conservation area is an area

> (a) of special architectural interest; or
> (b) of special historical interest;
> or, presumably, both. In any event, it must also be an area
> (c) whose character it is desirable to preserve;
> (d) whose character it is desirable to enhance;
> (e) whose appearance it is desirable to preserve; or
> (f) whose appearance it is desirable to enhance;
> or, again, presumably any combination of these.
>
> (Mynors 1984: 146)

However, there were no absolute criteria laid down for designation, save that, to be eligible, an area must be of special, not just some, interest. The effect of designating a conservation area is that 'special attention shall be paid to the desirability of preserving or enhancing its character or appearance' (Section 277(8), 1971 Act). Although local authorities were encouraged to determine which areas may be suitable for designation (MoHLG 1967a, 1967b), there was no over-riding obligation upon them to make designations. In practice, by 1990 each local authority had made at least one designation. The process of designation was, and remains, simple. Neither the 1971 Act nor its Circular (nor successive Circulars) contained further guidance as to what should be included, beyond that which was contained in paragraph 7 of Development Control Policy Note 7 (MoHLG 1969). It was noted that areas might vary in size between town centres and squares, terraces and smaller groups of buildings; they may centre on listed buildings; they may also feature other groups of buildings, open spaces, trees, historic street patterns and village greens (DoE 1987a: paragraph 54). Designation is made by the local planning authority.[1] Once a local authority has identified an area, the designation takes effect from the date of its resolution. Designation is implemented through (a) a notice in the *London Gazette* (or *Edinburgh Gazette* in the case of Scotland) and at least one local newspaper; (b) designation must be registered as a local land charge; and (c) the Secretary of State for the Environment and English Heritage must be informed (see Suddards 1988: 48–50). It is thus a simple, local and potentially rapid procedure: one from which there is no appeal.

There is no specific provision for the amendment or de-designation of an area, although this is implicitly allowed following the reviews 'from time to time' of areas. However, many areas have been amended through enlargement or amalgamation, and at least three areas have been de-designated. St Peter's Place in Birmingham was de-designated in 1976 following demolition of the church in its centre (Granelli 1973: 14) which, owing to another legislative quirk, being in ecclesiastical use was exempt from many planning restrictions and demolition control. Other de-designations are known in London and Liverpool, and it is known that a number of conservation officers have considered recommending de-designations. Yet the validity of amendment or de-designation is thus questionable until tested in the courts or clarified through further legislation (Ross 1991: 122).

Mynors (1984: 147) discussed the anomalies of designations being made by both district and county authorities.[2] He suggests that it would have been more sensible to allow only county authorities to make designations, which would have ensured some measure of consistency of designation quality, at least within counties. He noted that county councils were also more likely to be remote from merely local political pressures to designate (or not to designate) particular areas.

Extra publicity was to be given to planning applications affecting conservation areas (1971 Act, Sections 28 and 29(1)). What little provision for demolition control there was could be found right at the end of that Act, under the unpromising heading of 'Miscellaneous and Supplementary Provisions'.

These powers of designation and management are all to be exercised by the Local Planning Authority acting within the planning system as set up by the Town and Country Planning Act 1947.

Enhancement and the 1972 Act

More control was introduced by the Town and Country Planning (Amendment) Act 1972 (DoE 1972b), Section 10 of which introduced grants for those conservation areas designated as 'outstanding'. However, the method of designation of an area as 'outstanding' was not made explicit. The 1972 Act stressed the policy that 'conservation should always where possible be carried out on a self-financing basis, particularly by realising the enhanced value of improved property values' (DoE 1972b: paragraph 11). How conservation could be self-financing, particularly with regard to area-based conservation action, was not made clear. Section 8 of the 1972 Act also extended some control to the demolition of buildings in conservation areas, where the local authority had directed that the buildings in question should be subject to such control and where such an order had been confirmed by the Secretary of State. This was a clumsy mechanism, slow in operation and subject to possible abuse.

Article 4 Directions

Appendix D of Circular 12/73 (DoE 1973) dealt with Directions under Article 4 of the General Development Order, whereby a local authority can remove Permitted Development rights from certain types of developments, requiring instead a planning application. These have been used, for example, to control external painting and other finishes such as cladding, replacement of traditional windows by uPVC, and the conversion of walled front gardens to hard standings for vehicles (English Historic Towns Forum (EHTF) 1992 gives a good résumé of this argument).

> Although, in general, the Secretary of State will be favourably disposed towards approving an Article 4 direction relating to land in a conservation area, it must be emphasised that the existence of a designated conservation area is not, in itself, automatic justification for a direction. A special need for it must be shown.... The boundary should be drawn as selectively and tightly as possible, and should not automatically follow the boundaries of the conservation area.
>
> (DoE 1973)

This guidance was repeated as Paragraphs 4 and 5 of Appendix II to Circular 23/77 (DoE 1977a) and in Circular 8/87 (DoE 1987a). But the use of these Directions remains very limited. The EHTF (1992) suggested that there are only some 200 Directions in force in England, yet recent research revealed over 700 Directions, the great majority of which were conservation-related, used by only 89 LPAs (in a postal survey of a 50 per cent sample of LPAs: Chapman *et al.* 1995) and the policies of the DoE Regional Offices towards requests for Article 4 Directions are known to vary over the country and

Table 4.1 Resource implications of Article 4 Directions

	High	*Acceptable*	*Low*
Officer time to prepare	47	37	2
Officer time in enforcement	23	48	14
Officer time in public education	24	45	16

Source: Postal questionnaire of 50 per cent of local authorities (discussed in Chapman *et al.* 1995). Note that these are not absolute times but value judgements based on each responding authority's priorities through time. Several authorities have used quite a number of designations: Peterborough has the largest known number, with 102, while Glasgow has 42 and Liverpool 28. Yet many LPAs consider them to be time-consuming to prepare, difficult to secure the necessary approval and difficult to enforce (Jones and Larkham 1993) (Table 4.1).

The 1974 Act

A more substantial change was the Town and Country Amenities Act 1974 (DoE 1974) which replaced the clumsy and limited control of demolitions within conservation areas. This was another Private Member's Bill, passed between the two general elections of that year, albeit with government support (Ross 1991: 126). It significantly enlarged the scope of control within conservation areas by bringing the demolition of unlisted buildings in a conservation area within the remit of listed building consent requirements; although the prospect of applying for listed building consent relating to an unlisted building was a problem for subsequent years! It further noted the failure of many local authorities to set up Conservation Area Advisory Committees, despite the pressure to do so contained in the 1968 Act. Ray (1978) studied these committees, and found a variety of reasons why they were unpopular; an unpublished survey for the National Council for Civic Trust Societies in 1992 reveals a similar story. The 1974 Act also strongly suggested that local authorities should consider further conservation area designations, and put pressure on authorities who had made no designations to begin doing so, as

> the Secretaries of State [for the Environment and for Wales] appreciate that some authorities still have staffing difficulties following local government reorganisation, and must limit expenditure. Accordingly they have decided not to issue a direction under section 277 now. However, they believe that some authorities could have proceeded with designation more quickly and widely.
>
> (DoE 1974: paragraph 4)

The 1974 Act also set out general policy for conservation areas:

> Where any area is for the time being designated as a conservation area special attention shall be paid to the desirability of preserving or enhancing its character or appearance in the exercise, with respect to any buildings or other land in the area, of any powers under this Act.
>
> (Section 277(8), 1971 Act, as amended by 1974 Act)

This policy is considered important by the courts, as the case of *Richmond Borough Council* v. *Secretary of State* ([1978] 37 P&CR 151) shows. However, there is no definition of what constitutes 'special attention' (this point is discussed further below). Moreover, there has been considerable recent debate on the nature and scope of 'enhancement': early successful schemes removed accretions and clutter, with more recent schemes adding 'heritage bollards' and suffering from over-design (Booth 1993).

1977 Circulars and Regulations

Two official publications of 1977 consolidated earlier conservation guidance into a Circular (DoE 1977a) and the explanation of the operation of the listed buildings system, including applications within conservation areas (DoE 1977b). Paragraph 33 of Circular 23/77 (DoE 1977a) re-emphasised, in greater detail, the methodology of area designation:

> sometimes designation has been preceded by detailed and time-consuming surveys and by the preparation of a conservation policy. This work must be done but not necessarily before designation. A broad survey designed solely to identify an area as suitable for designation can be sufficient for this purpose. Authorities are asked to consider whether they could, within their present resources, designate further conservation areas, if necessary after a brief preliminary survey.

Yet this approach would appear, at least from an academic point of view, to be fraught with problems. The findings of broad and brief preliminary surveys may differ from detailed study, and area boundaries might require refining. The broad survey may omit items of merit, discoverable only during detailed study. The paragraph appears to castigate attention to detail, calling it 'time-consuming', without consideration that an early detailed survey may save time at a later date. A recent report for the Royal Town Planning Institute is also critical of this point, suggesting a comprehensive character appraisal as a prerequisite of area designation (Jones and Larkham 1993).

Changing attitudes in the 1980s

The Local Government Planning and Land Act 1980 (DoE 1981a) abolished the 'outstanding' category of conservation area introduced by the 1972 Act for the purpose of Section 10 Grants. By 1980 some 551 areas had received the 'outstanding' classification. However, Historic Scotland still uses this appellation: by 1994, of 574 conservation areas in Scotland, 204 were designated Outstanding. This Act required that any grant-aided conservation work should make a 'significant contribution towards preserving or enhancing the character or appearance of that area' (1980 Act, Section 10). There is no suggestion of what this 'significant contribution' might be (Civic Trust 1980b).

Two measures in 1981 affected, albeit in a relatively minor way, the previous increase of control within designated conservation areas. The Special Development Order (DoE 1981b) relaxed Permitted Development (as defined under the General Development Order) within those conservation areas designated after April 1981; control remained intact within areas designated prior to this. The Local Government Planning and Land

(Amendment) Act 1981 removed some powers, albeit dormant, of the Secretary of State regarding conservation areas. Critics of the Bill during its passage protested that this might lead to local authorities feeling that conservation areas were unimportant. Later proposals were to reduce planning controls outside conservation areas, rather than to see those designated areas as having more planning controls than was usual. This may seem a pedantic point, but 'it would be unfortunate if areas not blessed with designation came to be thought of as expendable' (Cantell 1981).

A further consolidation Circular (DoE 1987a) set out the current general principles for conservation areas. Designation is seen not as an end in itself but as a means to an end. The emphasis is placed firmly upon enhancement (DoE 1987a: paragraph 7). The manner in which new buildings blend with the area is seen as most significant (DoE 1987a: paragraph 61). An important point made within this Circular concerns Directions under Article 4 of the General Development Order. This freedom of LPAs to request Directions must, it is explicitly stated, be used sparingly, and the Secretary of State will not confirm such Directions in respect of conservation areas under normal circumstances, although he or she will be sympathetic provided that a special need for them can be demonstrated (DoE 1987a: paragraph 64 and Appendix II). This Circular also discussed the new conservation area consent requirements, which were a 'tidying-up' of the unsatisfactory requirement to obtain listed building consent for the demolition of unlisted buildings within a conservation area (the process set up by the 1974 Act).

Changes in the 1990s

The Town and Country Planning Act 1990 and Planning (Listed Buildings and Conservation Areas) Act 1990 again primarily 'tidy up' the current planning legislation. Little has been changed, and there are no substantive changes in the thrust of conservation area legislation; indeed, much of the wording of earlier Acts is preserved.

Responsibility for conservation policy was moved in 1992 to the newly created Department of National Heritage. The DNH thus has responsibility for general oversight of the conservation area and listed building systems, although their application in practice through appeals and so on still rests with the DoE and the Secretary of State for the Environment (DNH 1992). How this uneasy division of responsibilities will work in practice is yet to be resolved, and the lengthy gestation period of the Planning Policy Guidance Note on conservation and listed buildings, caused in part by this inter-departmental problem, is perhaps a bad omen.

In 1994, the then Minister made a surprise announcement that, in future, LPAs would be able to designate Article 4 Directions controlling certain types of change in conservation areas without the need to obtain consent from the Secretary of State. This went some way to addressing the vociferous concerns of conservationist bodies such as the EHTF (1992). The new measure was brought into operation in 1995 in the *General Permitted*

Development Order, but the scale and nature of LPA usage of this means of control cannot yet be assessed.

RECENT COURT CASES

Several recent court cases have had a significant impact upon the interpretation of the law relating to conservation areas and listed buildings, in particular with the statutory duty of local authorities (or other 'decision-makers') to pay 'special attention' to conservation areas (Section 277(8) of the 1971 Act as amended by the 1974 Act; now Section 72 of the 1990 Act). This duty had not previously been defined, and Suddards felt that

> many local planning authorities might say that a conservation area policy is not required because it would in any event pay special attention (ie attention over and beyond that which it would normally pay to any other area) to the sort of area which would be designated as a conservation area.
>
> (Suddards 1988: 51)

Steinberg

In *Steinberg and Sykes* v. *Secretary of State for the Environment and Another* ([1989] JPL 259), two residents in a conservation area were aggrieved at a grant of planning permission at appeal for a two-storey house in Camden. Steinberg and Sykes challenged the Secretary of State's decision in the High Court on the ground that the Inspector had failed to take into account the statutory requirement of Section 277(8). Lionel Read QC agreed that this obligation had not been fulfilled, stating that

> nowhere in his decision letter . . . does he mention this subsection or his duty thereunder, either in terms or by the use of any language from which it might, in my judgement, reasonably be inferred that he was intending to refer to it. . . . The obligation imposed by the statute is to pay special attention to the desirability of preserving or enhancing the character of the conservation area.
>
> (Lionel Read QC, in *Steinberg*)

The appeal decision was therefore quashed (Millichap 1989a; Stubbs and Lavers 1991). This decision was seized upon by conservationists, with the Civic Trust stating that

> it is not enough to say that a particular development will do no harm. The positive tests required by Section 277(8) [of the then 1971 Act] must be applied. Of course, opinions will vary as to whether a particular development will preserve or enhance the local scene but the Steinberg case will serve to enable the proper issues to be addressed in the future.
>
> (Civic Trust 1989)

Yet it is evident that the courts discourage the examination of the wording of appeal decision letters in a legalistic manner. It is not a legal requirement that a decision-maker should 'slavishly rehearse that he is aware of his duty and is discharging it by reference to consideration of preservation or enhancement' (Sir Graham Eyre, in *London Borough of Harrow* v. *Secretary of State for the Environment* (1989)). Yet many Inspectors use such wording.

After Steinberg

The apparent importance of *Steinberg* is the interpretation of the phrase that 'the concept of avoiding harm is essentially negative. The underlying purpose of Section 277(8) seems to me to be essentially positive' (Lionel Read QC, in *Steinberg*). In other words, a proposed development should not merely cause no harm, it must positively enhance the area. A series of cases have tested this proposition (Stubbs and Lavers 1991). Most importantly, in *South Western Regional Health Authority* v. *Secretary of State for the Environment* (1989), Sir Graham Eyre noted that Section 277(8) contained no exclusive test that could be applied by decision-makers to determine the outcome of every application within a conservation area.

> The sub-section does not set out a test, nor does it support the proposition that in relation to a consideration of development proposals they must themselves in every case preserve or enhance or serve to preserve or enhance the character or appearance of a conservation area.
>
> (Eyre, quoted by Stubbs and Lavers 1991: 12)

Harm may mean that the character or appearance will not be preserved or enhanced, but this is not necessarily so. It may be that a proposal would preserve or enhance some aspects of an area, but harm others (in fact, many applications, if strictly examined, would fall into this category). The decision-maker then has to balance preservation and enhancement against potential harm: it should not be supposed that all harm must be avoided.

Ferguson (1990) reports the case of *Unex Dumpton Ltd* v. *Secretary of State for the Environment and Forest Heath District Council*. Here, the Inspector had arguably considered whether the proposed development would preserve or enhance the conservation area, but not (in the post-*Steinberg* way of thinking) whether it would harm the area's character or appearance, let alone whether this harm would be demonstrable. Roy Vandermeer QC gave the following useful determinations.

1. As a general rule Inspectors cannot avoid the need to consider whether development proposals cause harm to interests of acknowledged importance and they should grant permission if they do not.
2. Inspectors considering proposals in a conservation area have, by Section 277 of the [1971] Act, a special duty imposed on them to

pay special attention to the desirability of the proposals enhancing or preserving the special character of the conservation area and it must be apparent from the decision that this duty has been discharged otherwise an error of law will have occurred.
3. If Inspectors find that the proposed development will not preserve or enhance the conservation area it is very likely that he will conclude that harm would be caused to the conservation area.
4. What falls to be considered is the appearance of the conservation area, not simply each individual component within it. Accordingly, it is possible that a proposal to replace one building with another in a conservation area will not harm the conservation area.

Millichap (1989b) examined a number of post-*Steinberg* appeal decisions, noting the problems of identifying 'demonstrable harm' and the tendency to treat 'preservation' and 'enhancement' as alternatives, although many decision letters refer to preservation *and* enhancement rather than preservation *or* enhancement, as does the legislation. The problem of distinguishing 'character' and 'appearance' is also raised, despite the lucid examination of 'character', 'appearance', 'preserve' and 'enhance' in Millichap (1989a). He concludes that applicants and decision-makers should beware of the ability to promote the interpretation of the section that favours their aims (Millichap 1989b: 504), a suggestion that further official guidance is necessary: a point explicitly made in his previous paper (Millichap 1989a: 240).

Bath Society

In *Bath Society* v. *Secretary of State for the Environment and Another* (*Journal of Planning and Environment Law Bulletin* 1991), the Court of Appeal held that the failure of an Inspector hearing a planning appeal relating to a conservation area to consider recommendations for the appeal site contained in the Local Plan constituted a failure to pay 'special attention' and resulted in a flawed decision. Glidewell LJ, in the Court of Appeal, thus set out the proper approach in considering an application for planning permission within a conservation area.

1. The decision-maker had two statutory duties to perform, imposed by Section 277(8) as well as Section 29(1) of the Act.
2. In a conservation area the requirement under Section 277(8) to pay 'special attention' should be the first consideration for the decision-maker. It was to be regarded as having considerable importance and weight.
3. If, therefore, the decision-maker decided that the development would enhance or preserve the character or appearance of the area, that had to be a major point in favour of allowing the development.
4. There would, nevertheless, be some cases in which a development could simultaneously enhance the character of an area but cause

some detriment. That detrimental effect was a material consideration.

5. If the decision-maker decided that the proposed development would neither preserve nor enhance the character of the area, it was almost inevitable that the development would have some detrimental effect on it. Then, the development should only be permitted if the decision-maker concluded that it carried advantages outweighing the failure to satisfy the Section 277(8) test and such detriment as might inevitably follow.

(*Journal of Planning and Environment Law Bulletin* 1991: 4)

Stanley (1991), commenting upon this case, found that in making this detailed approach, Glidewell LJ had not stated whether the Section 72(1) duty was to be considered first in order or in importance, although the latter was more likely. Yet the key point of this case, it was felt, was the judicial approval given to the use of presumptions, a concept with no statutory basis in the planning process despite the precedent of the 'presumption in favour of development in general' contained in Paragraph 15 of Planning Policy Guidance Note 1 (DoE 1992). Section 54A of the Town and Country Planning Act 1990 appears to introduce a presumption in favour of development that accords with the development plan and a presumption against development that does not: there also appears to be a trend to incorporate presumptions preventing the demolition of unlisted buildings in conservation areas into local plans. Yet the Section 72(1) duty requiring 'special attention' to be paid would seem to render such presumptions superfluous, although it is feared that they might lead to a further round of litigation.

South Lakeland

Shortly after the *Bath Society* case, the Court of Appeal also ruled on *South Lakeland District Council* v. *Secretary of State for the Environment and Carlisle Diocesan Parsonages Board*. This case dealt with the relationship between Section 72(1) and 'neutral' development, which did not harm the character or appearance of a conservation area. In doing no harm, it was argued that such neutral development acted to 'preserve' character or appearance. In accepting this argument, the Court overruled the narrow interpretation of preservation adopted by the High Court in the *Steinberg* case in 1989 (Stanley 1991: 1014).

The case has now been heard in the House of Lords. The judgement of Lord Bridge of Harwich, in dismissing the appeal by South Lakeland DC, agreed with the Appeal Court's interpretation of Section 277(8) and, therefore, its re-interpretation of the *Steinberg* judgement.

> It not only gives effect to the ordinary meaning of the statutory language; it also avoids imputing to the legislature a rigidity of planning policy for which it is difficult to see any rational justification . . . where

> a particular development will not have any adverse effect on the character or appearance of the [conservation] area and is otherwise unobjectionable on planning grounds, one may ask rhetorically what possible planning reason there can be for refusing to allow it.
>
> (House of Lords 1992)

Yet authoritative commentators have suggested that this is a problematic decision (cf. Hughes 1995). It is suggested, following legal and academic debate, that 'enhance' necessarily has a positive meaning, that 'preserve' does not, although there are both positive and neutral ways of preserving something, and that there is no requirement that a development proposal in a conservation area *must* either preserve or enhance that area (Hughes 1995: 689). In the light of the legal requirement for LPAs (and other decision-makers) to pay 'special attention' to preserving area character or appearance, there is a strong argument for suggesting that

> 'preserving' connotes the absence of change and not merely the absence of harm. The decision appears potentially damaging. If building within the curtailage of an existing building does not, in itself, cause harm, will building within the curtilage of another building cause harm; and at what point will the character or appearance of a conservation area be altered, or destroyed, if the process is continued?
>
> (Trafford-Owen 1991: 16)

This rhetorical and unanswered question is fundamental to the issues discussed in this book.

Canterbury

In *R.* v. *Canterbury City Council ex parte Halford*, a decision by Canterbury City Council to extend the designated Barham conservation area was quashed by the Court. This raised the question of the definition of conservation area boundaries, and established that, in the absence of a right of appeal against conservation area designation, a judicial review may be sought by an interested party.

A key issue in this case was the concept of the 'setting' of the conservation area and, in particular, whether a conservation area could be designated (or, in this case, extended) to protect particular views of the village and church. The High Court accepted that conservation areas should not be enlarged (and thus, arguably, originally defined) merely to include 'buffer zones'. But the judge also argued that, as Section 69 of the Act omitted specific reference to the 'setting' of a conservation area, Parliament must have 'intended that the interesting features and their setting were together to be treated as part of the "area"' (McCullough, quoted in *Journal of Planning and Environment Law* 1992: 852). It would appear, from this case, that land devoid of intrusive historical or architectural interest may be included within a conservation area so long as it comprises a relatively small proportion of the total area's extent:

it seems unjustified to expand a conservation area specifically to include such land (Jarman 1992; Millichap 1992). It should be noted that the Barham conservation area extension was quashed not on these grounds, but on the narrow ground that tree preservation had not been considered; this technicality has now been overcome and the designation made. Nevertheless, the airing of these issues in open court is significant. In particular, the judge noted that an LPA was not entitled to place a buffer zone around features of architectural or historic interest simply by great outward expansions of conservation areas. He gave the example of the famous and important distant views of Ely and Lincoln Cathedrals, but 'one could hardly put a ring five miles around each cathedral and designate it as a conservation area' (McCullough, quoted in *Journal of Planning and Environment Law* 1992: 852; see also Hughes 1995: 684).

Wansdyke

Wansdyke DC v. *Secretary of State* suggests that a development which clearly harms a conservation area may be sanctioned if, in so doing, other significant conservation interests would be promoted. A proposed sports development for a rugby club in the Bathampton conservation area was found, on appeal, to be damaging to the area. However, allowing this proposal would mean the likely development of a park-and-ride area and superstore on a site vacated by the rugby club to the direct benefit of a nearby conservation area: Bath. Since Bath is generally accepted as an 'internationally important' area, the Inspector ruled that the benefits of decreasing vehicular congestion through providing the park-and-ride facility, and attracting shoppers away from the city centre to the new superstore, would outweigh the accepted damage to Bathampton. This is a clear case of an exception being made to the general policies on conservation and development, following that of the *Bath Society* case: even if a given proposal is not in accordance with the legislative criteria, there may be other planning arguments in its favour leading to acceptance (Millichap 1993).

Chorley

The Chorley case focused on the 'character or appearance' part of the conservation area definition in an area where an applicant proposed to demolish a concrete and asbestos shed and replace it with a house. The LPA felt that the shed's demolition could improve the appearance of the area, but that the proposed house, although visually attractive, could adversely affect the area's character by extending residential uses further into the countryside. The Deputy Judge in the case discussed the apparent mutual exclusivity of the statutory phrasing of 'character or appearance'. He felt that 'they were words used separately but he did not discount the possibility that there could be cases where they meant, effectively, the same thing' (Vandermeer, quoted in *Journal of Planning and Environment Law* 1993: 930).

Subsequent commentary suggests that these issues and the potential conflict between the terms can be resolved if the nature of the particular area has been studied and the reasons for designation fully explained. 'In the case of an area designated for aesthetic reasons, an enhancement of *appearance* is likely to be considered to outweigh a change in character' (Hughes 1995: 681, his emphasis).

No.1 Poultry

The saga of Peter (now Lord) Palumbo's attempts to develop his site at No. 1 Poultry was finally decided on 28 February 1991 in the House of Lords (Watson 1991). Five Law Lords decided that the appeal decision by the former Secretary of State Nicholas Ridley was correct in granting permission for James Stirling's new building on the corner site now occupied by a group of eight listed Victorian buildings. However, they did not endorse Ridley's views that the new building was a 'possible masterpiece', being of such quality that it would contribute 'more both to the immediate environment and to the architectural heritage than the retention of the existing buildings'. Instead, Lord Ackner stated that 'in allowing this appeal, your lordships are in no way either expressly or implicitly concurring with the views of the Secretary of State' (quoted in Bar-Hillel 1991). Furthermore, the House of Lords regards this ruling as an exception, rather than forming a precedent. Yet, although dealing explicitly with listed buildings, this case has implications for development in conservation areas. SAVE Britain's Heritage, which fought the case against Palumbo, based its arguments on the fact that Circular 8/87 (DoE 1987a) makes no mention that the possible quality of a proposed replacement building could be a material consideration. SAVE's legal adviser said of the original appeal decision that

> everyone will be trying to get through this loophole ... unless the decision is challenged, future listed building inquiries will be bogged down in arguments about taste and aesthetics of new buildings instead of presuming, as the circular says, in favour of old ones.
>
> (David Cooper, quoted in Bar-Hillel 1991)

It is difficult to argue, as the Law Lords did, that this case will not set a precedent. The validity of Circular 8/87 and its presumption in favour of retaining listed buildings remains official policy, and Lord Bridge was correct in alluding to the 'special circumstances' of this particular site. Nevertheless, although each planning application should be treated on its own merits (see, for example, DoE 1991: paragraph A3), the law operates on the basis of precedent and case law, and the Government accepts that 'the Courts are the ultimate arbiters' (DoE 1991: paragraph 19). Future appeal decisions and court cases regarding development in conservation areas must be expected to raise the case of No. 1 Poultry.

CONCLUSIONS: LEGISLATION AND CASE LAW

In 1984, Mynors stated that 'the law as it now stands is somewhat complex – it is no wonder that there are several inconsistencies in the drafting of the later amending legislation!' (Mynors 1984: 145). Since then, there has been a specific conservation areas and listed buildings Act, bringing together and tidying the many changes and developments in conservation legislation. There has also been the creation of the Department of National Heritage, having specific responsibility for conservation policy (but none for the planning system relating to conservation issues).

Yet confusions remain in abundance. This examination of legislative development and legal quibbles suggests many. Most cases deal, implicitly or explicitly, with how an area's 'character' or 'appearance' are defined. But must all development 'enhance' a designated conservation area, as *Steinberg* suggested? Or does development which does no demonstrable harm thus act to 'preserve' it, as in *South Lakeland*? How far can the interests of one conservation area outweigh the causing of damage to another (*Wansdyke*)? As so many conservation areas lack clear character appraisals, how many cases will still confuse 'character or appearance' (*Chorley*)? And how many powerful developers will seek judicial review of conservation area designations (*Canterbury*)? These are key questions with implications far wider than the UK's own conservation planning system. Yet it is quite plain that they cannot readily be resolved through the UK's long-established and complex planning and judicial systems.

The policies and procedures laid down by statute are interpreted by official guidance, usually in the form of Circulars. These interpretations may change from time to time, as the political complexion of government changes, or even as influential individuals within the system change (as with the evident differences of opinion held by three Secretaries of State for the Environment during the No. 1 Poultry case). Yet both the statutes and the guidance are interpreted by users and, as a last resort, by the courts. The latter evidently place great reliance upon precedent.

The implications for conservation areas are plain. There are many areas of uncertainty contained in the statutes, Circulars and even in successive court decisions. These uncertainties often revolve around varying interpretations of specific words (for example the 'special' duty in Section 277(8), or the interpretations of the words conserve, preserve, enhance and so on). Various practices are hardly made explicit, as with the provisions for extending or de-designating conservation areas, and the lack of guidance on what precisely may constitute a conservation area. In one case, which gained wide publicity, it was accepted that a group of listed buildings could be replaced with a new building, designed by a prominent architect, at least in part on the grounds of the possible quality of the replacement.

Many of these problems stem from the basis of subjectivity implicit in conservation, and in the concept of amenity in planning in general. Some of the terms used are hardly amenable to casting in terms of 'objective', quantifiable standards. They will, therefore, continue to lend themselves to a variety of interpretations. Mynors's statement of 1984 remains true in spirit, despite the 1990 Acts. It seems inescapable that the whole area of planning law pertaining to conservation areas and listed buildings, and the types of development permitted within them, requires speedy clarification.

Indeed, the conclusions of Mynors's paper are worthy of quoting at length, since they are as relevant today as they were more than a decade ago. The continuing increase in public concern, the continuing confusion of legal precedents and the continuing lack of concerted legislative action since Mynors wrote merely emphasise his relevance.

> The problem with the legislation as it now stands is paradoxically that it provides at once too much and too little protection for conservation areas. Too much protection, because more and more land is being designated, with the emphasis being less on '*special interest*' and more on 'the familiar and cherished local scene'; and with limited financial and staff resources it is impossible (and it would in any case be unreasonable) for authorities to devote their energy to working equally towards protecting the whole of *all* of their conservation areas.
>
> There is too little protection, though, for areas which really are of special interest, since the legislation is largely toothless. There is no urgency to produce enhancement proposals, and no statutory force to them once produced. The scope of permitted development means that many small changes, which together will have a disastrous effect on an area's character, are outside any form of control. The control over demolition is seriously undermined by the various exemptions. The protection of areas of tree-cover, as opposed to individual specimens, is cumbersome and awkward.
>
> (Mynors 1984: 247)

It is clear, therefore, that there is a requirement for the overhaul of UK law as it is applied to conservation areas in particular, but to all aspects of conservation in general. The implications of the increasingly numerous 'exceptions to the rule' (Millichap 1993) are becoming confusing. However, experiences of numerous planning departments would suggest that major change in procedures is not required (Jones and Larkham 1993), rather a tightening-up of guidance, definitions and some procedures. The aims of the policy, for example, could be elaborated to take into account changing attitudes to conservation, thus answering the critics who complain that continued designations 'debase the coinage' of the concept. The use of Article 4 Directions has also proved problematic in practice, with the DoE reputedly reluctant to confirm draft Directions, and relatively few currently in use: further detailed guidance would be beneficial. The increasing reliance of powerful interest groups on the courts is not something which should drive policy formulation and implementation, but this appeared to be the case post-*Steinberg*.

The precedents of individual case law, compelling in law, are less so in the individual and local circumstances of conservation planning. It is for this reason that much of the British system is locally based, and this has been an evident strength over the years. Yet when an individual, as the owner or occupier of a listed building or a building within a conservation area, becomes embroiled within this quasi-legal planning procedure which is driven by changing legal precedent on the semiotics of legal phraseology, it is understandable that the individual reaction becomes one of confusion and complaint. The system is far from being readily comprehensible and does not operate in an open manner (cf. Reade 1991, 1992a, 1992b). What individual members of the public, and pressure groups purporting to represent them, want to know is the impact of decisions on their own immediate environments; often, in terms of purely visual impacts rather than on any theoretical criteria of 'preserve and/or enhance'. The following part explores, in detail, cases of actual changes to the urban landscapes of some conservation areas in an attempt to portray the workings of this system in practice.

NOTES

1 Mostly district planning authorities and London boroughs. The Secretary of State has the power to make designations, but in practice this power has never been exercised. County Councils may designate, but they must consult the relevant district council. In London, the GLC's designation powers have been taken over by English Heritage (Ross 1991: 121). Only in 1995 did English Heritage designate its first conservation area, after the London Borough of Wandsworth had refused to take action.

2 There are exceptions in National Parks, where only the county can designate, and in urban development areas, where the designating authority is the Urban Development Corporation (Development Corporation (Planning Functions) Orders, 1981, Article 3b and Schedule 1). The Secretary of State also has the power to designate (1971 Act, Section 277(4)), but this power was to be used in exceptional circumstances only (DoE 1977a: paragraph 34).

Part 2

THE CHANGING CONSERVED TOWN

5

AREA-BASED CONSERVATION

I hate Bath. There is a stupid sameness, notwithstanding the beauties of its buildings.

(Benjamin Robert Haydon, 1809)

INTRODUCTION

In many industrialised countries, as Chapter 2 suggested, a first phase of conservation legislation was concerned merely with identifying and listing monuments (and sometimes buildings). This was generally succeeded by a more sophisticated phase, concerned with the delineation and protection of wider areas. Over time, the workings of this second phase have become increasingly widely criticised: it is suggested that attention and resources are not targeted sufficiently sharply; too many areas are designated and for the wrong reasons; and perceived socio-economic disbenefits may outweigh generalised societal benefits. Yet area conservation remains popular. In England it is suggested that public support for this planning policy is second only to Green Belts. It is also a form of conservation activity in which the LPA can be seen to be proactive, a 'direct agent' in its shaping and guiding of change through planning policy, guidance, specialist advice and grant availability, and in the formulating and implementing of enhancement schemes. This is significant, given the depiction of LPAs elsewhere in this volume as being largely 'indirect agents' in their reactions to individual planning proposals. The proactive and reactive roles are combined within conservation areas, perhaps to a greater extent than elsewhere.

COMPARATIVE AREA-BASED CONSERVATION

Given the wide spread of area-based conservation approaches, a number of examples will briefly be studied. Both the development of ideas and practices are relevant; although individual national circumstances do vary and will dictate the availability of powers and the socio-cultural response to them, the differing reactions to the common problem of coping with heritage through the developed (and, increasingly, developing) world are instructive. Attention is, however, focused on practice in the UK; in particular, highlighting designation practices and innovations. Chapters 6 and 7 further discuss area enhancement and the relationship between policy and decision-making, while Chapters 8 and 9 will then develop detailed case studies of activities occurring in two types of UK conservation areas.

The Netherlands

In the Netherlands, the Monuments and Historic Buildings Act 1961 deals with both architecture and town planning, in order that both valuable individual buildings and planned townscape units (*beschermde stads-en dorpsgezichten*, literally 'protected town and village views') might be preserved. It introduced a system of licences, and no alterations were permissible to monuments without the Minister's permission. The preservation of whole areas was possible, but this concept has only slowly been adopted; monuments were the prime concern of the 1961 Act. This Act was a considerable advance on previous protection measures, but was seen as inadequate even by the late 1970s (van Voorden 1981). One of its principal practical defects was that the conservation legislation was quite separate from mainstream town planning legislation (Skea 1988: 17).

Both towns and villages may be designated as conservation areas, but designation required a joint ministerial decision by the Minister for Cultural Affairs, Recreation and Social Work and the Minister for Housing and Physical Planning (under the Physical Planning Act 1962); advice from other bodies, including the local council, was also required. A formal proposal was prepared, consisting of a written report and plan; but several years may elapse between initiation of a proposal and actual designation, as the legislation put no time limit on this procedural period. Appeals against designation may also be lodged. Detailed descriptions of the areas are usual, dealing mostly with the scale and appearance of buildings and the arrangement of public spaces (van Voorden 1981: 444–5). Although this may sound a superficial approach to what is *de facto* the examination of area character, and in reality only secondary sources are consulted – there is little or no physical survey – nevertheless quite detailed assessments can be compiled (van Voorden 1981: 445–9). In some cases, particularly where complete urban or village cores are designated, the protected area has been extended to include the adjoining undeveloped rural area: explicit recognition, in examples such as the town of Elburg (Figure 5.1), of the importance of the setting of historic districts.

In practice, the Dutch approach has been coloured by the considerable freedom left to local authorities by the 1961 Act. Changing local political approaches have, in some cases, played a significant part. The social democratic government of the town of Groningen, for example, moved away from concentration upon individual key buildings to a radical planning vision including large-scale pedestrianisation and urban renewal.

> Monument protection did not fit easily into this vision of the city and its central area. Nevertheless, an urban conservation policy ... could

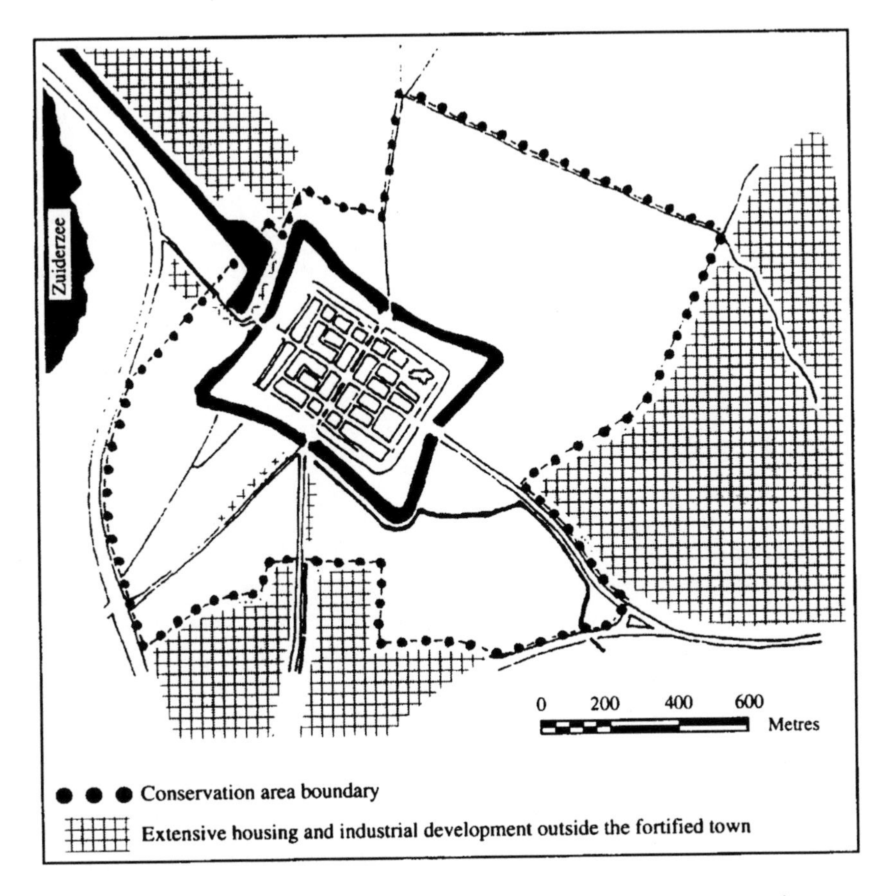

Figure 5.1 Conservation area around Elburg, The Netherlands (redrawn from van Voorden 1981)

> become a logical part of this reorientation in the city's planning. Historical buildings [and areas] were not to be preserved and renovated for their own sake, with a possible use being sought after the event; rather they were to be conserved because they were needed as part of the restructuring of the city.
>
> (Ashworth 1984: 611)

Nevertheless, although Dutch protected areas have been compared to UK conservation areas (e.g. by Dobby 1978), Skea (1988: 18) suggests that, because of their high quality, cultural importance and financial arrangements (since 1986 at least), they are closer to UK Town Schemes than ordinary conservation areas. The priority of the *Rijksdienst voor de Monumentenzorg*, the relevant government department, has been to select only the very best townscapes, and there is a lengthy process of evaluation, designation and plan preparation before full protection is afforded (Skea 1988: Figure 1). The administrative complexity of the Dutch system, Skea (1988) and van Voorden (1981) argue, has led to a slow rate of full legal protection, with many municipalities being unable or unwilling to prepare the necessary detailed development plan.

France

The 1913 law consolidating monument listing procedures also provided for 'protected perimeters' around them, and is still in force. 'As there can hardly be a town which does not have a listed church or *hôtel particulaire* at its centre, this means in effect that the French government has negative control over virtually the whole of the country's architectural heritage' (Kain 1981: 200). The then Minister for Cultural Affairs André Malraux introduced the radical concept of *secteurs sauvegardés* in his Act of 1962 as areas for positive enhancement, together with other measures for the protection of the architectural heritage (Sorlin 1968; Stungo 1972). In comparison with Britain, the number, scale and distribution of *secteurs* are limited. Within these designated areas, detailed restorations of small, closely defined areas, each known as a *secteur operationnel*, have been undertaken. Yet this once-innovative system has been widely criticised on the grounds of cost, slow progress and the injustice of focusing attention on very small areas. Estimates even from the early 1970s gave figures of 350–400 years being necessary to treat the *secteurs operationnel* then designated (e.g. Bourguignon 1971). Social problems have also been caused through displacement and gentrification (see, for example, Coing 1966; Bourguignon 1971; Soucy 1974). Since the mid-1970s attention has been turned away from the spectacular but limited schemes to more general *opérations programmés d'amélioration de l'habitat*, the Malraux Act itself has been amended, and social criteria have begun to match more closely those of architectural and historical values in French conservation (Kain 1981).

Kain identifies the French *Loi d'Orientation Foncière* (1967) as being potentially significant for conservation, in that under the *Plan D'Occupation des Sols* (POS) every settlement of over 10,000 inhabitants was required not only to zone land uses, set densities and plot ratios, but also to 'delimit areas, streets and buildings for protection and enhancement, and can specify regulations to govern the location, size, mass and external appearance of new constructions' (Kain 1981: 203) (although he admits that, in practice, developers are frequently allowed to contravene elements of a POS: Racine and Creutz 1975). A recent example where area-based conservation issues have been addressed through a POS is the Commune of Asnières-sur-Oise, 35km north of Paris. This was drawn up by a group of urban design specialists from Oxford Brookes University, particularly Dr Karl Kropf, and is an unusually detailed and thorough example of its type. The commune is divided into several regions, within each of which a series of regulations have been formulated to control future development. These have been based on minutely detailed surveys of local development history in terms of plot

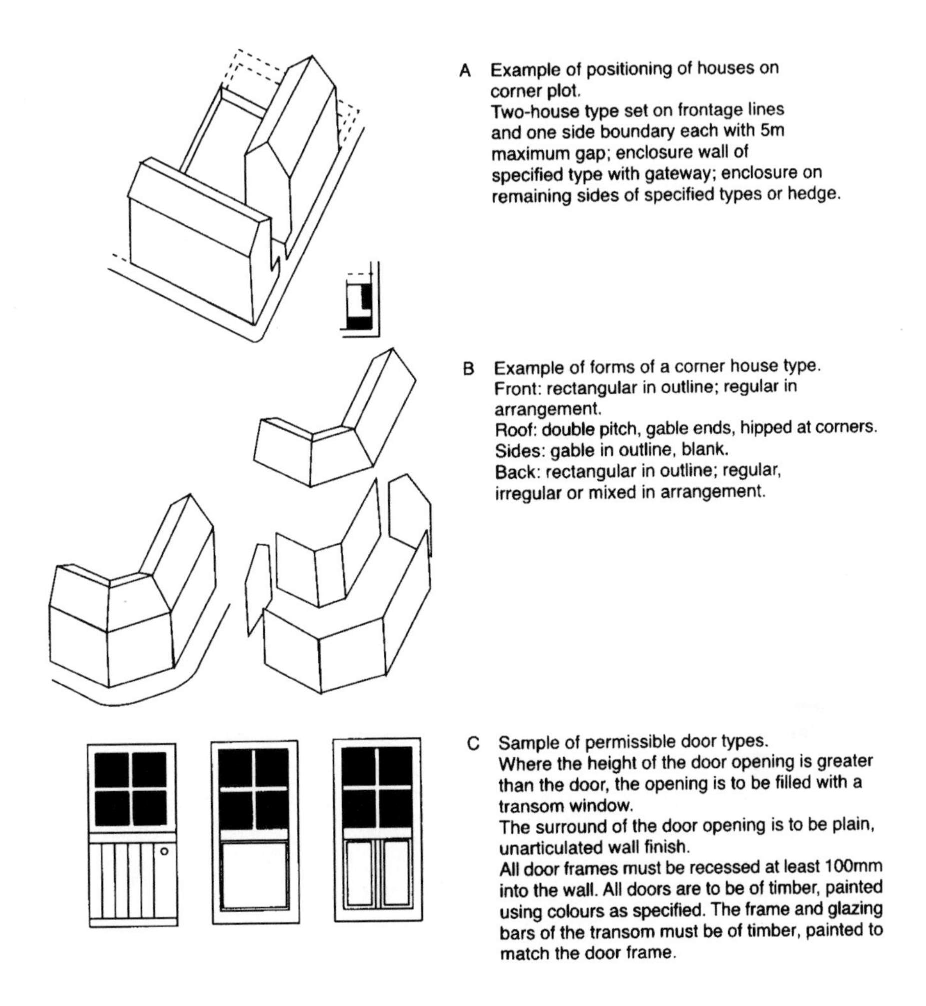

Figure 5.2 Prescriptive guidance based on morphological study (reproduced from Kropf 1993)

shapes and building coverage, and local architectural forms. An English-language draft, and commentary, is given in Kropf (1993: 339–64 and Appendix D); see also Samuels (1993). The Asnières POS shows how minute survey of settlement-wide morphological characteristics can be used in district- or site-specific contexts for the regulation of new construction and design (Figure 5.2). Particularly in areas of identifiable character, where streets, plots and buildings form a substantial component of that character and which it is deemed important to retain, such large-scale studies offer appropriate context to support site-specific interventions.

The United States of America

In the USA, local legislation is, perhaps, more significant than the national conservation legislation examined in Chapter 2. The historic district technique has been used since pioneering legislation in Charleston, South Carolina, in 1931, although take-up was slow until the 1960s (Reed 1969), and the total number of designations stood at just over 1,200 in the mid-1980s. Such designations are popular among preservationists, but they vary in their effectiveness at protecting landmarks, urban spaces and townscapes, depending upon the strength of local legislation and the level of local political support for their ideals (*A Guide to Delineating the Edges of Historic Districts* 1976). Many property owners and large sections of the business community resist these designations, fearing undue restrictions on land use and adverse economic effects (Listokin 1985). Particularly in residential districts, opposition has been based on the assertion that designation could raise property values, thus increasing tax liabilities and rents, and leading to the displacement of low-income and elderly households. However, an analysis of Washington DC found little actual support for the displacement threat (Gale 1991).

The Historic Landmark and Historic District Protection Act of 1978 for Washington DC is, perhaps, the strongest of the local Acts and deserves some attention. This Act provides a mandate for the establishment of a historic preservation review board, which may designate and maintain an inventory of historic landmarks and districts, and reviews all applications to demolish, alter, subdivide or build on individual landmarks or on properties in historic districts. The board is not democratically elected, nor is it made up of experts. Instead, it is composed 'with a view towards having its membership represent, to the greatest possible extent, the adult population of DC' (Bar-Hillel 1988b). The majority of decisions are taken only after public hearings. Criminal penalties of a fine of up to $1,000 and/or up to 90 days' imprisonment enforce the Act, but these are supported by much more severe civil penalties. Anyone who demolishes, constructs or alters a building in contravention of the Act 'shall be required to restore the building and its site to its appearance prior to the violation' (Bar-Hillel 1988b).

However, even the key historic cities, designated early in the inter-war period, may find themselves dependent largely upon voluntary organisations for any impetus to practical conservation, and for significant funding. Neighbourhood rehabilitation programmes, and commercial centre revitalisation, have had considerable detrimental impacts upon historic areas such as Charleston (P.B. Smith 1979) despite the rise of the heritage-tourism industry. Indeed, the country's second-oldest designated historic district, the Vieux Carré of New Orleans, has undergone considerable physical preservation on an area-wide basis, 'yet preservationists and long-time residents of the neighbourhood feel that, over time, it has lost a rich and valuable part of itself' (Sauder and Wilkinson 1989: 41).

Singapore

As has been discussed, conservation is a well-developed facet of planning in many Westernised developed countries, but is often regarded as less significant in the Third World (particularly with respect to academic studies: Ali 1990: 74). Singapore, however, clearly demonstrates that the rapidity of built form transformations and the difficulties of agreeing priorities of conservation and redevelopment are most significant (Kong and Yeoh 1994).

Little concerted planned renewal of any form existed in Singapore in the immediate post-war years. Only in 1966 was formal recognition given to urban renewal with the establishment of the Urban Renewal Department of the Housing and Development Board. Most early renewal projects concentrated upon slum clearance and comprehensive traffic planning schemes. Conservation *per se* received little, if any, attention: 'the imperatives of a rapidly developing economy (to provide housing, serve the transport and other social service needs of the population, facilitate employment and hence further economic development) dictated the planning agenda' (Kong and Yeoh 1994: 249). A new authority, the Urban Redevelopment Authority (URA), was constituted in 1974, and in addition to acquiring the existing roles of the Urban Renewal Department, the preservation of Singapore's historical and architectural heritage was explicitly added (URA 1975: 2–3). Two years later, in an evident major policy shift, the URA initiated conservation/rehabilitation studies of several entire areas, the most prominent of which was Chinatown. This was the first step in conceptualising an area-based conservation approach dealing with identity and character (Kong and Yeoh 1994: 249). Little was done on the ground, however, until a *Conservation Master Plan* was published (URA 1986), covering Chinatown, the central 'civic cultural district' and three other residential areas totalling some 100ha (Kong and Yeoh 1994: Figure 1). In 1989 amendments to the Planning Act recognised the URA as the national conservation authority. Ten areas were then officially designated as conservation areas, with another ten designated in September 1991. Despite this, members of the public have protested over the level of protection offered to areas and buildings, the gentrification of conserved areas and the subordination of cultural heritage to economic dictates (letters in the *Straits Times*, 1991–2, quoted by Kong and Yeoh 1994: 262). As redevelopment continues in Singapore, and as tourist income remains a significant part of the economy, it is interesting that Kong and Yeoh conclude their review of urban conservation in Singapore with the suggestion that

> state and public views about conservation converge on two counts. There is agreement about what conservation means; the 'improvement' and 'enhancement' of buildings and areas through refurbishing and landscaping, as well as preventing the demolition of existing buildings and areas. There is also agreement as to the purposes of conservation, whether as a legacy for the future; as a means of learning from or

> experiencing the past; as a source of character and identity for the city; or as an economic resource to attract tourists. [Yet] a critical public with a sufficiently distinct collective memory is often censorious of the final products of conservation efforts, even while the state is explicit in its aims to achieve a high degree of authenticity.
>
> (Kong and Yeoh 1994: 263)

Malta

In brief, Malta is instituting a completely new planning system, based upon the UK model, in a four-year period in the early 1990s. This includes the designation of conservation areas. The Development Planning Act 1992 empowers the Planning Authority to prepare, and from time to time review, a list of areas, buildings, structures and remains of geological, palaeontological, archaeological, architectural, historical, antiquarian or artistic importance, which are to be scheduled for conservation (Section 46, Paragraph 1). It is the intention that the World Heritage Site of Valletta together with the Grand Harbour and Marsamxett Harbour will be defined as the Valletta Harbours Heritage Conservation and Improvement Area, for which special policies will be developed; the historic core of every other island settlement will be designated as a conservation area. To give interim protection, all settlement cores identified from large-scale maps of the 1920s have already been designated as 'village' or 'historic cores', and these receive protection as *de facto* interim conservation areas, while more detailed survey work to define area character and refine boundaries is undertaken by the Planning Authority. Simultaneously, detailed surveys of every building within these historic cores are being undertaken to produce graded lists of structures of importance. Grade 1 buildings have important historical value or outstanding architectural importance; Grade 2 have 'interesting façades and architectural features that are worth preserving', and buildings which are part of a streetscape having a visual integrity and harmony are also included. Grade 3 buildings have no intrinsic historical importance and minor architectural interest, and may be redeveloped provided that the replacement is 'not out of harmony with the surroundings'. Draft guidelines on development control within urban conservation areas have been produced, from which the above quotes are taken (Maltese Planning Authority 1993). The intention of area studies and the listing exercise is to identify streets of scenic value: those which

> 1. have a considerable number of Grade 1 and Grade 2 buildings;
> 2. have a rhythmical effect composed by the location of doorways, window openings, balconies, or any other architectural features;
> 3. compose an interesting serial vision;
> 4. form part of a traditional processional route to villages; or
> 5. streets which are traditionally the routes to adjacent villages.
>
> (Maltese Planning Authority 1993: 23)

Although much local vernacular architecture and culture has strong relationships with the Islamic world, it is clear that the newly imposed conservation areas are designated according to West European ideals; the system has many of the shortcomings of the UK system upon which it is closely modelled, save that the statutory definition of a conservation area clearly encompasses features other than purely 'architectural or historic interest'. It will be instructive to see the development of this new policy and, in particular, whether the developing legal and appeals procedures become clogged with the same number of cases and arguments over minutiae of definitions as has been the case in the UK since the mid-1980s (see Chapter 4).

Summary of comparisons

Most, if not all, countries designate far fewer buildings and areas than does the UK. Designation is much more tightly, often centrally, controlled; the practical, aesthetic or other criteria for designating are rarely made explicit. The slow evolution of legislation and action has led to problems, including the issue of gentrification or displacement; although the actual impact of gentrification is variable, and this is still a matter of some political and academic controversy. The low level of financial support from central and local governments is a perennial concern, with local voluntary groups playing important roles in filling this funding gap. Even where it is argued that State and public views on conservation coincide, it is evident that controversy exists over the final products of conservation efforts. The increasing importance of heritage tourism is often detrimental to concern for 'authenticity'. Thus, whether the conservation system has been long established, as in France or the Netherlands, or is comparatively recent, as in Malta or Singapore, there is some commonality of practical and theoretical issues.

CONSERVATION AREAS IN THE UK

The process of designation

It is clear from many surveys (e.g. Gamston 1975; Jones and Larkham 1993; Morton 1994) that the manner in which areas are designated, and the attitudes of individual local authorities to boundary alignment, vary greatly. In response to the postal survey of all LPAs reported in Jones and Larkham (1993), some 66 per cent of all responding authorities include 'wide areas of historical significance' in contrast to just under half which draw boundaries tightly around 'groups of historical buildings'. Only 23 per cent of respondents indicated that they utilised both approaches depending upon the individual character of particular areas selected for designation. There is no apparent regional or urban/rural difference in the form of designation method used. An exception is the high level of authorities in London and the South East which designate areas with wide boundaries (almost 80 per cent), rather than tightly around historical buildings (only a quarter in London and less than half in the wider South East). This method is used in areas where authorities indicate that some designations stem from the pressures for development which have been particularly high, and where conservation area status is used as a method of development control.

Areas are inevitably designated following a formal report to the Planning Committee (or other appropriate body of the Council). However, the standards of such reports have raised concern in the professional planning press (Morton 1991). Jones and Larkham (1993) found no recent examples of inadequate short reports, although their extensive consultations suggested that these still do exist, especially for designations which relate to extensions of areas. There are, however, a wide range of Committee reports which examine the issues in some detail. Wolverhampton MBC provides brief historical and architectural summaries of area character and importance, for example Castlecroft Gardens, which

> today retains the character of a small and self-contained hamlet within the heart of the built-up area. Comprising many houses of considerable architectural quality and historic interest in a distinctive tree setting, the estate represents an early and highly distinctive experiment in town planning which merits every effort being made to preserve its special quality and character.
>
> (Wolverhampton MBC, Report to Planning and Transportation Committee, July 1988)

Such a brief report (three paragraphs) is sufficient for a small, well-defined and closely drawn area. More complex areas require much more detail. The report to Norwich City Council on the proposed extension to the Central Conservation Area (February 1992) is exemplary. This 12-page document discusses the background to the proposal, the historical importance of the city centre, the proposed new boundaries (with maps) and character assessments of distinct areas within the designated area. There was also a consultative paper on the rationale underlying the area's reassessment. Much of the text of this document is reproduced in Appendix 1.

It is clear that quality, not quantity, is important in these documents. They are a significant stage in policy formulation. It is unlikely that the reported two-line Committee report would be sufficient in any circumstances but, as has been shown, acceptable reports may be produced ranging from a few paragraphs to a dozen pages, depending upon the size and complexity of the area concerned. Nevertheless, with Morton (1994) suggesting that his extensive professional practice and academic research show that specific character statements or appraisals exist for only about 10 per cent of the UK's 9,000-plus conservation areas, the designation report is clearly important. In many cases, it may be the only written record which an LPA could use in policy-making, development control decision-making, appeals and court action which may explicitly refer to 'character' and/or 'appearance'.

The motives for designation

The vast majority of designations are officer-led. Predominantly they aim to control development, control demolition, to protect the 'very special' and to encompass familiar and cherished townscapes. There are also a range of 'knock-on' effects of designation. Not only does it enable the local planning authority to keep a watching brief on the effects of change, but also it is valuable to national bodies who would not be consulted about developments outside designated areas. The Council for British Archaeology, for example, considers that area status provides valuable protection for sites which are not scheduled and are not necessarily of national importance.

In some cases, designation is the result of a systematic initial townscape appraisal by officers. However, this is certainly not a commonplace activity and most appear to be a response to actions which challenge what is 'very special' or familiar. In such cases, designations are stimulated by threats of demolition: a situation that may become less commonplace as measures to control external demolition become clear.

Some designations, however, are criticised for not being historically worthwhile. Rather, they are thought to be designations driven by political expedience or under extreme pressure from residents. Such designations fuel the argument that the protection of further areas of townscape is devaluing the 'special' quality of the older, more historic areas.

In a very small number of cases, new conservation areas appear to be designated for political reasons rather than for the benefit of the historic townscape. Developers complain of instances where rapid designations have been made following the submission of large-scale development proposals for previously unprotected areas. These designations are considered by members of the British Property Federation (as reported in Jones and Larkham 1993) to be deliberate blocking tactics to delay or attempt to prevent developments taking place. In some of these instances 'political' designations result from changes in local authority administration during the planning stages of development. An example of a designation considered to be politically motivated by developers affected by the protection is in central Hammersmith. London Regional Transport and Bredero Development obtained outline planning permission to redevelop a large site in Hammersmith. This involved a planning agreement and exchange of lands agreement with the Borough Council. Following local elections, the balance of political power had changed, and the authority introduced policies opposing major office schemes. The developers state that the local authority wrote to them indicating that they would use every means available to block the scheme. One month after the letter, the Borough designated the site and its surroundings as a conservation area. There were no consultations, and there had been no indication previously given to the developers to indicate that any part of the area was a valuable townscape or was in need of consideration. The effect of the designation was to prevent the developers from demolishing any of the buildings without listed building consent (now conservation area consent). The only listed building on the site already carried consent for demolition. The developers, concerned at there being no consultation or mechanisms for appeal to the Secretary of State, were angered that designation would render the development less viable because of the delays involved. Having been refused the necessary listed building consent, the developers appealed to the Department of the Environment. The Inspector dismissed the appeal but this decision was overturned by the Secretary of State. The developers then submitted a revised scheme and negotiated consent with the local authority. The scheme was fundamentally the same as the original application, and included no measures for the enhancement or preservation of the character of the conservation area. (This case was reported at length in Jones and Larkham 1993.) Cases of what appears to be poor practice similar to this example suggest that conservation area designation may be used as an obstructive or delaying tactic – especially where the local authority indicates as such in writing (although such explicit statements are rare).

In many situations, local residents will be anxious that conservation area status is approved; if for no other reason than the frequent assertion by estate agents that property values are higher in designated areas. Indeed, pressure may be placed upon the planning authority by influential or active local groups and parish councils to designate particular localities. Many local amenity societies have exerted such pressure on their local authorities; indeed, a number of societies have been formed expressly for this purpose (mostly in middle-class suburbs and commuter villages: Larkham 1985). The example of the Edgbaston conservation area in Birmingham is useful in this context. First designated in 1975, and extended in 1984, this area comprises much of the high-class suburbia of the Calthorpe Estate. Although the Birmingham Unitary Development Plan (UDP) (October 1991 modifications) made a commitment to regular review of the city's conservation areas, this review was immediately spurred by the proposal in May 1992 to demolish four large Victorian houses, together with anticipated substantial developments elsewhere on the conservation area's boundaries. Officers reviewed the area and proposed further extensions. Leaflets were delivered to all affected properties, and leading local landowners (including the University of Birmingham and the King Edward VI School) were invited to comment. Two public exhibitions were mounted, at which a total of 16 people attended. Eight questionnaires were returned to the LPA; six in favour of the extension – although two of these thought that the area should be even larger – and two against. A developer involved with the four Victorian houses objected that there was no need to preserve and enhance what was not considered to be an area of special architectural character or historic interest, and complained that the proposed extension was not made within the context of the UDP and that the four buildings had also been locally listed. The major educational establishments felt that their records as landowners were sufficient protection for the area, and did not wish to compromise their requirements to alter uses or properties on their campuses in the light of changing educational circumstances. Despite such objections, the extension has been recommended; although the villas have since been demolished and redevelopment is proceeding. This case clearly shows a boundary review initiated because of an immediate threat to adjoining buildings, and the limited responses to what was as comprehensive a consultation exercise as could be desired. Local residents wished for a larger conservation area; large landowners did not wish to be constrained, and the developer affected by the proposal was vehemently against it.

Dissenting views of designation

Not all views on designation are as favourable as those usually put forward by residents. In a few cases, a vociferous public group, usually a minority, have organised campaigns against designation. In one example, the pre-1974 Structure Plan for Eardisland (Hereford and Worcester) suggested conservation area designation, and proposals were made to the LPA's Planning Committee in 1975, 1981/82 and 1990. In each case, despite some local support, members of the Parish Council remained vehemently opposed to designation. Their objections were not well articulated, but revolved around fears that designation would restrict potential development (W.N. Bloxsome, Principal Planning Officer, pers. comm.).

Likewise, designation of an area in Aslackby (South Kesteven District), although initially supported in a public meeting, was opposed by the Parish Council. Reasons for opposition were as follows:

1. The creation of a conservation area would create two communities through the application of different planning criteria in the same village.
2. The Local Plan's Conservation Policy C9 would be sufficient to protect and enhance the village.
3. The imposition of additional controls over unlisted buildings would not be fair without the grant aid resources available to listed buildings.
4. The conservation area would restrict growth.
5. The conservation area would inhibit modern development.
6. Conservation area status is only wanted by newcomers to the village.

These objections mirror some of the general fears of the restrictive nature of conservation in general. There appears to be a misunderstanding of the constructive interpretation of 'conservation' in UK law, policy and practice, now enshrined in PPG15 (DoE 1994). The reference to Local Plan Policy C9 was technically wrong, since this policy referred only to buildings protected within designated conservation areas. However, the reference to designation pressure from newcomers mirrors findings elsewhere (cf. Cloke and Park 1985). Despite this local opposition, the area was formally designated in November 1991 (Edwards 1995: 41–3).

Large local landowners may object, fearing the curtailing of current or future activities. This occurred in the Edgbaston case just discussed, where the educational establishments were the largest local landowners and did not wish conservation to 'interfere' with their property management policies.

More important, perhaps, than such localised objections are those from national bodies. For example, the British Property Federation has raised the issue of politically motivated designations undertaken at speed and without consultation (as has been suggested earlier). The Outdoor Advertising Council, when contacted by Jones and Larkham (1993), responded that it felt that the designation process should be amended, as it opposes the wide spread of 'ordinary shopping and business areas' included, and the controls over certain advertisements afforded by conservation area status. The OAC suggested the following:

1. No new conservation areas should be designated without the prior approval of the Secretary of State.
2. LPAs should be required to advertise proposed conservation areas more widely.
3. The public and interested organisations should be given reasonable time to lodge objections.
4. If objections are lodged, a local public inquiry should be held.
5. All conservation areas should automatically lapse after five years unless reconfirmed by the Secretary of State after going through the same procedures as for the designation of a new area.

The OAC justified these proposals on the grounds that they would 'make LPAs act in a more responsible manner in considering whether to put forward designation proposals'.

Likewise, in his acerbic critique of conservation area planning, focusing on his home area of Upper Bangor, Reade (1991, 1992a, 1992b) clearly suggests that there is no value in designation for the sake of designation. Such action is of real value only where significant policies are developed and resources allocated for their completion. Without this, designated areas will inevitably stagnate or even decay, rendering designation eventually fruitless and (by implication) de-designation inevitable.

Changing designation criteria: changing types of areas

It is clear that designation practices – i.e. exactly where the line is drawn on the map – have changed over time. Some anomalous designations have been found, many of which resulted from early surveys and evidently poorly developed conceptions of the area concept. Chipping Norton (Oxfordshire), for example, is essentially a two-row mediaeval planned town, with market-place and regular narrow and deep burgages, yet the original area boundary cut through the plots on the east side of the market, rather than following the back lane (Robinson 1982: 127). In this instance, standing buildings dominated the concept of area value. Town plan, and particularly street and plot patterns, were evidently not regarded as important. Some of Wolverhampton's designations of the early 1970s also had a peculiarity, in that small areas of modern development were specifically excluded from the designation, producing 'doughnut-shaped' areas. It was apparently felt too incongruous to include within the protected area groups of houses or a block of flats which were then less than two decades old. Again, this seems a particularly narrow vision of the area, and one wonders what would happen if radical redevelopment was proposed within this undesignated island. These views contrast sharply with those of some London Boroughs, particularly those covering the expanses of Georgian estate development, which have in the region of 80 per cent of their area covered by conservation designations. Bath has for some time had only one giant area of some 3,000 acres. Indeed, the larger area appears to be a recent trend, with Norwich proposing the amalgamation of several areas within its mediaeval core into one area in 1992 (see Appendix 1).

Concern has been expressed over the nature of recent designation, and whether this has been 'debasing the coinage' of the conservation area concept (by Morton 1991 and others). English Heritage research suggested that the most common recent designations are village centres and suburbs (Pearce *et al.* 1990; B. Hennessy, English Heritage, pers. comm.). Of these suburbs, some are Victorian bye-law terraced housing, and an increasing number are 1930s speculative suburbia. Conservation officers suggest that an impetus for designation of the latter is the current concern for inter-war and, increasingly, post-war architecture, and the '30-year' rule generally preventing buildings younger than 30 years from being listed. A further impetus affecting both types of area is the rapidity of change and, therefore, the increasing rarity of both late nineteenth-century terraces and inter-war local authority and speculative suburbia in near-original plan and architectural forms. Officers do not accept that they are acting against the wording of the Act in these cases, but argue that the 'special' nature of these areas lies in their typicality and familiarity, and that it is the best such areas that are sought and designated.

Designation reviews

Formal and regular reassessment of designation is not particularly common. None of the LPAs responding to Jones and Larkham's (1993) survey make annual reviews of conservation areas, with only 8 per cent having any set time period within which to reassess each area. The vast majority are 'reassessed as and when necessary' or when resources are made available. This non-committal response accounts for over three-quarters of responding authorities. Many LPAs had large numbers of conservation areas, differing greatly in character and development pressure. Such authorities had insufficient resources to review each area within a set period of time, and instead relied on *ad hoc* reviews or would review an entire area when a particular issue arose in, or even adjoining, it. The example of Edgbaston, Birmingham, is typical in this respect. Stratford, however, with over 80 areas, hopes to review 10 areas per year, although these are carried out in batches by consultants after a competitive tendering process. It could be argued that the use of a range of consultants, and seeking low tenders, may lead to a diverse range of survey approaches and thus different influences on review decision-making.

One in nine responding authorities have never reassessed their areas. This is more common in the most rural areas, including 27 per cent of Welsh authorities and 23 per cent of non-metropolitan districts in the North West. None of these authorities are part of cities or are based around major towns. All describe themselves as rural or mostly rural.

Local authorities suggest that reviews either result in adjusting boundaries (almost always an extension), in comprehensive additions of streets or quarters, or occasionally in no change being thought necessary. Of the authorities interviewed who had undertaken reviews, all either had made, or were in the process of making, additional designations. Two had used students on placement to undertake comprehensive area surveys and propose boundaries (with varying degrees of success). The duration of the review process was extremely variable, with one authority undertaking reviews continually over a period of years, as and when officer time permitted. The legislative processes behind review are the same as those for new designations. Consequently, it is entirely dependent upon individual authorities. Authorities committed to the consultation process do appear to undertake public discussion with newly affected residents and organisations.

The impetus for boundary reviews appears to vary greatly. Even comprehensive authority-wide boundary reviews must be the result of changing circumstances, otherwise it could be argued that the new area to be included should have been incorporated within the original designation (although one planning officer suggested that the process of a new officer becoming familiar with an area also led to the recognition of conservation-worthy areas). However, a very limited number of the designation reviews appear to be no more than extensions beyond areas of historical significance to incorporate fringe areas of built development or open landscape. Where designation is utilised purely as an opportunity to stifle pressures for development, or as a 'buffer' to the inner area, as has occurred in a number of cases (and see the Barham case detailed in Chapter 4), this is clearly poor practice in conservation, especially where no justified reasoning is provided for the review. More usually, boundary reviews result either from comprehensive surveys of already designated areas and their surroundings perhaps during the local plan review process, or the appreciation of more modern architectural periods, including incorporation of groups of recently listed buildings.

Although reviews of designated areas and, indeed, the most recent new designations, are incorporating a new variety of townscapes, it is inevitable that further regular reviews will be required to incorporate selected post-war townscapes or localities which are the subject of new detailed study.

The range and types of designated conservation areas

Several recent studies have examined the spread and typology of conservation areas, with Pearce *et al.* (1990) providing a comprehensive listing as of *c.* 1988. Annual updates and commentaries are provided by Max Hanna's annual reports in *English Heritage Monitor*, produced by the English Tourist Board. Jones and Larkham (1993) produced an overview from the early 1990s which, although less comprehensive than Pearce *et al.* (1990), usefully provides a geographical overview of several key types of designated areas; this information is used in the following section.

The types of areas designated by those responding to Jones and Larkham (1993) began with those areas traditionally viewed as 'conservation areas': historic city centres, town centres, village centres and some tightly drawn around specific buildings, churches or monuments. It was not until the mid-1970s, and especially the early 1980s, that much attention was paid to residential areas. Since then, the variety of areas designated has grown to embrace almost all townscape forms. This expansion in the variety of areas designated may result from a broadening of ideas about what is 'historic' and worthy of protection, following completion of a 'first phase' of designations in traditional historic town centres. It may also be a response to a new *laissez-faire* attitude to development outside conservation areas in the early 1980s (see Chapter 3). This attitude followed especially from publication of DoE Circular 22/80 (DoE 1980) which determined that the LPA should have no role in the design of development outside conservation areas (Punter 1986a gives a useful critique of the influence of this guidance).

Designation of historic towns

Commercial centres receive the most intense pressures for redevelopment and piecemeal change of any type of townscape (cf. Whitehand 1987a; see also Chapter 8). It is consequently not surprising that these were the first type of area which received attention from LPAs. In most cases, protection of the urban landscape was clearly the prime motive behind designation, but it is also evident that, in certain cases, conservation area status was (and still is) used to promote regeneration and is based upon the potential of the area in heritage and development terms: heritage references are currently very fashionable in retail environments (see, for example, the case of the Nottingham Lace Market, discussed by Crewe and Hall-Taylor 1991; Tiesdell 1995). Much of the research work for designating many historic town centres had been undertaken in advance of the 1967 Act, and the pattern of 'historic' towns and associated threats in Britain was identified in a Council for British Archaeology report published in 1972 (Heighway 1972). English Heritage has stated that, by the end of 1986, the centres of 94 per cent of pre-industrial towns had been designated. In contrast, only half of the centres of towns established after 1750, presumably considered to be 'less historic', had been designated (Pearce *et al.* 1990: xxii). Although the number of areas continues to increase, the designation of post-1750 towns remains incomplete.

As the number of historic towns is limited, it is unsurprising that, on average, authorities contain a small number of such conservation areas, often only one. There are slightly higher levels in the south and east of England, which may result from almost constant pressures for redevelopment in town centres and the need to designate areas to prevent continued erosion of the built heritage. Furthermore, the sheer number of towns in this area, with comparatively little open landscape, may be a reason for the higher number of conservation areas. Authorities with an above average number of historic town conservation areas include high-profile 'honeypot' towns such as Alnwick, Chester, Monmouth and Stratford-upon-Avon. However, a considerable number of such conservation areas are small, country towns with predominantly rural hinterlands. For example, Huntingdonshire, the New Forest, Horsham, South Somerset, Angus and East Devon District Councils all stated that they have five or more historic town conservation areas.

Designation of town centres

A majority of commercial centres could not readily be described as 'historic' in the same way as towns in the preceding category although, of course, they all have a history of at least local merit. This is certainly true of the definition given to historic towns in the English Heritage survey (Pearce *et al.* 1990). The designation of such 'non-historic' urban commercial centres followed those of historic towns in the early 1970s. Once again, the number of such areas is limited to approximately one per authority in all regions, except for the former West Midlands Metropolitan County (specifically the Black Country boroughs). Many of these West Midlands designations comprise previously free-standing towns in the Black Country which are now part of the West Midlands County conurbation. Their historic townscape is predominantly Victorian and derives from rapid economic growth as a legacy of the Industrial Revolution. Of note is the late introduction of designations in comparison with other metropolitan areas or the more rural Midland hinterland.

Designation of residential areas

The type of residential area designated varies greatly. These have typically been well-established residential areas, and mainly pre-Victorian. However, as has previously been indicated, increasingly more modern areas, even post-war examples, are being designated. Certainly, continuing development is leading to the inclusion of very recent residential development within the boundaries of older residential conservation areas, a problem discussed further in Chapter 9.

Twenty per cent of authorities have now designated twentieth-century residential townscapes, including a number of areas first developed in the post-war period. Conservation area designation and listed building guidance suggest an emphasis upon 'historic' townscapes. However, it is evident that some authorities are designating modern areas of high townscape quality, presumably to prevent them from architectural and aesthetic deterioration over time.

The majority of residential suburb conservation areas are concentrated in the South East and the metropolitan counties, with the exception of Tyne and Wear. In authorities such as the City of Birmingham, Wolverhampton, Bradford, Trafford and the City of Liverpool, residential conservation areas are the most numerous and were among the first such areas to be designated. Similarly in Scotland, the Cities of Edinburgh and Dundee stand out as having a high number of early designated areas. The highest concentrations of designated residential suburbs are in the inner boroughs of London. Kensington and Chelsea, Wandsworth and Islington have been particularly active in their designations. It is in London where suburban expansion in Britain has been most extensive and varied (cf. Barrett and Phillips 1987). The boroughs are charged with protection of some of the finest residential townscapes, including a number of laid-out estates, Georgian squares, and experimental garden suburbs as well as less affluent areas such as prime examples of inner area Victorian and Edwardian terraced housing, inter-war 'Metroland' ribbon development and model London County Council municipal housing. Although London authorities acted early to designate residential areas, presumably assisted by their status as local planning authorities from 1965, the adoption of residential conservation areas in regions outside the metropolitan sub-regions has been far more recent.

Designation of villages

English Heritage has identified the largest single type of conservation area in England to be the rural village (Pearce *et al.* 1990). A limited number of village centres, subsumed within the urban fabric of conurbations, are greatly outweighed by the number of free-standing village conservation areas designated in the rural areas of Britain (although the former may be the fastest-growing category of new designations: B. Hennessy, English Heritage, pers. comm.).

The highest concentrations are within the areas which have the most developed rural settlement patterns: lowland areas of East Anglia and the East Midlands and the metropolitan areas of Yorkshire. The rural upland and heath areas of Wales and the North have notably lower concentrations. Designations include industrial villages, and reflect the high proportion of quasi-rural commuter villages within the boundaries of the metropolitan counties.

As with the designations of town centres, many village conservation areas are established with the purpose of environmental enhancement in mind. Rural decline in many peripheral areas of the country has led to considerable deterioration of the built fabric and the setting of conservation areas. One aim of protection is to prevent further damage resulting from poor upkeep. However, without capital investment from local authorities it is unlikely that these already depressed areas will be significantly regenerated.

Many authorities in rural areas have been quick to designate villages which contain the most extensive areas of buildings and streetscape which have escaped the comprehensive changes of the Industrial Age. Consequently, the built form and plan will often form a high priority for conservation. However, in common with the designation of conservation areas in towns suffering industrial decline, authorities designating village conservation areas in depressed rural areas aim to prevent a further decline, inevitably sensitive to the possibility of warding off new residents, investment and ultimately economic growth in these settlements.

Designation of industrial archaeology and transport areas

Most recently, conservation areas have been designated around industrial archaeology and transport. There are relatively few examples, even in areas with significant industrial history, but these areas do reflect the diversity of recent designation practice. They are also a product of the growing importance of the 'heritage culture' including 'heritage tourism' and marketing.

Interest in industrial heritage has resulted in the establishment of open-air museums and heritage centres, as well as a considerable number of conservation areas in industrial quarters, such as Birmingham's Jewellery Quarter. Here, designation as a conservation area formed part of the rejuvenation of an historic quarter of the city, which had experienced considerable deterioration of its townscape (Slater 1994). Conservation area protection has been followed by streetscape improvements and investment in small workshops and businesses which constitute the character of this part of the City. Further large-scale enhancement proposals are being formulated (Tibbalds *et al.* 1990; Llewelyn-Davies 1993) and a conservation-oriented shopfront design guide has been published.

However, conservation area protection has been found inappropriate in a number of cases where large-scale industrial features or industrial landscapes merit some form of protection, but where the features are not sufficiently concentrated in one area to enable a coherent conservation area boundary to be drawn around them (Alfrey and Putnam 1991).

Transport-linked areas include those based around canal locks and basins, as well as those based on an increasing awareness of the quality of railway architecture and the valuable urban spaces that these land-uses involve. The contribution of rural and urban canals to public amenity space is well recognised nationally. It is evident from the increase in water-based holidays, the use of towing paths and the popularity of canalside sites for commercial and residential development. Redundant railway lines are also being re-used as cycleways and linear parks.

Both the buildings and open spaces associated with canals are highly characteristic of early industrial architecture, and convey a strong regional identity. Although the locks themselves will be preserved and maintained for operational purposes, designation is often used to conserve the buildings which are gradually becoming functionally obsolete, and to enhance the paved and planted areas alongside the canal. However, despite the apparent popularity of canals as a form of linear conservation area, the rationale behind designation of long lengths of rural canal of relatively little intrinsic merit and under no direct threat of development remains little explored.

A further type of industrial conservation area is large, often disused, central railway stations. The sites often have a high potential value for redevelopment. These extensive inner-city sites are attractive to developers. Many are functionally obsolete but aesthetically, culturally and historically valuable. Careful consideration of their character has led to a variety of preservation and enhancement schemes, including conversion of Manchester's derelict Central Station to the G-MEX exhibition and conference centre. Furthermore, many stations still in use retain a considerable range of historic buildings, often of local materials and reflecting local, regional or pre-nationalisation company styles.

Few railways have been designated as linear conservation areas, unlike canals. The Settle–Carlisle line is an exception. This may have been partly prompted by the public outcry when British Rail proposed to close the line. This unique conservation area is 76 miles long, crosses through several LPAs, but was a single designation on the initiative of the Yorkshire Dales National Park Committee with the co-operation of Craven and Eden District Councils and Carlisle City Council. English Heritage has agreed to the establishment of a Town Scheme for the entire area.

> The circumstances [of this area] are thought to be unique. The Settle–Carlisle Line is the only piece of railway in Great Britain to have conservation area status throughout its whole length. It is also unusual in its blend of distinctive heritage and working railway. It may be that in order for this complicated [grant-aid] scheme to work ... relaxations or dispensations of principles may need to be examined by all concerned, including the local authorities, English Heritage and British Rail.
>
> (Darlington 1993: 62)

Designation of other types of conservation areas

The wide variety of miscellaneous conservation areas includes those surrounding specific monuments, follies, historic parks and gardens and open spaces. They are frequently townscapes, and often open landscapes, which have some unique feature or character. Across England and Wales they are few and far between.

Historic parks and gardens

It is generally accepted that open spaces, including parks, gardens, coppices and orchards, are important to the landscape and, where appropriate, conservation measures should be used for their preservation. The Department of the Environment originally indicated that the conservation area legislation should not be used in the protection of historic parks and gardens. Despite this, a number of authorities have used it for this purpose. The Department's concerns presumably stemmed from the emphasis of the legislation upon urban conservation, of historic and architecturally valuable buildings and associated urban or quasi-urban open spaces. Yet it is clear that, from the mid-1980s, rural historic landscapes are becoming more studied, more used for recreation and thus marketed; rural vernacular architecture reappraised (cf. Horne 1993). Such trends may conceivably extend the traditionally urban focus of the conservation area as a policy instrument.

There are no planning measures specifically designed for protection of historic parks and gardens. To overcome this lack of protection, planning authorities employ a variety of means including Tree Preservation Orders, listed building law, green belt status, and registration as historic monuments as well as conservation area and other development plan designations. According to Stacey's study of the protection of historic parks and gardens in the planning process (Stacey 1991, 1992), conservation areas occur consistently among the four most popular methods used. A policy of designation of historic parks as conservation areas was pioneered by Derbyshire and Staffordshire and is now recommended by the Garden History Society (Lambert 1993). English Heritage is also compiling a second edition of the *Register of Parks and Gardens of Special Historic Interest*. It is envisaged that the final total of sites on the register will be no more than 2,000. This register will include only sites of national importance.

Some local amenity organisations, such as the Chiltern Society, are calling for more designations of gardens and open space conservation areas, and the extension of existing areas to include areas of landscape. There remains no comprehensive method for their protection. Although satisfactory for the protection of small areas of landscape and open space, conservation legislation appears to be inappropriate for areas where the majority of the proposed designation consists of soft, and especially informal, landscaping.

Agricultural archaeology and dispersed settlements

A final, and rare, example of the 'miscellaneous' area is the designation of areas which derive their special character from their extensive agricultural history. Many particular locations of Britain have been significant in agricultural advances (similar to those at the forefront of early industrial development). One such area, the Moss Valley, south of Sheffield, has been designated not only for its small and dispersed settlements but also for the complex mediaeval transport network and the complex agricultural and industrial patterns that exist.

Another innovative and successful (but atypical) conservation area is the Barns and Walls Area, a 72km^2 expanse of Upper Swaledale and Arkengarthdale in Yorkshire, designated in 1989. Co-operation between the National Park Committee who made the designation, English Heritage, who has made Town Scheme funding available (totalling £126,000 in 1992/93), and the Countryside Commission, who funded a two-year post for a Project Officer, has ensured the success of this designation despite initial opposition from local farmers. To date, 116 field barns and 4.5km of dry-stone walling have received grant aid, and discussions are proceeding to extend the scheme (Darlington 1993).

These examples are an imaginative form of designation, protecting field patterns, other aspects of the historic agricultural landscape and the characteristic dispersed settlement pattern. However, it is easy to see why, without clear advice and guidance, this type of area can be criticised for being a misuse of the legislation. With a description of why the area has been designated, and what is special about the landscape of this discrete area, its historic and architectural character is equally easily defended.

CONCLUSIONS

This chapter has explored practices of area-based conservation in a number of contrasting countries, together with a detailed study of the UK based on research undertaken for the Royal Town Planning Institute. The differences between practice in the UK and abroad are, in general, readily understood. The principal differences appear to be that the UK system is generally operated at a very local level, that of the LPA; in other countries, for example the Netherlands, control is largely at the level of the State. The UK, in contrast to such countries, appears to have greatly over-designated; others have arguably under-designated. In both cases, what actually happens on the ground is largely decided at the local level, although some degree of initiative is usually required in terms of adapting other forms of legislation or control to the purposes of conservation, as occurred with the Asnières POS in France. Yet the level of control exercised in this example is unusual, and it has yet to be seen whether this novel approach will become more widely adopted in France, let alone developed for other planning systems.

The UK system of area-based conservation has been subject to detailed examination, in particular through consideration of designation motives, dissenting views, changing criteria and the lack of designation review. This, together with the description of the types and distribution of UK conservation areas, raises important general questions about the designation process and the overall effectiveness of the area-based approach. Particular aspects of the built environment are over- or under-represented; there are clear regional variations and rather less clear temporal variations; motives for designation vary widely although the mechanism is sinple and effective; and approaches to boundary delineation also vary. Some summaries of conservation have tended to suggest that the UK has a uniform approach, based on the application of a uniform legal system (e.g. Great Britain 1993). It is clear from the discussion and examples given in this chapter that the application of area-based conservation in the UK is far from homogenous. The local variation is simultaneously a strength and a key weakness of the UK system.

The following chapters will explore some of the strengths and weaknesses. The UK local experience of area-based enhancement, and the relationship of development control in conservation areas to stated conservation policy, is reviewed in Chapter 7, following an introduction to the study of agents and processes of built fabric change which will be used throughout the remainder of this part. Chapters 8 and 9 then focus on a number of micro-scale examples, demonstrating to what extent (if at all) some of these larger-scale concerns operate at local levels.

6

DECISION-MAKERS AND DECISION-MAKING IN THE CONSERVED TOWNSCAPE

> An architect must perform in the dual role of designer for the future and defender of the past.
>
> (Richard England)

INTRODUCTION

Changes in the conserved townscape may be studied for their own sake alone, and this can provide information about rates, scales and types of change or non-change, as did Sabelberg's study of the persistence of historic features in the urban landscape of Tuscan towns (Sabelberg 1983). Yet this provides relatively little insight into *why* changes occur. Within the discipline of geography, urban morphologists have increasingly widened their activities from narrow conceptions of urban form itself to include consideration of the individuals, organisations and processes shaping that form (Slater 1990a). This line of inquiry is one of three prevalent in current geographical urban morphology which stem both directly and indirectly from the ideas of M.R.G. Conzen (Whitehand and Larkham 1992: 7). Yet relatively little is known of the operation of these agents of change.

> The reason is clear: much of the information necessary for a well-founded discussion of the role of private enterprise in the built environment is located in the files of individual people and organisations or, where it is available in a more consolidated form, as in the case of local authority building and planning records, it is dauntingly time consuming to extract.
>
> (Whitehand 1992b: 418)

Using the detailed data sources held by LPAs, reconstructions of urban development processes of unparalleled detail and completeness have been pieced together, sometimes for quite lengthy periods (ibid.). Such reconstructions greatly aid a second strand of Conzen-inspired research, concerned with the planning and management of the urban landscape. Detailed processes of decision-making are reconstructed and management procedures and policies examined. These two aspects of research are the concern of the present part of this volume.

In order to understand the actual changes occurring on the ground, it is necessary to consider the decision-makers involved. These may be divided into two groups: first, those directly involved in changes in that they initiate, design or implement development; secondly, those exercising external, less direct influence, mainly through the statutory system of development control.

DIRECT AGENTS OF CHANGE

The initiator of a change and the architect designing it are probably the agents having the most direct influence on the townscape. The architect also frequently acts as an intermediary between the LPA and the initiator. The initiator is often, but not invariably, the landowner. Specialised contractors and consultants are also of significance. These agents form a chain, or perhaps more correctly a web, of decision-making that begins with the initiator. The initiator makes the initial decision to begin the process of change, and it is almost invariably the initiator, rather than any agent acting for him or her, who is described as the 'applicant' in the application submitted to the LPA. Not only does this decision set in motion a train of events that leads to a change in the physical fabric, but also the initiator frequently exercises considerable influence over the choice of other firms and organisations that participate in the later stages of the development process. If the proposed change is substantial, the initiator is usually the building owner, but in the case of minor changes the initiator may be a tenant. The reasons underlying the initiation of a change are numerous, but one factor that is often involved is the obsolescence, in one respect or another, of the property (Chapter 3). Concern to improve the townscape is rarely a motivating factor.

The initiator usually employs an architect to design and draw the plans which almost invariably accompany an application. Selection and engagement of an architect is usually the first link in the web of decision-making, particularly where structural work is involved. The architect generally has considerable influence on the later selection of consultants, specialised contractors, builders and suppliers. The several other agents directly engaged in urban fabric change, notably consultants, builders and specialised contractors, are not linked in any set sequence. Standing relationships between these agents, and between these and the other direct agents, are quite common – where an initiator shows a marked preference for a particular architect or other agents, using them again and again in different development schemes (Larkham 1988a).

INDIRECT AGENTS

As with the direct agents of change, those having less direct influence have received little detailed study until relatively recently. The most valuable recent studies are those by Simmie (1981) on the detailed operation of the planning system in Oxford; by Short, Fleming and Witt (1986) on the planning system and its effects on housebuilding activity in Berkshire; and by Punter (1985, 1990b) on aesthetic planning control and office development in Reading and Bristol.

The first major group of indirect agents to be involved in the development control process is the LPA planning officers. They are professionally qualified and have responsibility both for large-scale planning (Structure and Local Plans) and for the direct and detailed control of development. Virtually all significant proposed changes to the built fabric, and material changes of use, must receive the permission of the LPA. When a formal application for permission is submitted, not only may it be preconditioned by the initiator's prior knowledge of the planning officers' attitudes to certain types of development, but subsequent negotiations between the applicant and the planning officers may be of considerable importance in changing an initially unacceptable application to a form found acceptable by the LPA.

The actual decisions on planning matters are made by the Planning Committee of the local authority, which is composed of elected public representatives, amongst whom any aesthetic or technical knowledge or training is rare. Indeed a Past President of the Royal Institute of British Architects (whose bias should be taken into account) has criticised these committees as being

> cultural pariahs operating out of their depth, with extraordinary arrogance, and frequently destroying the efforts of some of the more imaginative architects. Their role should not be the making of aesthetic and planning judgements.
>
> (Manser 1980)

These committees discuss the formal recommendation of the chief professional planning officer on most applications (in some, usually minor, cases, authority has become delegated to the chief planning officer to make decisions on the Committee's behalf). It is therefore the professional planners who determine, to a large extent, the nature and detail of the information upon which the Committee acts. Nevertheless, the Committee makes the final decision. Although in practice heavily dependent upon the advice of their professional officers, with one study suggesting that Committees reversed the officers' recommendations in only 1.8 per cent of 6,709 cases studied (Fleming and Short 1984), Committees may act in a variety of ways, suggesting that other pressures (including local and national political concerns) have acted upon them. They are reputed to spend more time debating detailed design changes than the principles involved in large projects, and some are particularly active in making detailed modifications to applications (e.g. Witt and Fleming 1984: section 5.4).

However, councillors' attitudes to any planning discussion, including that on aesthetic matters, are conditioned not only by the planning officers' recommendations and other lobbying, but also by factors such as ideological and political position, attitude towards the role of the elected representative, and perception of the role of planning (Witt and Fleming 1984: Chapter 6). Their position is of paramount importance in the democratic decision-making process of planning. However, their role as agents of change in the townscape is, in practice, probably considerably less influential than that of other agents, including the professional planning officers.

The last significant indirect agent is the general public. The influence of the public in UK planning has increased considerably since the mid-1960s. The increase in official regard for public participation seems to be a consequence of a new paradigm in planning (Long 1975: 73). This began with the critique of the existing planning system, based on the Town and Country Planning Act 1947, which was prepared by the government-appointed Planning Advisory Group (1965). The Report of the Committee on Public Participation in Planning (Skeffington 1969) is popularly upheld as being another step towards greater public participation, and a number of studies point to this report as being an important landmark in the rise of the voluntary amenity movement (e.g. Barker 1976). Yet the report has been widely criticised, and its recommendations have been considerably diluted by subsequent government action.

Although individual members of the public rarely respond to invitations to comment on planning applications, local amenity societies or pressure groups respond more frequently and, because they often possess considerable local knowledge and, in many cases, some professional expertise, they may be able to present the public's viewpoint with force, even eloquence. However, the impact of local amenity societies is difficult to assess. They have few sanctions that they could bring to bear on a recalcitrant planning authority, save for the possible stirring-up of adverse publicity: 'the persuasive power and status of any group is a function of the size of the problem which it would create by its non-cooperation' (Moodie and Studdert-Kennedy 1970: 65). Having little leverage to exert, the societies usually rely on persuasion and, increasingly, upon becoming incorporated within the planning system through various consultation processes. Informal contact between societies, councillors and planners is high (Barker 1976; Buller and Lowe 1980), and there often seems to be a striking similarity between the views of local planners and those of the amenity societies consulted (Barker 1976; Gamston 1975; Larkham 1985). It is true that many LPAs see such local groups as potential allies (Jennings-Smith 1977: 245). Co-operation is thus important, whereas direct confrontation is often seen as self-defeating because it may prejudice, to the detriment of future consultations, a close working relationship that has been built up. The number of documented cases where the views of local groups and LPA have diverged, and the amenity group prevailed, is small in comparison to the total number of planning decisions being made. Thus, however much consultation is urged by the DoE, the position of amenity societies is somewhat anomalous within the planning system. Their influence is limited, indirect and reactive.

A last factor that should be considered is planning policy. At a national scale, the DoE issues guidance to local planning authorities which, together with the official attitude towards listed buildings and conservation areas, could be said to form a national conservation policy. This is now contained in Planning Policy Guidance Note 15 (DoE 1994). As studied in Chapters 2 and 4, however, this policy is confusing and sometimes contradictory in implementation. At the local scale, LPAs may act as direct agents through the formulation and implementation of policies for the enhancement of historical townscapes. Such policies do exist, and enhancement schemes are considered further in Chapter 7, but their effectiveness may be questioned, since there is little finance (either from local authorities or central government) to carry out such proposals. The achievement of other aims and policies of conservation depends upon the effectiveness of the development control system, which is more a negative form of control preventing what is perceived as 'inappropriate' change rather than encouraging favourable change, and this is also examined further in Chapter 7. As this is essentially a short-term system, and since aggrieved applicants may appeal to the Secretary of State against a local authority's decision, the UK planning system is in some difficulty in attempting to cope with conservation problems (Blowers 1980; Larkham 1986: Chapter 8).

Yet policy, as Bruton and Nicholson (1984) suggested, is often non-statutory, and even unwritten. Powerful individuals or groups within an LPA can *de facto* make 'policy' through concerted action, often over a lengthy period. Such 'policy' may not reflect formal local or national policy documents. Thus, when one planning officer arrived in the historic city of Peterborough, he soon became aware

> of an unholy trinity – city engineer, chief building inspector and deputy town clerk – who completely ignored listed building legislation requirements by serving dangerous structures notices and had successfully removed several 'awkward' buildings in that way.
>
> (Peterborough planning officer, pers. comm.)

Likewise, much of Manchester's Cathedral conservation area has been demolished. Despite designation in 1972, and firm policy statements since, the LPA continued to use dangerous structures notices made under the Manchester Corporation Waterworks and Improvement Act 1867 to secure demolition of numerous key buildings. SAVE Britain's Heritage has been critical of this use of outdated powers to overcome national listed building requirements and local conservation policy.

> Manchester Council's use of the 1867 Improvement Act to deal with supposedly 'dangerous' structures represents a considerable extension of the normal dangerous structure procedures. . . . The Manchester Act states that the building is 'to be taken down either wholly or in part, or to be repaired or secured *in such manner as the Corporation shall think requisite*'. . . . Its use in the 1980s is an anomaly. The suspicion must be that the Council is prepared to use its powers to clear the way

> for developments which it sees as desirable – in some cases even before planning permission has been sought.
>
> (Powell and Fieldhouse 1982, their emphasis)

In general, though, indirect agents have little visible effect upon the form of development. In some cases, however, both the public (as represented by amenity societies) and the planners may have a decisive influence on the appearance of new buildings. Ealing Civic Society, for example, appears to have been influential in rejecting a tower block, office and shopping centre in favour of a town square with colonnaded shops and an absence of high-rise blocks (Griffiths 1985). Similarly, in two major retail developments in the heart of the historic city of York, the Planning Department has been a vital influence. The store for J. Sainsbury PLC was made possible after long negotiations both with prospective developers concerning commercial viability and design, and with the elected Planning Committee to reverse a previous policy opposing retail development of the site (Larkham 1988c: this example is further discussed in Chapter 10). For the Coppergate shopping scheme, a design and detailed tender competition based on a planning brief led to the selection of a scheme incorporating a massive multi-storey car-park, flats, shops, and the Jorvik Viking Museum, within buildings clad in traditional brick and roofed with pantiles. This scheme won the 1985 Silver Jubilee Cup awarded by the Royal Town Planning Institute.

CHARACTERISTICS OF AGENTS AND THEIR TOWNSCAPE IMPLICATIONS

Recent morphological research suggests that two of the most significant characteristics of agents of change – most particularly the direct agents – are their provenance and type (Whitehand 1984). The ways in which these characteristics have changed through time and across the country has recently been examined (Whitehand 1992b). Decisions made by local agents, especially initiators and architects, appear generally to result in different types and styles of alterations or additions to the building stock than decisions taken by agents based far from the site of the proposed change.

> It seems inescapable ... that boardroom decisions taken in the metropolis against a background of national scale operations would have produced different results from those taken by local individuals with a field of vision ending abruptly at the edge of their town's sphere of influence.
>
> (Whitehand 1984: 4)

It may be felt that national firms based outside a given area may be insensitive to local circumstances and traditions, and 'the extent to which firms have local roots and are imbued with a sense of place takes on a special significance' (Whitehand and Whitehand 1984: 245). A 'sense of place', or *genius loci*, is usually seen as a very individual response to a familiar locality

(Chapter 1), and it is expected that this would be more pronounced in local agents.

There is a tendency for large national concerns to be based in London, and thus London-based agents have a large impact. In the case of retail developments, a major aspect of this is the adoption by each chain of its own house style, which is reproduced in many different towns. Minor changes to commercial property are more often dealt with by local agents on behalf of the chains. Small changes to dwellings are often initiated by owner-occupiers, often caused by a change in the family life-cycle (cf. Slater 1978), while large residential developments are undertaken by local authorities or major housebuilders, the latter still retaining some local characteristics (Ball 1983). Whether an agent is a speculative developer or is building for owner-occupation may well affect the architectural style used. Commercial owner-occupiers may adopt new architectural fashions more rapidly than speculative developers, as has been shown in the central areas of Northampton and Watford in the inter-war period (Whitehand 1984). Circumstantial evidence suggests that this also applies to residential development. However, this stylistic conservatism of speculators is not so marked in the post-1950 period for either town centres (Freeman 1986) or residential development (Larkham 1986).

Bentley (1983) similarly suggested that the type of developer influences the style of development. He argued that the agents whom he termed the 'institutional patrons' of architecture, principally pension funds and insurance companies, have adopted a particular style of development. This is characterised by an emphasis on new buildings rather than old ones, which reinforces the move towards a more restricted range of uses in town centres; a desire for large projects rather than small ones, with a consequent decrease in the number of routes available through the area concerned; the repetition of standard, adaptable building designs; an emphasis on tall, free-standing buildings; and a separation of buildings from their street settings by zones of open space, often filled with planted barriers (Bentley 1983: 7). In short, that style is Modern. Smaller developers have never needed the economies of scale and other benefits that the Modern style gave to the large 'institutional' developers. Larger developers are themselves poorly represented in small towns and hardly at all in villages. Smaller, regionally-based firms may have adopted particular styles more for reasons of prestige than for the reasons suggested by Bentley, as the importance of fashion (architectural or other) as an image-enhancer should not be discounted.

The inter-relationships of agents: an example

It is instructive to consider the general points made above in the light of actual examples. A detailed study of four Midlands conservation areas (Larkham 1986, 1988a) has allowed identification of the majority of agents of change active over the period 1967–84, and their spatial inter-relationships may be analysed. The areas are the town centre of Solihull; the core of

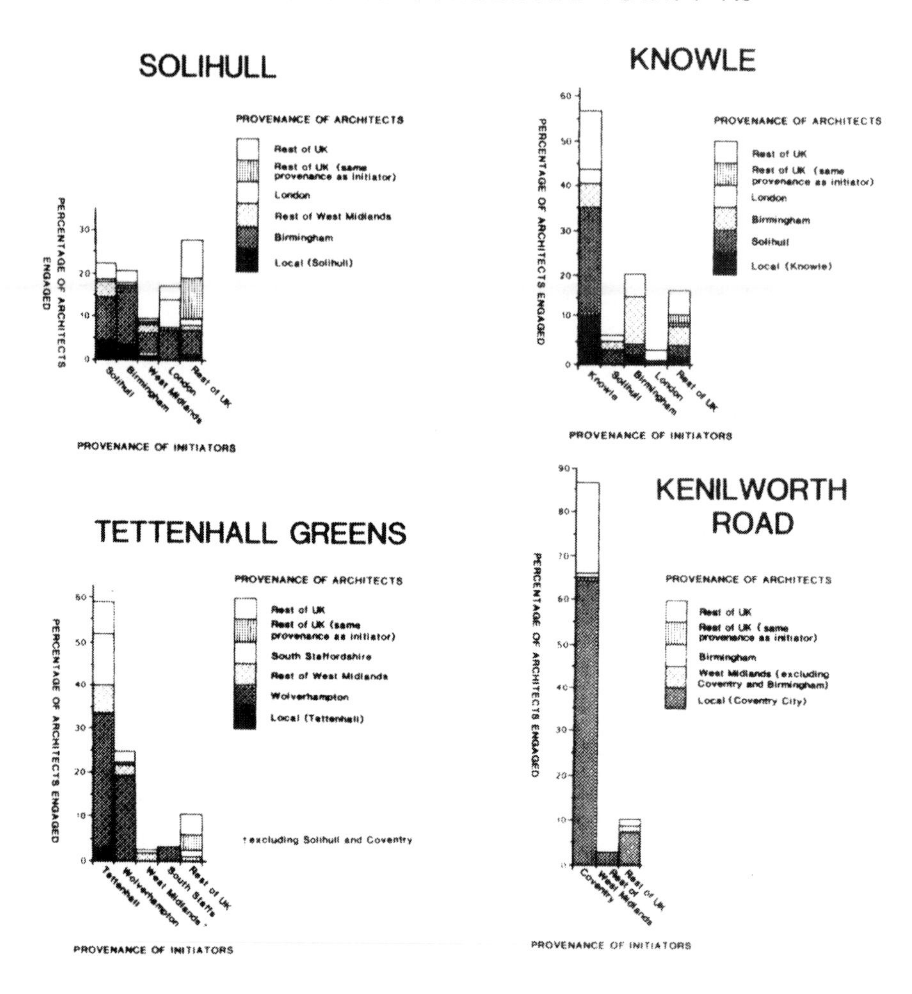

Figure 6.1 Spatial relationships between initiators and architects in four West Midlands conservation areas

Knowle, a commuting village close to Solihull; Tettenhall Greens, a suburb of Wolverhampton; and Kenilworth Road, a major radial route into Coventry. Analysis of the provenance of the initiators of change and the architects used gives very different results for the four areas (Figure 6.1). This may be related to the characteristics of the areas and of the agents.

Knowle and Tettenhall show a basic similarity, with local initiators engaging just under 60 per cent of all recorded architects. However, the provenance of architects engaged by these local initiators varies. Only 3 per cent in the Tettenhall case are local, compared to 10 per cent for Knowle. But the addition of architects based in the nearest large towns of Wolverhampton (for Tettenhall) and Solihull (for Knowle), each of which is only 2–3 miles away, brings the total in both cases to some 35 per cent. The overall pattern seen for Tettenhall, where Wolverhampton-based initiators are next in importance to local initiators in engaging architects, with a decline in distance thereafter, is not seen in Knowle. Here, the local centre of Solihull is considerably less important than Birmingham, which is of major importance in its number and variety of architectural practices, and has a distinct 'metropolitan' influence.

In both Tettenhall and Knowle, it is clear that architects from the study areas themselves are of little importance whatever the origin of the initiator. This appears to be a result of the low numbers of architects practising from local addresses: Tettenhall having but one, and Knowle only two until 1983. In general, initiators from close to, but not actually within, the study areas appear to employ more architects from the same location as themselves than architects of any other provenance, and this may simply be a result of familiarity. Familiarity with, or personal introduction to, architects was an important way of making contact throughout the study period, particularly for local small initiators, as interviews with various local architects and developers showed. It was not until March 1984 – virtually the end of this study period – that the RIBA Code of Practice allowed architects registered with the RIBA to advertise, and non-registered architectural practitioners were restricted in advertising by the Architects' Registration Act. However, of those initiators from the 'rest of the UK', many are 'corporate clients who are well equipped to find their own way to capable performers of architecture' (Harrison 1984), if they do not actually employ their own architects in-house. Other initiators active in Knowle are strongly concentrated in the West Midlands region, if outside the county itself: they may be expected to know, and employ, regional architects; indeed many of the architects employed by this category of initiator are based in Birmingham or Solihull.

Data for the Kenilworth Road area show a marked difference from either Tettenhall or Knowle, the two 'village' areas. Coventry-based initiators account for some 86 per cent of the architects engaged, and this includes some 64 per cent where the architect is also from Coventry. Initiators from the rest of the county are few, and use Coventry architects, as, largely, do those initiators from the rest of the UK. There is minimal employment of architects based elsewhere in the county, but just over 20 per cent originate from elsewhere in the UK. Thus, in this almost totally residential area, there seems to be a very strong link between local initiators and local architects, seemingly influenced by the types of changes for which planning applications were made, and the variety of Coventry-based architects interested in this type of small residential work. For Solihull, there is a pattern quite different from any of the previous areas. Local initiators are employing only about 22 per cent of the named architects, with Birmingham-based initiators close behind with some 21 per cent. In this commercial area, therefore, local initiators are of less consequence. Solihull's local architects are likewise of little importance, while Birmingham architects are employed quite widely by all groups of initiators. Particularly for Birmingham, and slightly less so for the 'rest of the UK' group, there is a tendency for both initiator and architect to have a common origin. This is only partly explained by company architects operating from the same address as a company initiator.

The Coventry-based architectural practice of R.A. Geden ARIBA is a useful example to illustrate some of the trends outlined, particularly in relation to the residential Kenilworth Road area. This practice was set up in 1962, at a time when other local architects 'openly stated that they were not interested in residential work'. The practice originally grew on the basis of 'one-off' houses and very small estates, mostly local, and became known mostly by word-of-mouth recommendation. The practice has won four awards for this type of housing, including the Daily Mail Award for the Best Speculative Home, 1978. The practice 'does not concentrate particularly on this area [Kenilworth Road], and quite a large proportion of our work does take place in conservation areas throughout the West Midlands and, to a small degree, in other areas'. Although these types of housing remained important to the practice into the 1980s, it had diversified into public housing, factories, offices and a variety of other buildings. According to the practice's broadsheet, most work was carried out within the West Midlands and Warwickshire (R.A. Geden, pers. comm.).

One architectural practice may thus be seen building up a reputation for one particular type of residential construction, which happens to be well-suited to the conditions found in Kenilworth Road, in addition to diversifying into other areas and building types. Although private recommendation seems important to the practice, publicity from competition successes must also help it to become known. Its success may be judged from the 47 applications, for all types of changes, that were submitted in the Kenilworth Road conservation area during this study period. Although not all were approved or implemented, the significance is that this particular practice was chosen by a large number of initiators, mostly local, and indeed was used more than once by several initiators.

Another relevant example is provided by the former Anglia Building Society, which had engaged the Solihull-based architects Warr, Barnett and Sadler on several occasions for applications relating to premises in High Street and Poplar Street, Solihull. In general, the Society dealt with most of its design work in-house. Large projects or overflow work were usually dealt with by outside consultants, selected from a panel with whom the Society had built up a working relationship over a number of years. On less than 10 per cent of the occasions where outside consultants were employed were they engaged for their particular local knowledge, or where the Society had negotiated contracts with a developer habitually employing a local architect (Design Manager, Anglia Building Society, pers. comm.). The Anglia Building Society would thus usually be an external initiator (based in Northampton) employing an external architect of the same provenance, its own Design Department. The use of a local architect in several cases in Solihull was unusual.

AGENTS AND TOWNSCAPES: SOME GENERAL PRINCIPLES

Based upon a body of recent research in a number of historic town centres and residential areas, a number of generalisations may be made about the influences of direct agents of change on the adoption of types and styles of development in recent decades. These are particularly relevant to town centres owing to the numbers and scale of new developments.

1. The initial adopters of new architectural styles in any given area, especially of styles that are novel (for example, Art Deco and Modern) as opposed to historicist (for example, neo-Georgian and neo-Tudor), tend to be external architects and/or initiators.
2. Following a stylistic innovation, local architects and initiators are often very quick to adopt, adapt and diffuse the style (as was the case with Art Deco).
3. In town centres particularly in the inter-war period, innovators have tended to be either local initiators seeking to present a new image by using an external architect, or external chain stores redeveloping sites and often using in-house architects to apply a house style.
4. During that same period, innovation was largely for bespoke (i.e. owner-occupied) building. Speculative developers have tended to be conservative in their styles.
5. There tend to have been significant time-lags between the initial adoption of a new style and its later widespread adoption in the same area.
6. Speculative developments are particularly important numerically in the diffusion of styles following their introduction often as bespoke developments.
7. Although one architectural style may dominate a period, there are few sharp divisions between architectural-style periods.
8. For both bespoke and speculative developments, external agents are more likely to build on a large scale. Most recently, larger developments in town centres may be hidden in whole or in part by retained façades or by the subdivision of long façades to give the superficial appearance of several buildings.

Superimposed upon these general principles, and affecting their detailed operation, are considerations of national and regional economic trends and fluctuations, building cycles, the position of towns in the urban hierarchy (which may, for example, determine demand and the number of local agents available), the degree of Metropolitan influence, and changes in architectural fashion on the part of both architects and their clients. The operation of these factors will ensure that each area remains different in detail, although the general principles will always be of importance in explaining the form that development actually takes.

This perspective on the decision-making process informs the remainder of this part. Chapter 7 examines the LPA both as a direct agent through consideration of conservation area enhancement, and as an indirect agent through studying the relationship of stated policy to the development control decision-making process. Chapters 8 and 9 examine in detail changes to the conserved urban landscapes of town centres and residential areas, and the processes and agents involved.

7

AREA ENHANCEMENT, DEVELOPMENT CONTROL AND POLICY

> Stop this useless longing for the past. Pass by, we are working for the future since the threads of history are in our hands!
>
> (George Bernard Shaw)

INTRODUCTION

There is a clear expectation in UK legislation, and on the part of local people, that it should be the responsibility of the LPA to 'manage' conservation areas, with regard to enhancement, guiding development, and development control. Authorities which are not seen to be carrying out such actions are subject to sharp criticism, as in Reade's commentary on the problems of management and the lack of investment and enhancement in the conservation area of Upper Bangor (Reade 1991, 1992a, 1992b). Yet, equally clearly, in practice there are problems with priorities and funding for direct intervention in conservation areas, for example where an individual LPA has a large number of designations. Not every area can thus expect large amounts of public investment and enhancement. All areas, however, are subject to the nationwide processes of control over development, and LPAs should exercise their guiding and controlling functions in each designated area. The function of this chapter is to examine both facets of LPA activity: the direct intervention in terms of 'enhancement', and indirect action in terms of controlling development, in particular with respect to the relationship between development control decisions and stated conservation policies.

ENHANCING UK CONSERVATION AREAS

Introduction

The formal aims of UK conservation, as embodied in the legislation, place considerable weight on the value of 'enhancement'. Indeed, authorities are now required, under Section 71 of the 1990 Act, to prepare proposals for the enhancement of each conservation area. Consequently, management of the townscape, whether in conservation areas or outside them, is most often represented by a variety of environmental enhancement schemes. These are usually highly localised and often do not result from co-ordinated enhancement proposals following from the protection of areas through designation.

Yet, although this legislative emphasis upon enhancement does explicitly recognise that designated areas can and should change, and should not be 'preserved in aspic', the direction of that change is problematic. Enhancement is another legislative term never defined in statute. It is 'a wildly indefinite term' (Dobby 1978). Guidance does suggest that enhancement could appropriately include

> (a) the removal of all that presently harms the character of the area (for example unsightly street furniture to be removed or upgraded, vacant sites to be landscaped); and
> (b) the promotion of positive improvements (such as design policies for specific sites or new buildings generally, paving schemes, grant-aiding the replacement of missing architectural details).
>
> (Mynors 1984)

In practice, driven often by the need to justify actions and decisions to DoE Planning Inspectors and in court, Vallis (a practising Conservation Officer) has resorted to dictionary definitions, suggesting that enhancement means 'to heighten or intensify: to boost or strengthen' (Vallis 1994: 18). Guidance from English Heritage prefers the single word 'reinforce' (English Heritage 1993).

Such approaches raise the question of whether enhancement is *de facto* changing the character or appearance of areas, or merely reinforcing existing characteristics. In either case, one should know exactly what the character and appearance of a given area are, which strengthens the requirement for detailed character appraisals, as is argued by Morton (1991). Vallis notes with regret that the *South Lakeland* decision has relegated 'enhance' to a secondary position, owing to its emphasis on 'preserve'.

> But it is still relevant in terms of its proper meaning. However, there is a tendency to misuse the word, using it to mean merely 'to make more attractive'. If it is used in this way, subjectivity is being introduced, and I feel that this is quite unjustified.... Strictly, the dictionary defines enhancement as 'action taken in the treatment of an *existing* situation'.
>
> (Vallis 1994: 18)

Yet there have been several critiques over what LPAs are actually doing under the label of 'enhancement'. Even in 1971, the noted urban designer Gordon Cullen commented that 'we have seen a superficial civic cycle of decoration using bollards and cobbles' (Cullen 1971). Oliver suggested that the manner in which small-scale streetscape enhancements were being carried out in the 1970s was destroying individual character and place-identity by promoting a form of nationwide anonymity and uniformity (Oliver 1982). More recently, Booth (1993) suggests that early enhancement schemes could be seen as positive, in that they took steps to remove unsightly items which detracted from the appearance of areas; particularly overhead power and telephone wires and street sign clutter. More recent schemes, he contends, add new and alien elements, such as 'floorscape enhancements' using inappropriate materials such as coloured block pavers in herringbone pattern and the proliferation of 'heritage street furniture', purchased from catalogues, not specific to the area, and appearing throughout the country's historic (and other) towns. Montgomery (1992) criticises this as being merely 'bland "mail order" street furniture'. Newby (1994) forcefully makes this point, in a discussion of the tourism dimension to conservation and enhancement of historic districts.

> Enhancement is a further challenge to individuality. The installation of street furniture to support the heritage dimension of the townscape is now widespread. Throughout Britain there seems to be a feeling that 'nineteenth-century' cast iron goes with all periods of townscape. The message that this is a quality environment is expressed by having the lettering picked out in gold.... At first this approach was individual, but its repetition throughout our historic towns serves more to destroy its individuality than enhance it.
>
> (Newby 1994: 220–1)

One large LPA not only uses heritage replicas in some areas but adds cast heritage-style decorative elements to standard lamp columns in others, although, increasingly, it is using colours other than the standard black! The proliferation of 'heritage bollards', often (but not always) in an attempt to control intrusive car-parking, is also a problem. A rare deliberate attempt to move away from standard cast-iron 'heritage furniture' was made in Derby Street, Leek, in July 1995. This enhancement scheme used 16 hand-carved oak bollards, gritstone horse troughs as planters, and brightly-painted benches. Yet, in general, it seems, as Gamston noted over two decades ago, that there are

> evident dangers in the idea of enhancement becoming too closely associated with our own culture-bound notions of 'prettification'.
>
> (Gamston 1975: 19)

Recent guidance from English Heritage (1993) reinforces Booth's message to do less but to do it better. Vallis (1994: 21) also noted a growing tendency for use of the term 'enhance by contrast', particularly relating to the insertion of new, large-scale developments, usually of alien style and materials. He feels strongly that this term is a nonsense: 'contrast does not mean enhance'. Nevertheless, this is a common tactic, for areas and individual buildings (Figure 7.1). Public reactions to these contrasts are often initially negative. This concept is discussed further in Chapter 10 in relation to new building styles.

In order to direct 'enhancement' towards the definition of 'strengthening existing qualities', Booth has suggested several guidelines to avoid the

Figure 7.1 'Enhancement by contrast': Kirby House, Coventry (author's photograph)

insidious mediocre or heritage replication schemes of recent years. These include

> selection of natural, not imitation, materials;
> departing as little as possible from originals;
> limiting the range of materials to those functionally necessary;
> observing local detail in surfaces and in street furniture;
> and, above all, to resist gilding the lily.
>
> (Booth 1993: 23)

LPA enhancement schemes

Regardless (often quite literally so) of the above debate, LPAs often carry out a wide variety of 'enhancement schemes'. In this way they can be seen both to be discharging their statutory duties to conservation areas, and also to be actively managing processes of change. Yet not all schemes are initiated nor wholly funded by the LPA. Figure 7.2 illustrates the variety of schemes and the agents involved. Clearly, the most common form of enhancement scheme is pedestrianisation, the success and failures of which have been well documented in the planning press. In particular, there are issues of planning and design. In planning, the resolution of pedestrian/vehicular conflict often takes priority over the character and appearance of the area. In design, the

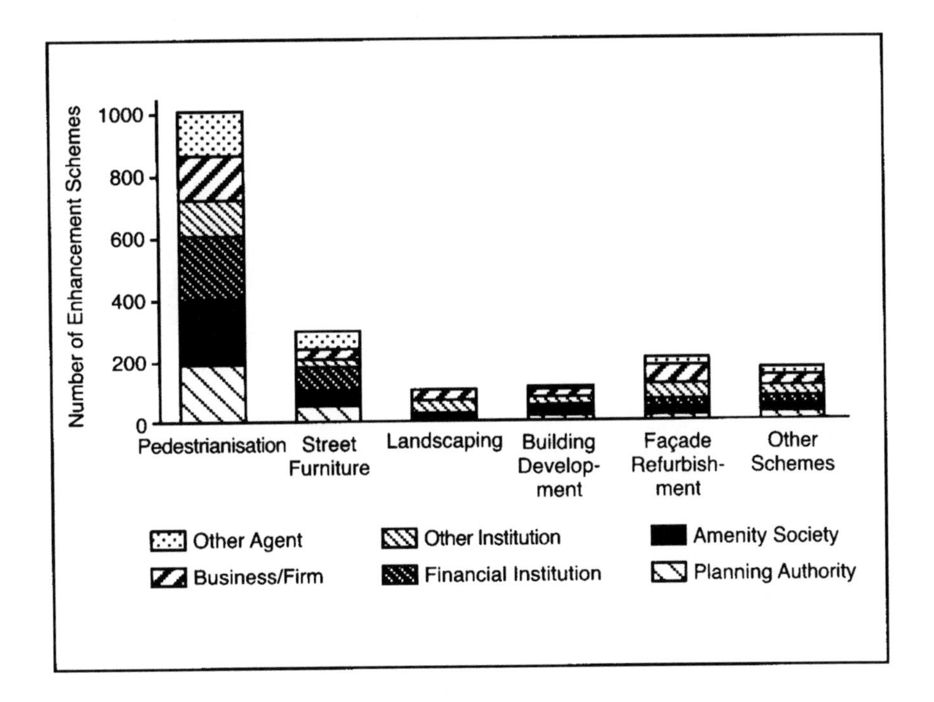

Figure 7.2 Types of enhancement schemes suggested from a survey of LPAs (adapted from Jones and Larkham 1993)

traditional patterns and proportions of building:pavement:street:pavement: building have often given way to an undifferentiated sea of paving stretching from wall to wall. Nevertheless, pedestrianisation remains a very popular strategy. A variety of agents has been active in initiating and especially providing capital for these schemes. They include the local authority, amenity societies, various institutions and local businesses and development agencies.

Other enhancements include street furniture projects but, again, the local authority, amenity societies and other institutions have leading roles to play. For other schemes, institutions (excluding banks and property investment companies) have played a disproportionately active role, especially in landscaping and façade refurbishment. However, discussions with local authorities and amenity groups emphasise their role in suggesting, although not in carrying out or paying for, such schemes.

Development grants and funding

Funding of enhancement activity is central to the success of conservation. In the opinion of the majority of LPAs, the availability of grants from local authorities, central government or from other grant-giving bodies has not been high since the late 1980s (Figure 7.3). This deficit has, in the past, been acknowledged by the DoE. The limited funding available appears to be highly targeted, with historic town centres and villages receiving the larger share (Figure 7.4). Only 15 per cent of authorities suggest that funds are available for environmental enhancement in residential conservation areas, 7 per cent of areas based around industry and archaeology and 11 per cent of other areas. Funding from English Heritage is likely to become far more restricted in the mid-1990s as the previous grant regimes are replaced by a much more sharply targeted 'partnership' scheme (*Planning* 1994).

Funding sources for enhancement are, however, not limited to the more formal channels and, despite limited budgets, many authorities are able to fund (or provide the stimulus for funding) a variety of enhancement schemes. Clearly, with issues such as planning obligations and compulsory competitive tendering high on the current political and planning agendas, the role of the private sector in funding enhancement is set to increase. However, it is likely that the impetus for private sector investment will still be stimulated only by local authority or other public body pump-priming. Jones and Larkham (1993) report that many officers remain concerned about this responsibility where grant funding may be directed at one private scheme and not others, thus raising the issue of equity. A number of local authority officers express great concern that conservation budgets, already small, are usually first to suffer in local authority expenditure cutbacks. Although such allocations are low, their importance in 'pump-priming' should not be under-estimated: cuts here have a disproportionately large effect on conserved townscapes, particularly so during periods of economic recession.

Direct capital programmes

Local authority budgets often include works such as street paving, tree-planting and street furniture and often supplement grant funding from English Heritage or Historic Scotland. Reductions of local authority budgets throughout the mid- to late 1980s have resulted in many authorities reducing the quantity of money specifically directed at conservation, often completely. Wolverhampton, for example, at one point reduced specific conservation allocations from some £30,000 to zero; even historic towns such as Stratford-upon-Avon suffered significant reductions, while Chichester not only reduced its budget but also reduced its staff. These economic constraints have necessitated a more imaginative approach by officers to achieve conservation enhancement.

Many of the more successful programmes, whatever the capital programme funding, appear to rely upon a consensus of support for conservation. Often this support runs with a commitment to the environment, both built and natural, with environmental improvement and enhancement being at the core. Public consensus is more difficult to achieve. Officers frequently complained of public concern over the grant aid of schemes to improve

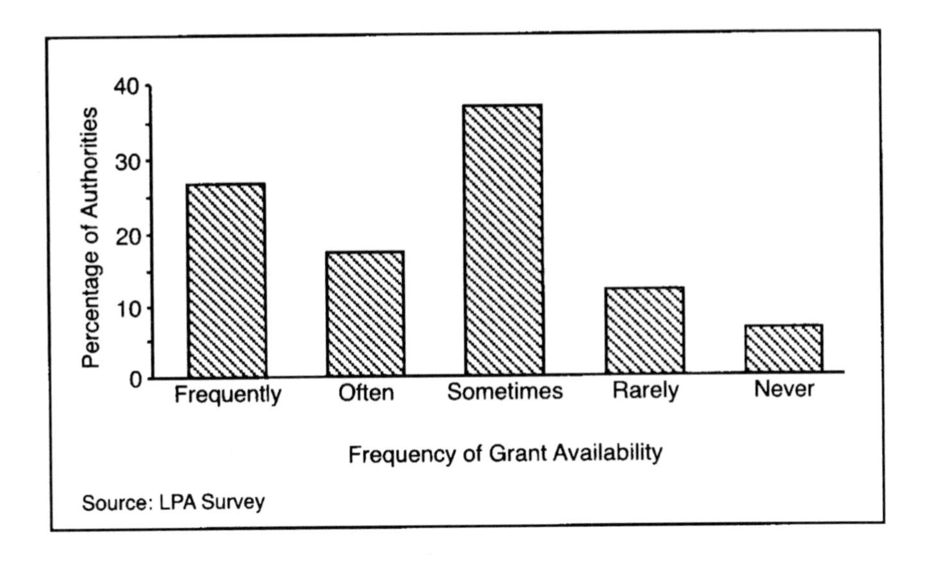

Figure 7.3 Availability of enhancement-related grant aid (from Jones and Larkham 1993)

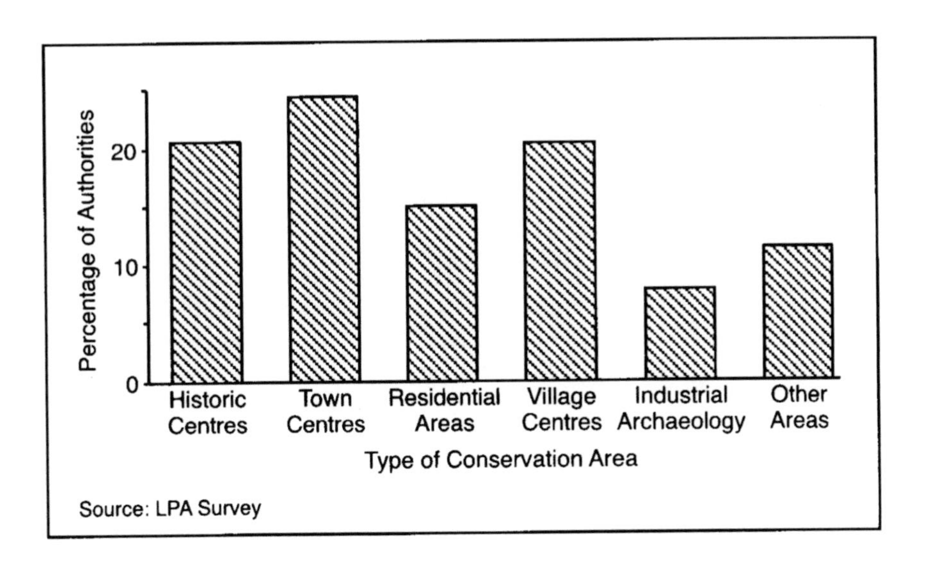

Figure 7.4 Targeting of grant aid (from Jones and Larkham 1993)

'eyesores' when individuals cannot gain funding for private reinstatement or enhancement.

There is concern that the largely politically driven requirements for compulsory competitive tendering (CCT) are not assisting the goal of enhancement. Low tenders can lead to low-quality results, but there is considerable political and financial pressure to use low tenderers. Some companies have also been reluctant to tender for projects where high-cost, high-quality materials (e.g. York stone paving) are specified. CCT and European Union legislation have also led to foreign (i.e. not local to the area) materials being used in enhancement schemes, purely for financial reasons. In some paving schemes in Scotland, for example, Spanish granite setts have been laid by Italian craftsmen, rather than using indigenous original materials and local labour (K. Murray, Tibbalds Monro, pers. comm.).

Stretching limited conservation enhancement budgets

Many authorities have little or no formal budget to spend on enhancement in conservation areas, or to provide grant funding for enhancement works. However, imaginative collaboration with colleagues in planning departments and in housing, environment or highways departments can ensure that conservation and enhancement of designated areas remains on the funding agenda. Much of the success of this collaboration relies upon the commitment of individual councils to the historic built environment and the influence of conservation sections with colleagues elsewhere in the LPA. Where it is successful, the detail of rehabilitation projects and improvements made possible by external funding sources can be greatly influenced by conservation staff.

Highway 'improvements'

The issue of traffic in conservation areas is often central to schemes for enhancement. Traffic flows and related issues such as car-parking are crucial to both the continued functioning, and the character and appearance of the majority of conservation areas. Indeed, pedestrianisation appears to be the most prevalent of enhancement activities (Figure 7.2), highlighting the importance of a traffic-free environment in conservation areas. Its popularity stems from the first such scheme, in Norwich, in the late 1960s: although small in scale, removing traffic and enhancing paved surfaces here constituted a significant improvement. Although such schemes are usually restricted to town centres, the problems of through traffic, on-street parking and the incorporation of traffic-calming measures are also important in many of the rapidly increasing number of residential and village conservation areas.

Where the highway authority is investing in 'improvements', these usually relate only to safety, parking and traffic-flows; this is often at the expense of visual character. This limitation can be highly damaging to the historic character and appearance of conservation areas. Many district conservation officers consider that a good working relationship with the highway authority, especially where this is at county rather than district level, is essential. In many large 'shire' towns the district authority has an agency status from the county: considered to be an invaluable control to guide enhancement activity and funding. Such co-ordination facilitates the development of a conservation-influenced scheme at an early stage, and appreciation of the historic character of the area by the engineers. It also enables all important highway authority budgets to assist in conservation. Many areas, however, report little or no contact with the highway authority. Some, including some key historic areas such as Stratford-upon-Avon, complain of poor workmanship, use of inferior and inappropriate materials and bad results even in nationally significant conservation areas (Conservation Officer, Stratford-upon-Avon DC, pers. comm.). Likewise, the reinstatement of surfaces by statutory undertakers has long been a particular problem, especially where high-cost, high-quality materials were originally used. The recent privatisation of many statutory undertakers, and the rise of other major undertakers such as cable television companies, is also reportedly leading to problems.

It would appear that, across the UK, local authorities are less active in this respect of proactive conservation management than they are in the reactive role of monitoring applications, probably in part owing to the restrictions on local authority spending over the last decade.

CONSERVATION PLANNING POLICY AND DEVELOPMENT CONTROL DECISION-MAKING

Introduction

The scope of the problem facing any attempt at positive conservation, or townscape management, is great. Conflicts occur at a variety of scales and levels, from proposed total redevelopments to the most minor fabric changes. The importance of planning policy in the regulation of development within designated conservation areas in particular appears problematic. These conflicts are examined using data from a sample of three of the same four conservation areas in the West Midlands used earlier in this book.

The scale of the problem

The sheer scale of the potential problem is immense. Some designated areas are small 'preservation areas', where very little pressure for development is anticipated. In other areas, however, there is demonstrably a considerable pressure. Conservation area designation far from precludes change. Nevertheless, an overriding consideration should be that designation should protect the area's character and appearance and thus its special architectural and/or historic interest, as these are the reasons for original designation. Yet it would be an unusually determined LPA that would attempt to restrict development to these necessarily low levels. Not only may individually insignificant changes have a great cumulative impact upon the visual appearance of the townscape, with façade changes and advertising signs being particularly important in commercial areas, and changes to original windows and doors significant in residential areas, but each change affects the historical and architectural value of the building concerned, and thus the historicity of the townscape. So significant are these minor changes, many of which are Permitted Development under the GPDO, that English Heritage has recently mounted a campaign to highlight their detrimental effect (English Heritage 1991); a new pressure group has conducted a survey of historic towns resulting in a high-profile publication arguing for more strict control (English Historic Towns Forum 1992); and surveys of local authority conservation and design publications show a high level of awareness of the issue in design guidance material (Chapman and Larkham 1992; Jones and Larkham 1993). Yet this is not to condemn out of hand all change. Indeed, changes are necessary to adapt buildings to suit modern requirements, without which the buildings may become disused, decay and may be lost. Nevertheless, cumulative change is a potentially insidious process, little-studied in detail (but note the case study in Waley 1983), whose long-term effects in the conserved townscape are unknown. An immediate step for positive conservation would be to monitor this process in some detail, using data readily to hand in LPA files, thus building up a rather better picture of precisely what is happening in these areas than the present brief recounting of a site's planning history may show to a Planning Committee when discussing a current application (Larkham 1990c).

Refusals of permission are relatively few: the great majority of all planning applications are granted. Indeed, for three conservation areas studied in the West Midlands (Solihull High Street and the village suburbs of Knowle and Tettenhall), a total of 1,022 applications deposited between the date of conservation area designation and 1984 gave an average percentage of refusals as 22 per cent, significantly higher than the average of 13.8 per cent for England and Wales between 1979 and 1986 (DoE 1985a, 1987b). This suggests that processes may be operating slightly differently within conservation areas than elsewhere, leading to more refusals of permission. Nevertheless, with just under 80 per cent of all applications granted, it may be thought that the legal presumption in favour of development may, in sensitive areas, be unwise. It is still too soon to say whether the 'primacy of the development plan', a key policy contained in the revised Planning Policy Guidance Note 1, will have affected this trend (DoE 1992).

Conservation and management as observed

The seven LPAs in the West Midlands varied widely in their attitudes towards conservation during the 1980s. This is reflected in the personnel appointed and the resources of time and finance allocated. Conservation as an issue has been dealt with by a variety of officers, often with little or no specialised training or knowledge. Trained conservation officers were rare at the start of the decade, as was the case elsewhere (Civic Trust 1981), and indeed it took one LPA some six years to appoint a replacement after the previous conservation officer left in the late 1970s. Conservation issues are most frequently dealt with as a matter for development control, in order to limit undesirable development within designated conservation areas. Other, more 'positive', aspects of conservation, such as the production of management and enhancement proposals, are more patchily dealt with: one authority even suggesting that its principal planning assistants worked on conservation issues only when there was a lull in Local Plan work. The availability of finance is extremely variable, despite the existence of various grant-aiding bodies. In the Midlands, and again nationally, LPAs are increasingly using for conservation purposes finance actually allocated to other budgets (B.A. Hennessy, English Heritage, pers. comm.). These other sources include highways, leisure, recreation, Derelict Land Grants for landscaping, Housing Improvement Grants, and finance from the British Waterways Board for work involving canals.

Thus in these areas, each designated as being intrinsically worthy of conservation and of special interest, but not of outstanding national significance, positive conservation is very limited and is constrained largely by a lack of finance. In addition, in an economically depressed area such as the Midlands, local authorities have other pressing problems that are locally deemed more deserving of the limited finance available.

Conservation and policy

It has been suggested that much 'everyday' conservation-related work is dealt with through the development control process. Although the decisions on individual applications recommended by officers should conform to local authority policy, a number of problems are encountered in attempting to analyse this relationship.

First, much LPA conservation policy often consists of standard phrases derived directly from the enabling legislation or DoE guidance that have been inserted into the appropriate local policy documents, which thus give little guidance as to the actual achievement of lofty aims or their direct local relevance (Jones and Larkham 1993).

Secondly, development control is constrained by DoE guidance which, throughout the 1980s, strongly suggested that aesthetic considerations and details of design should be left to the developer, it not being the place of the LPA to intervene (DoE 1980, 1985b; Punter 1986a). Although intervention in matters of detail has always been permitted within designated conservation areas, observation of LPA decision-making suggests that the presumption against intervention is so strong that it does influence conservation area development control decisions and that, despite the revised PPG1, this persists up to the present.

Thirdly, other factors inevitably influence the decisions made by Planning

Table 7.1 Reasons for refusal of planning applications in three West Midlands conservation areas

Refusal reason	*Times used*	*Percentage of total*	*Times used as sole reason*	*Percentage of times used alone*
Conservation area character	158	36.8%	9	5.7%
Conflicts with stated policies	132	30.8%	28	21.2%
Detrimental to neighbours' amenities	23	5.4%	2	8.7%
Physical planning constraints	37	8.6%	2	5.4%
Design	51	11.9%	7	13.7%
Technical	15	3.5%	—	—
Create undesirable precedent	6	1.4%	—	—
Other	7	1.6%	1	14.3%**
Deemed refused*	(6)		—	
Unknown reasons*	(10)		—	
Total	429	100.0%	49	11.5% (average)

Notes: *Omitted from totals and percentages
**Too few cases for meaningful conclusions to be drawn

Committees, including party politics, personalities, local pressure, and so on (Short, Fleming and Witt 1986). The result is a system whose results, when examined in detail at a local scale, appear to be *ad hoc* and whose relationship with stated policies is unclear (for example, see Whitehand, Larkham and Jones 1992).

It is thus instructive to examine LPA conservation policy, such as it is, in the light of actual development control decisions in conservation areas. It is most useful in this respect to study the reasons given for the refusal of planning applications, given that more are refused in these areas than the national average. The refusal reasons found in the three West Midlands conservation areas studied are given in Table 7.1.

The reasons stated for these refusals are easily analysed. It is instructive to note which refusal reasons are used alone, and which are used mostly in conjunction with others. It seems most likely that reasons used alone are those that the LPA considers would be most appropriate, and would be sufficiently strong to withstand challenge on appeal to the Secretary of State. It is important that the reasons given are capable of winning support in the case of any appeal, since successful appeals lead to development that is, in a sense, 'unplanned' and which may affect the LPA's strategy. A good Midlands example of this is the number of examples of 'façadism' won on appeal in the Colmore Row and Environs conservation area of Birmingham (Barrett and Larkham 1994: see also Chapter 10). Reasons that are used in combination may reinforce each other, so that where, for example, one reason may prove weak, two or more may be sufficiently strong to halt the proposal, even at appeal. The recent spectre of costs being awarded against LPAs when individual refusal reasons are successfully challenged at appeal (even if the entire appeal is not successful) is an added incentive to produce robust refusal reasons.

Table 7.1 shows that the most commonly used refusal reasons are those referring to the character or amenities of the conservation area. However, such reasons are used largely in combination with others, and are rarely used alone as sole reasons for refusal. This may owe much to a feeling that the assessment of the character and amenities of any area is rather subjective, and that arguments resting purely upon their protection could be difficult to uphold at appeal. It is increasingly evident that many LPAs do not possess detailed assessments of the character and appearance of their conservation areas:

> The preparation of a character appraisal report and policy statement is, or should be, an essential pre-requisite for the designation of a conservation area. Many conservation areas do not have character appraisal reports at all. Other authorities take so long to find them in the town hall dungeons that even where they exist they are quite clearly not in regular use.
>
> (Suddards and Morton 1991: 1012; see also Jones and Larkham 1993)

Refusal reasons drawing upon the concept of character cannot easily be substantiated in the absence of studies of character, nor can adequate policy be developed. The lack of such studies is inexplicable given the fundamental nature of 'character' in the concept and definition of conservation areas. Yet refusal reasons from this grouping are widely used, particularly by one LPA, which shows a marked propensity for supporting a wide variety of refusal reasons with the phrase 'the application is out of keeping with the character of the conservation area and detrimental to its visual amenities'.

In a wider study of refusal reasons, Rawlinson (1989) also noted the high incidence of amenity-related refusals. Under the general heading of 'amenity', he considered the size of the proposed redevelopment, its density, layout, appearance, overlooking, noise and impact on the general visual amenity of the area. In his six sample LPAs, such reasons accounted for 323 of 508 refusal reasons in 1984 and 318 of 522 reasons in 1987. He noted that

> within the amenity group of reasons there is perhaps a distinction between size, intensity, layout and appearance as one group of factors associated with what the development will look like *in itself*, and a group comprising overlooking, noise, disturbance and adverse effect on the amenity of the area which are more to do with the impact of the

proposed development in its *surroundings*.

(Rawlinson 1989: 19–20, his emphasis)

From his wider sample, Rawlinson supports the popularity of refusal reasons on the grounds of amenity. Such refusals are not the sole preserve of conservation-related applications.

The next most commonly used set of refusal reasons refer to specifically stated policies, of which parking standards and land-use zoning are of greatest importance. The precise wording indicates that the refusal is directly because the proposal is counter to the LPA's standards and/or policy in those areas. Such reasons are slightly less common than the general character reasons, but are far more likely to be used alone; indeed, of the reasons used as sole refusal reasons, this group is by far the largest. A series of studies of development in mature residential areas points to the high degree of reliance placed by LPAs on such easily-quantifiable parking, density and highways standards; with the form of development tending to be an afterthought, when battles over density and access have been concluded (Whitehand 1990a, 1990b; Whitehand, Larkham and Jones 1992). This study also produced examples of refusal reasons related to putative policies, such as highway improvements and town centre plan preparation, although no highway improvements were carried out in the two-and-a-half-decade study period, and there is evidence that the town centre plan was never begun (Whitehand 1989b: 411).

Design criteria are next highest in numerical terms, although far behind the previous two categories. Their high ranking is of some interest, since it appears to reflect the apparent tendency for Planning Committees to become considerably involved in matters of detail and design (as has clearly been the case in the Midlands; see also Simmie 1981; Witt and Fleming 1984). The proportion of design reasons used singly is quite high, and this is surprising given the often subjective nature of design matters, which leads to their being relatively easy to overturn on appeal, and the strong arguments that could be advanced for the total abolition of aesthetic control by LPAs. Such refusal reasons often refer to the poor relationship of the proposed building to its surroundings, or to the creation of 'clutter' on a façade. An interesting refusal reason, never used alone, is that a proposal 'will create an undesirable precedent'. This is unusual in that an application is legally to be decided upon its intrinsic merits; the part played by arguments of precedent is, to an extent, counter to this, and has yet to be researched in detail.

The aesthetic and tidiness arguments so evident in these study areas and in Simmie's study of Oxford (Simmie 1981) form the most commonly employed logic basis for planning, and its most trivial but simultaneously major claim to autonomy from outside influences. Such aesthetic reasons for changing an application can be followed with relative ease by prospective developers, for whom 'at the planning stage it makes only a marginal difference ... what colour bricks or what shape windows they employ' (Simmie 1981). Such arguments mirror many middle-class tastes and values, and indeed are reflected in a large proportion of comments on planning applications made by local amenity societies, whose members are also largely middle class.

The examination of stated conservation policy in these study areas suggests that, in fact, there is no such thing as a simple stated conservation policy, since conservation includes the regulation of a wide variety of aspects of the urban landscape, each of which may require separate policy statements. 'Conservation policy' is, therefore, not as easily dealt with as a single entity as are more easily definable topics, such as office restraint policy. Indeed, this complexity may be seen in the grouping of refusal reasons, where 'conflicts with stated policies' involve a wide variety of LPA policies from land use to advertising, and this is totally separate from more general considerations of conservation such as the amenity value of the area. These considerations are controlled by the refusals related to the character of the conservation area and the design of the proposed structure.

Although stated conservation policy may be characterised as being particularly wide-ranging, it may be seen that many of the reasons given for refusal of planning applications are closely related to that policy. The 'conflict with stated policy' section may be expected to bear a strong relationship to the conservation policies, and this does appear to be the case; even though the stated policies on, for example, land-use zoning may not specifically be conservationist. Specific physical planning constraints are harder to relate to specific policies, more so as those decisions also show the influence of the County Council (to 1986 in this Metropolitan County, when it was abolished) which, as Highway Authority, had the power to direct that an application be refused on highway grounds. Design matters could be said to be strongly related to conservation and policy but, as already suggested, this is a rather sensitive area, and there are strong feelings that, even in these conservation areas, LPAs should not become involved in regulating design nor have policies upon it. Technical grounds for refusal are such that applications would almost certainly be refused whether they were inside a conservation area or not; they are certainly not conservation-specific.

Thus there does appear to be a strong link between the wide-ranging stated conservation policy and the reasons given for refusing planning permission. This is hardly a surprising finding, since LPAs cannot act in an arbitrary manner, and must give logical refusal reasons that can be substantiated on appeal. It is logical for such reasons to be supported by a statement of policy, even if this is an informal (non-statutory) statement. It may also be argued that, for various reasons, the stated refusal reasons may not be the actual reasons. Therefore, the analysis thus far is insufficient fully to explain why various development proposals have been refused.

Refusal reasons and the type of proposed development

The type of proposed development is often of some importance in the selection of reasons for refusal by the LPA planning officers. Some reasons will obviously be tied to certain types of development, for example advertisements to advertisement policy, or to the idiosyncrasies of individual applications; nevertheless, it is useful to examine the possible links between development type and refusal. It is known, for example, that one type of development – for offices – attracts a variety of refusal reasons, including non-compliance with residential retention policies, land-use policies and physical planning policies (McNamara 1985: 354ff). Whether each development type in the Midlands study areas attracted a similar variety of reasons, or whether refusal reasons were more restricted, can be examined.

Table 7.2 presents data from the three study areas. It can be seen that refusal reasons differ according to the type of development proposed. Commercial developments involving new building, major rebuilding and change of use are most frequently refused because the proposal is in conflict with the LPA's stated policies, usually those on land use. Parking and servicing space requirements are also of importance, since in all of these areas there is only limited rear access, and front servicing along busy main roads is unacceptable. Constraints of physical planning are a second group of reasons commonly used for the refusal of commercial development where the proposal would constitute overdevelopment of the site or an undesirable overintensification of use: it should be said, however, that definitions of 'overdevelopment' and 'overintensification' are vague and probably subjective. For proposed changes of use, offices proposed in previously residential areas are refused not only on grounds of land-use policy, but also because the necessarily increased disturbance would be detrimental to the amenities of residents. 'Amenity', however, is undefined in primary planning legislation (Cullingworth and Nadin 1994: 164–5), although guidance, appeal and court cases have provided numerous, sometimes inconsistent, definitions. Residential development, however, is more likely to be refused as being likely to have a detrimental effect on the character, appearance or amenity of the area, or for detailed reasons of design. By far the most frequently given reason for the refusal of signs and advertising is concern for the character of the conservation area, yet, as has been discussed, most LPAs cannot show proof of character appraisal.

This analysis permits greater insight into the relative strengths of policies. Design matters are rarely used to refuse commercial development even as a subsidiary reason, since land-use policies appear stronger: they usually appear in published Structure or Local Plans that have been through the process of public consultation, and thus considerable weight is attached to them at appeal. This also explains the predominance of land-use policies cited in the refusal of change of use applications, where other reasons are of quite minor importance in comparison. There being no land-use policy to refuse

Table 7.2 Reasons for refusal related to type of proposed development in three West Midlands conservation areas

	Development types							
Refusal reason	*(1)*	*(2)*	*(3)*	*(4)*	*(5)*	*(6)*	*(7)*	*(8)*
Conservation area character	2 (15.4)	15 (28.9)	4 (6.3)	13 (56.5)	8 (80)	101 (64.4)	9 (39.2)	5 (5.8)
Conflicts with stated policies	6 (46.1)	4 (7.7)	27 (42.8)	—	1 (10)	30 (19.1)	3 (13.0)	61 (70.9)
Detrimental to residents' amenities	1 (7.7)	6 (11.5)	3 (4.8)	—	—	—	3 (13.0)	10 (11.6)
Physical planning constraints	1 (7.7)	6 (11.5)	15 (23.8)	3 (13.0)	—	5 (3.2)	4 (17.5)	3 (3.5)
Design	3 (23.1)	13 (25.0)	9 (15.3)	4 (17.5)	1 (10)	16 (10.2)	3 (13.0)	2 (2.3)
Technical	—	7 (13.5)	2 (3.2)	2 (7.7)	—	—	—	3 (3.5)
Undesirable precedent	—	—	1 (1.6)	—	—	4 (2.5)	—	1 (1.2)
Other	—	1 (1.9)	2 (3.2)	1 (4.3)	—	1 (0.6)	1 (4.3)	1 (1.2)

Development types: (1) Commercial new building and major rebuilding
(2) Residential new building and major rebuilding
(3) Commercial floorspace additions
(4) Residential floorspace additions
(5) Façade alterations
(6) Signs and advertisements
(7) Miscellaneous minor
(8) Material change of use

Notes: Figures in brackets are percentages of column totals
Only the major element of any proposal is recorded in this analysis
Demolitions have been excluded as in only two cases was demolition the major element of the application

residential development in zoned residential areas, as most conservation areas are, the reasons given in such cases are more varied and, frequently, more numerous, referring to the major stated, but rather general, policies of the preservation of character and amenity. Despite the existence of strongly stated policies by one LPA on the control of signs – for example, it 'will refuse to consider projecting signs' – advertising is again more usually limited by reference to the general policies, rather than to specific advertising policy. This is somewhat unusual. Part of the same LPA's policy states that 'the council will not consider signs that it feels are not essential', and it may be felt that this is again a rather subjective assessment, which might be challenged at appeal.

CONCLUSIONS

This brief examination of conservation policy and its relationship to development control decision-making sheds light on precisely how the statutory planning system operates with regard to conservation. Since most action concerning conservation areas appears to be undertaken by development control sections, albeit often with the advice of a conservation officer, the refusal reasons issued by LPAs to unsuccessful applicants form a useful data source.

Local planning authority policy relating to conservation seems particularly wide-ranging and is a rather illusive subject of study. Much of it is non-statutory or informal in nature and may, indeed, be in the form of unpublished standards. This is similar to a wide range of LPA policy in general.

Local authority action regarding conservation as a whole – the everyday management of change within designated conservation areas – is carried out as a development control function, and is rather negative. It is largely reactive and in many instances it is clear that submission of an application is the first indication to the LPA that a site's future is being reassessed; officers often find that policy is unhelpful when seen at this site-specific level, leading to *ad hoc* decisions: hardly 'management' (Whitehand, Larkham and Jones 1992; Larkham 1990a). It is suggested, from a variety of studies, that the influence of these development control sections is negative because it is reactive only to outside development pressure, and weak because it is constrained by pressure from the DoE and it enters into the development process relatively late (Punter 1986c).

Conservation policy as it affects development control also, from this evidence, appears to be largely negative, and to be concerned with the protection of the existing character of the area. 'Amenity protection' reasons are frequently used, although most often to support other refusal reasons. Problems are widespread over the lack of ability to define character; the lack of legal definition of 'amenity' could also pose problems, with the increasing trend of appeal inspectors and the courts to arrive at precise definitions of terms. Nevertheless, these reasons are most often upheld at appeal in these study areas. Conflicts with other specific planning policies are also very common. Considerations of design are fairly frequent, as might be expected in these areas of special control where detailed planning applications are required, and outline applications discouraged. Refusal reasons appear to be carefully formulated with a view to possible appeals, despite their relative scarcity. This explains the frequent use of multiple refusal reasons and references to Council policy and standards.

Development control and conservation policy can and should act in a positive manner to facilitate appropriate development by providing a basis for negotiation (cf. Davies *et al.* 1984: paragraph 2.2). Nationally, the importance of such negotiations between LPA and the potential developer is realised. Examination of applications in the Midlands study areas suggests that such negotiations are also common in these conservation areas, particularly where the LPA actively solicits the views of statutory consultees and other interested parties (such as voluntary amenity societies) and relays these to the developer. Despite this negotiation, however, refusal rates in these conservation areas remain high.

In general, the conservation areas examined showed a somewhat higher rate of refusals than the national average. Many factors undoubtedly contribute to this. However, on comparing the data for these areas with those provided by more general studies (including Whitehand 1984; Freeman 1986), it appears that the characteristics of applicants and types of development proposed are not markedly dissimilar from those elsewhere in the country. This suggests that it is the fact of designation that leads to more refusals. The high use of refusal reasons based on the character of the area, and their success at appeal through the 1970s and 1980s, appears to support this suggestion.

The planning system does, therefore, seem to work slightly differently in these conservation areas than in other areas. In the application of special consideration and control, conservation policy, acting through the medium of development control, must be having some effect upon the built form of conservation areas. Yet the vagueness and generality of much of this policy (Jones and Larkham 1993) suggests that there is still little proper notion of townscape management.

8

AMOUNTS AND TYPES OF CHANGE IN THE CONSERVED CITY CENTRE

The policy of sweeping clearances should be recognised for what I believe it is: one of the most disastrous and pernicious blunders in the chequered history of sanitation.

(Patrick Geddes, *c.* 1915)

INTRODUCTION

Chapters 6 and 7 have presented the perspective on urban change used in the following three chapters; in particular, the importance and inter-relationships of the various agents of change in the built environment, and the practical implications of the LPA policy-making and development control systems in general. Chapters 8, 9 and 10 examine the key questions posed in Chapter 1 through micro-scale case studies using this perspective of urban morphological study.

The amounts of change to historical, supposedly conserved, urban landscapes, particularly since the end of the Second World War, can often strikingly be seen in even a cursory visual examination of a town or city accepted as having historical or architectural value. This immediate visual impact is probably most significant for the majority of users of urban areas, the inadvertent 'consumers' of architecture, urban design and conservation. Not only are new buildings evident, but there are more large-scale changes, areas of comprehensive redevelopment and ring roads among them. Heighway's survey on behalf of the Council for British Archaeology in 1972 examined 702 historical towns in England and Wales, and found current and proposed development posing serious threats to the historically significant town plan and buildings in 457 cases. A further 65 had already undergone damaging development (Heighway 1972). A trained eye will see even those more minor changes that seek to be 'historical' in nature, but which are usually given away by detailed deficiencies. For example, window replacement with glaringly new materials, particularly uPVC, has recently come to the fore as an example of minor, but often insensitive, change (English Heritage 1991; EHTF 1992). Yet town centres are popular subjects of conservation area designation, while village centres form probably the largest single category of conservation area (Pearce *et al.* 1990; see also Chapter 5). In many cases, the nature and scale of change occurring within supposedly conserved central areas appears, in visual terms, to belie the conservation designation itself.

But there are data sources available other than visual inspection that record, in scrupulous detail, all but the most minor changes. Since 1947 in Britain the registers and files of planning applications held by local planning authorities are invaluable in this respect. These allow the plotting of all types of change, and their accurate dating. Using this information, a picture can be built up which not only identifies and dates each change made, including the interior and rear elevation changes rarely visible, but also the proposed changes that were submitted as planning applications but were refused, or were withdrawn.

Of equal importance to this detailed recording of amounts and types of change is the attitude of the planning system to proposals for change. In many countries this is not determined by a strict codification of regulations, although Spain follows this path, and Vilagrasa (1992) documents the differences between the Spanish and British conservation planning systems. Instead, an accumulation of precedent determines the approach to change – in Britain this consists of development attempts, appeals to the DoE, and case law as decided in the High Court and the Court of Appeal. This was highlighted in Chapter 4 through consideration of particular cases that have tested the notions of conservation and development: Palumbo's repeated attempts to redevelop several listed buildings within a conservation area at Mansion House Square, London (finally resolved in the House of Lords) is one of many appropriate examples.

USING CASE STUDIES

The research upon which the next three chapters are based relies upon the use of detailed case studies to elucidate the number, types and forms of changes within selected study areas, working at the level of individual sites and individual planning applications to demonstrate key trends. There are many problems with such material, not least the problem of representativeness, the distortions caused by the researcher's perspective, and the considerable demands on time and resources. It is not claimed, for example, that the study areas chosen, nor the detailed examples of planning studied, are 'typical' in any statistically significant way. Yet the UK conservation area designation and management policies are constant across the country (with the exception of Northern Ireland: Hendry 1993) and the cases are selected from two major groupings of conservation areas, the town centre and residential area; they do appear to be representative of the pressures and processes occurring in historically significant urban landscapes. The case study method does, as recent discussion points out, have considerable advantages in planning research. These include

> firstly, the ability to retain a holistic and meaningful view of real-life events ... secondly, the possibility of in-depth analysis looking at a multiplicity of causal links. Thirdly, they are particularly relevant to 'how' and 'why' type questions, explanatory investigations and research into operation[al] links which need to be traced over time. Fourthly ... a more sympathetic approach [is offered], getting inside the process, or the minds of actors or individuals, as 'discovering' explanations.
>
> (Punter 1989)

Such research projects are not usually designed for quantitative analysis, and the data collected are not amenable to more than the most basic statistical manipulation. Instead, qualitative and interpretative techniques are used to supplement the basic quantitative data. This project is one of a growing group being carried out by members of the Urban Morphology Research Group using detailed case studies of current planning practice in an attempt to elucidate the workings of the planning system, particularly concerning the development and management of historical urban landscapes (cf. Whitehand 1987a, 1992a); a small number of studies of aspects of the planning system, principally dealing with aesthetic control, have also begun to use this approach (Punter 1985, 1990b).

PLANNING INFORMATION AND THE MEASURING OF CHANGE

Since the reforms of the Town and Country Planning Act 1947, LPAs in Britain have been given responsibility for controlling all physical development, defined by Section 12 of the Town and Country Planning Act 1962 and subsequent Acts as 'any building, engineering, mining or other operations in, on, over or under land'. This responsibility has been exercised through the regulatory apparatus of the application for planning permission, which must be submitted for all prospective development except those minor changes allowed by the General Permitted Development Order. In law, development must not take place without planning permission. The LPA's primary role is thus that of a gatekeeper, an indirect agent in the process of change, reacting to applications submitted by others and attempting to control poor applications by the refusal of permission, or to improve proposals through negotiation. The local authorities intervene actively, by initiating their own developments, only relatively rarely.

The data source provided by this bureaucratic process of application, consultation and decision-making is rich. A summary record is kept in the Planning Register of each application, and the files of application forms, correspondence and architectural drawings are also retained by LPAs from 1947 to the present. With the co-operation of the LPA, therefore, every single change to an area in the post-war period may be examined. Although this research is time-consuming, it allows absolute precision in dating and analysing types of change and in the identification of the agents – especially developers, architects and landowners – active in the area. Data are available not only for development proposals that received planning permission and were subsequently built, but for proposals that were refused permission, proposals withdrawn before the LPA's decision, and sometimes on informal discussions between the LPA and a developer that did not result in an application. But, in general, only those proposals that became part of the formal planning process through the submission of an application are considered: of course, there may be many more proposals under consideration that did not reach this stage and are unrecorded, for instance because economic conditions were unfavourable, or because LPA policy precluded certain types of development in some locations. In general, the Planning Register and the file of applications, plans and correspondence should yield the following information:

1. Name and address of the initiator of the application; initiator's agent; architect, sometimes also the builder, consultants, specialised contractors and other agents involved in the development.
2. Site address; block diagram of the site in relation to surrounding buildings and streets.
3. Initiator's interest in the land (as owner, lessee, etc.).
4. Brief description of the proposed works, which can be supplemented by examination of the plans.
5. Plans and elevations of the building as existing.
6. Plans and elevations of the building as proposed.
7. Technical annotations, often with details of materials and finishes.
8. Decision of the LPA, with reasons for refusals and conditions attached to grants of permission.

There are no alternative data sources from which to measure urban change in Britain as accurately as planning records permit from 1947 and as a closely related local authority source, building plans, permit often from the mid-nineteenth century. Without these comprehensive records, the researcher is dependent upon iconographical records (maps, plans, building elevations, fire insurance plans and so on) that may vary considerably in detail, frequency and coverage. Records of local businesses may be available in some cases, but these are of limited value unless, for example, a particular architectural practice or developer has been especially active in an area and will permit access to its records. Alternatively, Foote (1985) has examined the use of photographs as a source for examining urban change. His analysis opens up possibilities where photographic coverage is good and where other sources may not exist, but these preconditions may be rare. Local authority data sources similar to Britain's planning files have been exploited with considerable success in Spain (the *Licecias* or *Permisas de obras*: see Alió 1987; Vilagrasa 1984), and planning files, among other sources, have been used by Gad and Holdsworth (1988) for their detailed study of office development and occupancy in King Street, Toronto. Thus local authority sources of building and planning data should be examined as a first step in the study of built fabric change in all countries where local authorities exercise some degree of planning control.

CHANGES IN BRITISH CENTRAL CONSERVATION AREAS

Local authority development control information has been used in examinations of the types and amounts of change to buildings within conservation areas. For this analysis, the types of change found in the planning applications have been divided into several categories on the basis both of the written description of the proposed development in the Planning Register and inspection of the detailed architectural drawings. These categories are new building, major rebuilding, addition of floorspace (or extension), alterations to the façade, advertising signs, internal alteration, refurbishment, miscellaneous minor changes and demolition. Each individual planning application may contain more than one type of change or, if considerable alteration is proposed, several elements may be found. For the following analysis, each identifiable element is counted: the number of elements is therefore somewhat greater than the actual number of planning applications. But the use of these identifiable elements gives far greater precision in measuring the total change to the townscape, which is the aim of this analysis. For convenience, all 'additive' changes, those not involving demolition, have been added together to give an indication of the levels of activity present. Those applications that received planning permission, and which the evidence from site visits, the knowledge of planning officers, and the lack of subsequent similar applications suggests were implemented, have been separated from those applications that were not implemented owing to refusal of permission, withdrawal, or for other reasons. This separation gives an indication of the level of intervention of the planning authority in the development process.

Numbers and types of development

This form of analysis has been carried out for several entire designated conservation areas (Larkham 1988b). These include historic town centres of varying characteristics, from mediaeval towns showing a single phase of regular, apparently planned, plots (Henley-on-Thames) to towns having several distinct phases of mediaeval planning (Ludlow) and regularly planned Georgian and Regency towns (Leamington Spa). Figure 8.1 shows the example of Solihull, a small market town between Birmingham and Coventry. The conservation area here was designated in 1968, and its boundary is tightly drawn. This is one of the areas used for analysis in Chapter 7. It principally contains the High Street, largely still retail in character, where the majority of buildings are nineteenth-century in date but which stand upon the remnants of regular mediaeval plots. These plots have been truncated at

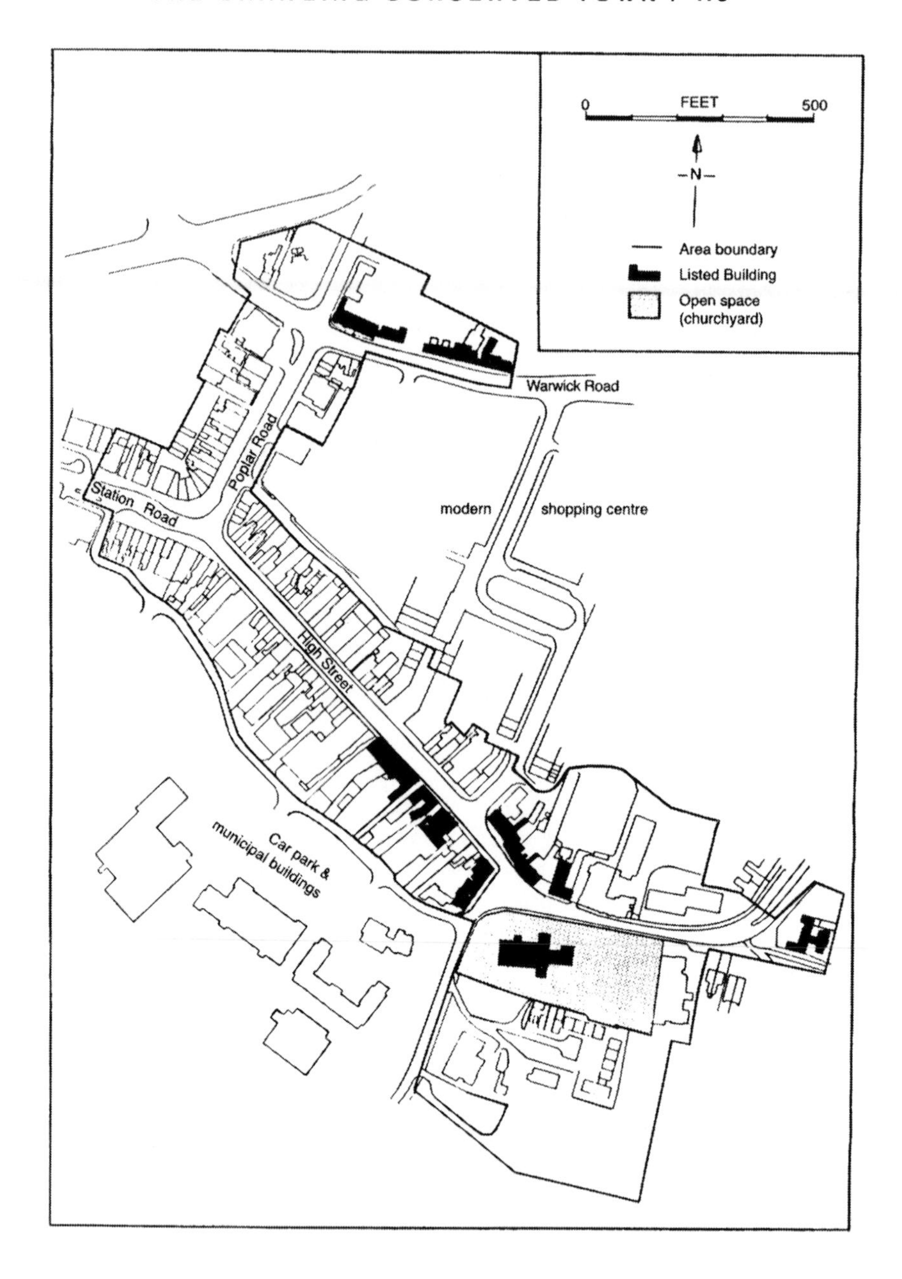

Figure 8.1 Extent of Solihull conservation area

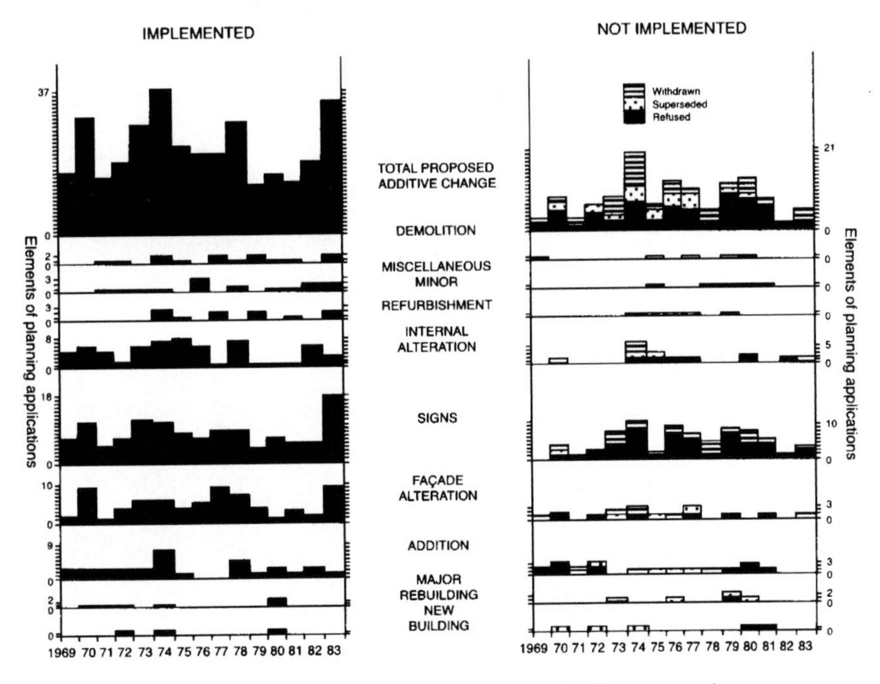

Figure 8.2 Analysis of built fabric change, Solihull conservation area

the rear for a car-park and a shopping centre of the 1960s. A considerable volume of total additive change, both implemented and not implemented, can be seen (Figure 8.2). However, major structural work, such as new building or major rebuilding, is quite rare. There are rather more additions, while the majority of the development activity is for interior alterations, signs and façade alterations. The significance of these last three groups is closely related to the levels of commercial activity within the area. For a considerable number of commercial and retail properties, applications are initiated by new or prospective occupiers who are intent upon adapting the premises to suit their own requirements of layout and corporate image, sometimes with little apparent regard for the historical or architectural significance of the building. In extreme cases, this has resulted in the cladding of an existing structure with timber or metal hoardings, entirely obscuring the character and appearance of the original building (Figure 8.3). It is instructive to note that the Main Street revitalisation programme in Canada has spent much energy and money in removing such excrescences and revealing the detail of the buildings preserved beneath them (Holdsworth 1985). Façade changes usually involve new window and door layouts, stallrisers and pilasters, often removing older shopfronts to give large expanses of plate-glass windows. Signs obviously display the name, and often logo, of the firm. The importance of both types of change in presenting the corporate image of the modern market-conscious firm must not be under-estimated. In fact, the incidence of façade and internal

Figure 8.3 Victorian shop with insensitive 1960s cladding, Leicester (author's photograph)

Figure 8.4 Example of LPA shopfront design guidance (reproduced by permission of Vale of White Horse District Council)

Key: Good practice:

A Chimney repaired: retains character
B Roof repaired and maintained: guards structure from decay
C Sympathetic restoration of original features
D Console bracket: helps to frame fascia board
E Pilaster: provides a frame to the opening and interest to the eye
F Stallriser provides base to display window

Bad practice:

A Leaking roof in need of attention: threatens whole building!
B Alteration to chimney: out of character
C Flat roofed dormer window conflicts with traditional building style
D Advertising 'clutter' obscuring design and proportions of building
E Neglect of unused upper floors: supporting income lost and decay passes unnoticed
F Over-large fascia sign alters balance of design and encourages moisture trap which may start decay
G Decorative pilasters removed or covered over: detail and proportions lost
H Large areas of glass are uninteresting and, without a stallriser, can be prone to vandalism

changes has been used as an index for the development of the shopping centre of Utrecht (Buissink and de Widt 1967). Such indices have proved sufficiently sensitive to allow categorisation of the quality of retail streets, the amount and type of commercial activity being paralleled by the frequency of building activity. The increased frequency of changes to signs, façades and interiors is due to changing commercial pressures and subtle variations in demand, as is suggested by Palser (1984a, 1984b). For example, the corporate image of Top Shop, part of the Burton Group, was changed several times, and the shops refitted, within a five-year period in the 1980s. An economic climate that seems to encourage frequent mergers and take-overs in the retail

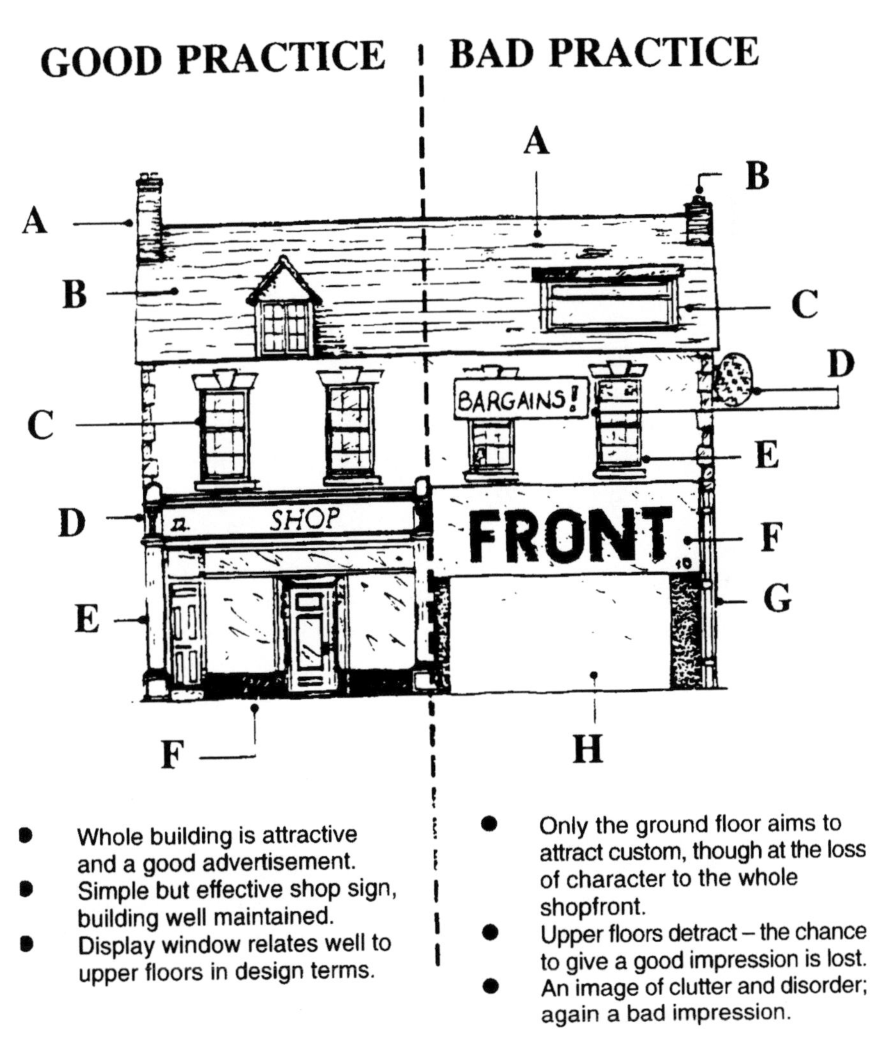

sector, as has been the case in Britain over the past decade, also serves to increase the amount of change as one corporate image is supplanted by that of the victor in the take-over battle.

Many conservation areas have local policies regarding the types of shopfronts which will be approved, the materials preferred and the size, illumination and style of signs. Solihull MBC has particularly strong policies, especially regarding signs, which account for the considerable number of rejected and withdrawn sign applications. Internally illuminated projecting perspex box signs, for example, are not favoured: externally illuminated fascia signs, traditionally signwritten, are preferable. The attempt by a leading building society to erect a new fascia sign in 1978 was refused on the grounds that the proposed sign was

> out of keeping with the character of the area and detrimental to the visual amenities of the street scene. The Council is particularly anxious to ensure that façade signs are appropriate to the building. Box signs are incongruous on a predominantly brick façade.
>
> (planning file 3334/78, Solihull MBC)

The poor design of shopfronts and signs can have a considerable adverse visual impact on conservation areas (Binney 1978; SAVE 1978), and many local authorities have issued design guidance on this topic to assist prospective applicants; indeed, this is one of the more popular forms of design guidance (Chapman and Larkham 1992) (Figure 8.4). Despite such firm policies and guidance, the majority of firms are particularly sensitive about their 'image installations' (a term used by the Midland Bank Premises Department in 1968), and the reaction of Bejam Freezer Food Centres Ltd is typical.

> Our company has a distinct policy on fascia design in as much as we have a corporate identity in terms of colour and lettering style and I take a very strong personal interest in any proposed deviation from this even in the more sensitive areas such as conservation areas. I fully believe in integration within areas of architectural merit, but I will fight strongly for the right to utilise our image, even in modified dimensional form irrespective of its location.
>
> (Property Director, Bejam, pers. comm.)

This comment was made in response to a proposed advertising sign for which Solihull MBC refused planning permission as it was 'incongruous and would create a garish, strident and dominating feature completely out of character in the High Street, and constitute a substantial injury to the amenities of the conservation area' (planning file 2603/75, Solihull MBC).

The scale of changes

It is very evident in town centres that development in the later twentieth century has been occurring on ever-larger scales. In most British towns

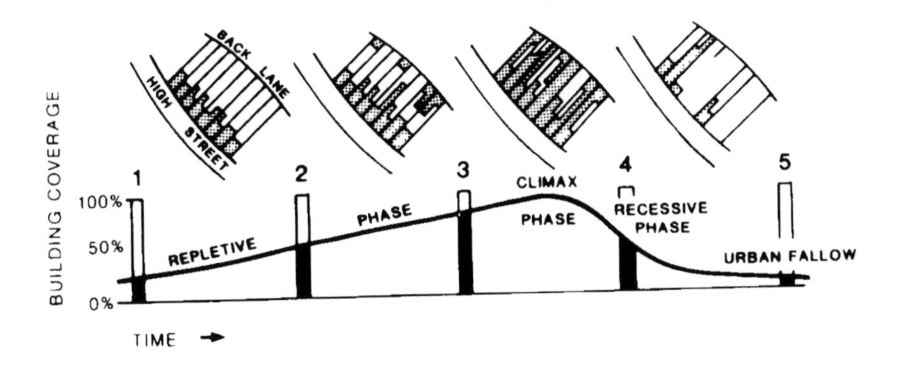

Figure 8.5 Phases in the burgage cycle (after M.R.G. Conzen)

founded or subject to phases of planning during the mediaeval period, plots were laid out by a landholder. To ensure regularity in rental income, plots were of uniform dimensions, so far as was practicable given constraints of terrain and pre-existing features such as routes and defences that would distort an ideal layout (Slater 1987). Plan analysis in different regions of England suggests that the statute perch and acre were used by mediaeval surveyors in planning plot series. Although a typical English mediaeval plot width of 33 feet (approximately 10m) has been suggested (Conzen 1960), some plots were 66 or 99 feet wide and up to 1 acre in area, as in Hedon, Yorkshire (Slater 1985). Over the centuries, a cycle of development has affected these plots (Conzen 1962). More and more building takes place, until the greater part of the plot is built over. This process takes place not only horizontally, by the addition of outhouses or new buildings in the rear of the plot, but vertically, by increases in the number of storeys. At some point, however, this process reaches a maximum; later still, plots or whole areas may be cleared prior to a new phase of development. As this development cycle proceeds, the original plots are also metamorphosed, as amalgamations and subdivisions occur. This is the expected pattern of development in English towns, and is referred to as the 'burgage cycle' after the dominant form of tenure which produced the English mediaeval plan layout (Figure 8.5).

Four city centres of largely mediaeval origin have been investigated in some detail. In Northampton and Watford, plot widths in the early twentieth century were predominantly of 10m or less, with some buildings having special functions, such as inns, often occupying larger plots. Watford comprised a single, long, mediaeval street, while Northampton has a much more complex plan reflecting a series of mediaeval adjustments (Whitehand 1984; Northampton's plan is discussed by Lee 1954). Aylesbury has Saxon origins, was a manor of the Crown in the Norman period and, although its growth was slow, its central position within Buckinghamshire led to the location here of the Assizes from 1218 (Freeman 1986: 29). It, too, has a complex plan denoting several phases of growth. Worcester, the largest of the settlements studied, became the site of an episcopal see in *c.* 680 and was fortified some two centuries later. As the lowest bridging-point of the Severn, this was a key strategic and commercial centre, and the town centre was heavily developed during the mediaeval period. A detailed analysis of Worcester's town plan has recently shown the numerous phases of development discernible by the streets and plot boundaries remaining into the late Victorian period (Baker and Slater 1992). Three separate studies have examined all new developments in these urban centres (Whitehand 1984; Freeman 1986; Vilagrasa 1990; Vilagrasa and Larkham 1995). If these new developments are divided into groups of frontage widths under and over 10m, the notional 'typical' mediaeval plot, it can be seen that such towns vary considerably in the speed with which the mediaeval plot lineaments have disappeared beneath twentieth-century redevelopments (Figure 8.6). In the post-war building boom from the early 1950s, the proportion of redevelopments that have taken place within plot frontages of 10m or less has diminished, except in the case of Worcester, where a larger proportion of redevelopments have remained of more traditional size. Buildings of this period were, according to one critic, monotonous, repetitious, squeezed out of a tube and cut to the desired length: this was the period of 'toothpaste architecture' (Aldous 1975: 57), typified by the Barclaycard headquarters building in Northampton (Figure 8.7). This scale of development has produced a reaction over the last decade and a half, particularly evident in Northampton, where attempts have been made to re-create the traditional scale of plot frontages. This has been done by designing long-fronted buildings to give the appearance of more than one building (Whitehand 1984). The visual appearance of structural bays of traditional proportions has been produced by giving stronger pillars at ground-floor level, breaking up the unifying shopfronts and fascias popular in the 1950s and 1960s, and giving the upper floors some relief and different materials. The success of these attempts has been indifferent, particularly on large-scale developments such as the Grosvenor shopping centre, Northampton (Figure 8.8). On less monumental buildings, lower and less in bulk, this tactic is often much more successful. Only during the 1980s have planning authorities begun to recognise the significance of the historical plot pattern, particularly in smaller towns where development pressures have been less, and survival of these features is consequently greater. The explicit concern of Taunton Deane Borough Council for the retention of the remaining burgages in Wellington (Somerset), and the walls marking the boundaries of these mediaeval plots, was unusual at the time (Taunton Deane Borough Council 1984: 17), but this

Figure 8.6 Plot widths of new buildings in four town centres (data on Northampton and Watford from J.W.R. Whitehand; data on Aylesbury from M. Freeman)

Plot frontages of new buildings in central areas

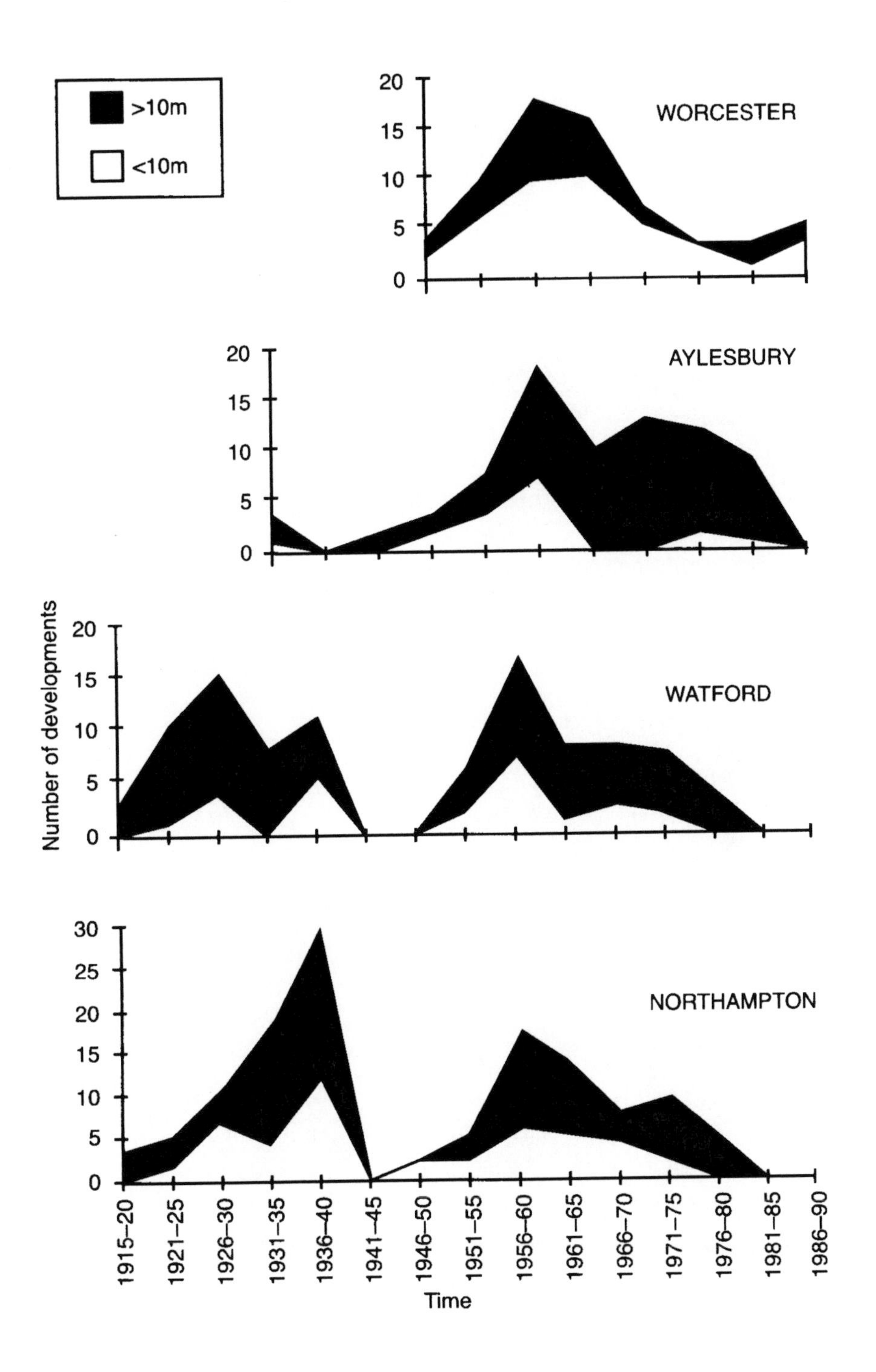

Figure 8.7 'Toothpaste architecture': the Barclaycard building, Northampton (author's photograph)

Figure 8.8 Attempted subdivision of large frontage: the Grosvenor shopping centre, Northampton (author's photograph)

is becoming less so as planning and conservation policies are revised and updated.

In towns of more recent origin, particularly Georgian and Regency towns such as Leamington Spa and Cheltenham, plots are generally much smaller than the deep mediaeval burgages. There is far less scope for more intensive development on the plot tail. Being of relatively recent origin, Leamington Spa being developed from *c.* 1810, for example (Arnison 1980), the buildings of these towns were less susceptible to change during the Victorian period than the collections of Georgian and Tudor buildings in towns such as Henley-on-Thames and Ludlow, where there were some notable Victorian redevelopments. These towns seem to have a different development cycle to those of mediaeval origin. Redevelopment is less common over time, and more changes occur within buildings to adapt them to modern uses, for example by amalgamating buildings to provide large areas of relatively unobstructed floorspace for the demands of modern retailing display. In both Leamington Spa and Cheltenham, town-centre shopping malls have been created by the amalgamation of landholding in an entire street-block, which is then redeveloped. The rear access alleys form the main mall, while the frontage buildings are modified to open both onto the street and onto the mall (Figure 8.9). In the Leamington example, one of the streets is relatively minor, and on this street frontage there has been some redevelopment for car-parking and service vehicle access.

The architectural styles of changes

As with other fashions, architectural styles change over time. A number of changes, some quite marked, are visible during the course of the twentieth century. Some have been overtly ahistorical, even anti-historical, while others have been increasingly sympathetic to the context of plot sizes, existing architectural styles and building materials in historical town centres.

Although the identification and naming of architectural styles is itself controversial, the styles discussed here have been identified after reference to numerous standard works of architectural description and criticism, and current architectural periodicals. The information is derived from studies of five town centres – Aylesbury, Epsom, Northampton, Watford and Wembley – using local authority planning and building control data (Larkham and Freeman 1988). The prevalent architectural styles can be classified as those with historical precedents, for example neo-Georgian, showing symmetry and some use of classical detailing, and the eclectic historicism of post-Modern (although there are many aspects to post-Modernism in architecture, historicism being only one: see Jencks 1991; Jencks and Chaitkin 1982). Novel styles have no such obvious antecedents, and these include Art Deco and Modern. As the data permit the accurate dating of buildings, the discussion of styles can be related to concepts such as the diffusion of innovations in architectural style, and the degree to which the dominant style of one period persisted into the next.

Figure 8.9 'Backland' shopping mall, Leamington Spa (author's photographs)

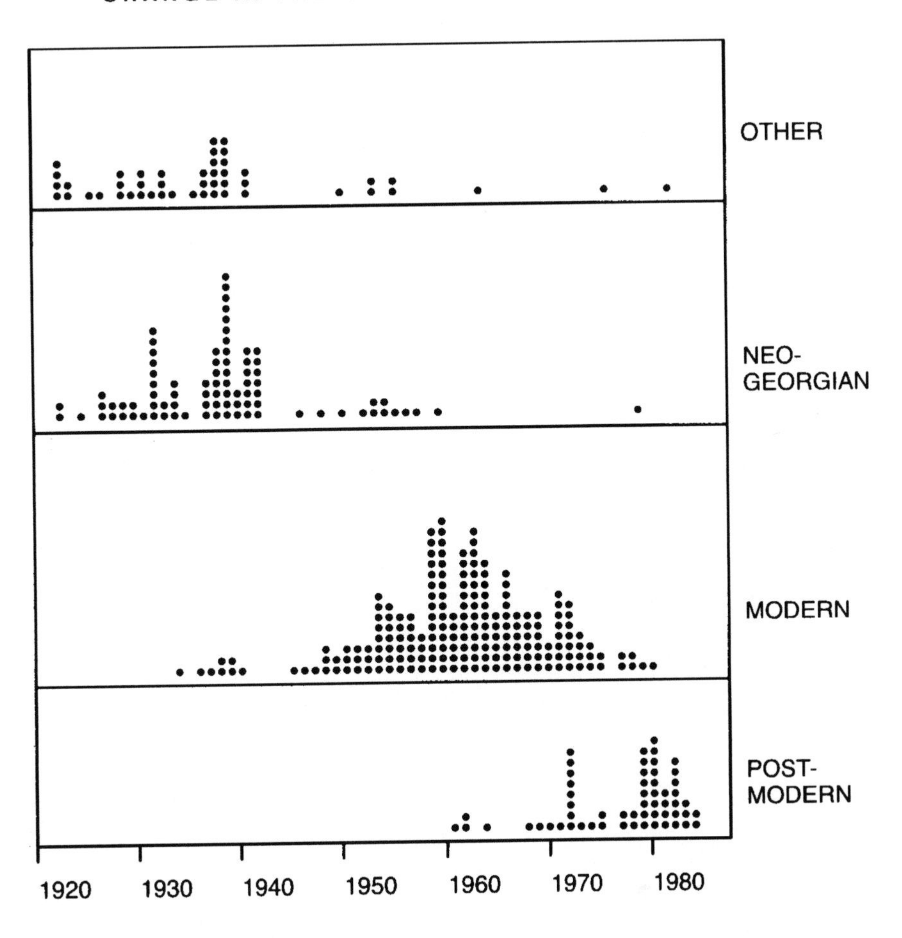

Figure 8.10 Changing architectural styles in five town centres (compiled from Figure 2 of Whitehand 1984, Figure 5.1 of Freeman 1986 and Figure 6.1 of Callis 1986)

These changes in styles are shown in Figure 8.10. Neo-Georgian styles were popular early on in the town centres examined. Although in Aylesbury and Wembley, the study period commenced too late to chart the adoption of this style, in Epsom, Northampton and Watford, neo-Georgian buildings were erected from the early 1920s onwards. The national retail and service chains were particularly important adopters of this style. These were, however, essentially conservative and historicist buildings. The radical Art Deco styling, following the Paris *Exposition des Arts Decoratifs* of 1925, was first introduced in Northampton in 1927 and in Watford in 1929. The only Art Deco building in Aylesbury was the Odeon Cinema of 1935, part of a nationwide dissemination of this style amongst the expanding cinema chains during the 1930s, with many of the Odeon cinemas being designed by Harry Weedon, a Birmingham-based architect. Some of the earlier examples of Art

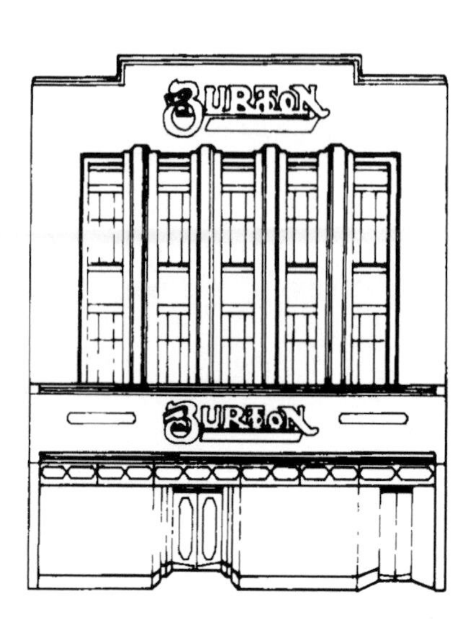
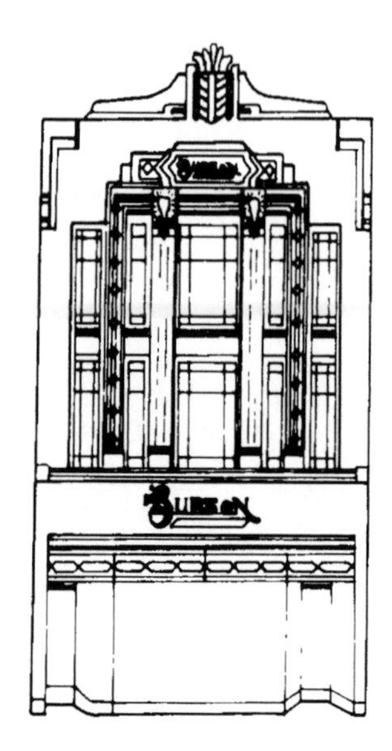

Figure 8.11 Examples of Burton's version of Art Deco (redrawn from the company's advertising material)

Deco in Britain show distinct traces of neo-Classical, with traditionally proportioned window openings and pilasters. The Art Deco styling on these examples is merely *'appliqué'*. This shows how the markedly novel Art Deco, epitomised in American skyscrapers such as the Chrysler building, was adapted to form what might be termed the 'English commercial' variant of this style, which is often a classical pilastered structure with Art Deco details, mouldings and glazing bars. As was the case with neo-Georgian, national chains were important in the diffusion of this style. In addition to the cinema chains already mentioned, the men's clothing retailers Montague Burton, which sometimes had more than one shop in larger town centres, had a house style typical of the 'English commercial' Art Deco (Redmayne 1950) (Figure 8.11).

But Art Deco was a short-lived architectural revolution. No sooner had it appeared in commercial townscapes than its position as the fashionable style was usurped by the sudden rise of Modernism. In Epsom, Aylesbury and Northampton, the first suggestion of the Modern style appeared in the 1930s. Indeed, Northampton was the site of 'the first consciously Modern house in England', commissioned in 1925 by an industrialist from Behrens, a German architect working in Vienna (Gould 1977: 10), and it is possible that this had some influence on commercial styles in that town. Following the hiatus of the Second World War, redevelopment in the post-war building boom (Marriott 1967) continued the Modern idiom in all five town centres studied, to the virtual exclusion of all other styles. This was the period of extensive developments and 'toothpaste architecture', and the new buildings introduced new materials as well as new styling, with glass façades, steel and concrete supplanting traditional materials such as brick (Whitehand 1984; Freeman 1986).

A suggestion of a reaction against the monolithic, functional designs of Modernism appeared in Aylesbury in 1961. Lincoln House, a small office development, had a mansard roof and dormer windows that clearly make its classification as Modern inappropriate: strict Modernism would not tolerate such historicist references. Indeed, this double-coding of styles, the combination of the Modern architectural language with another, was the first sign of post-Modernism (Jencks 1991). It has been suggested that the British *avant-garde* abandoned the Modern Movement in the 1950s, but only in the following decade is there any evidence for this in commercial building (Lyall 1980). This stylistic reaction to Modernism was, however, slow to become widely adopted, and Modern styles remained predominant throughout the 1960s and into the 1970s, although declining sharply in numbers. But in the 1970s a sudden and decisive architectural revolution took place. Only four further Modern buildings were erected in the commercial centre of Aylesbury between 1971 and the end of the study in 1983, for example (Freeman 1986). But it is evident that post-Modern developments in North America are as extensive and obtrusive as their Modern predecessors (Knox 1991); in Britain, as with Art Deco, the style has been adopted in a rather more restrained manner, as is suggested by the majority of the examples discussed in Glancey (1989).

EXAMPLES OF CHANGES

Worcester

As an important mediaeval trading centre and growing industrial area in the nineteenth century, Worcester entered the twentieth century with a city centre consisting of crowded shops, factories and housing, particularly between the High Street and the river, and along the Birmingham and Worcester Canal. Considerable thought was given to the replanning of Worcester immediately following the Second World War, in the period before conservation was a major concern. A survey showed that

> the City still retains in the central area its nineteenth-century plan, with its jumble of factories, shops and houses, patches of high-density, low-grade housing, and lack of open spaces and amenities. It has emphasised the need for rezoning in this central area, particularly along the river front, for a number of new public buildings, shops, schools, clinics and open spaces. It has recommended the encouragement of holiday and tourist traffic by the provision of improved amenities and services, and an extension of the City's hotel and catering facilities.
>
> (Minoprio and Spencely 1946: 13)

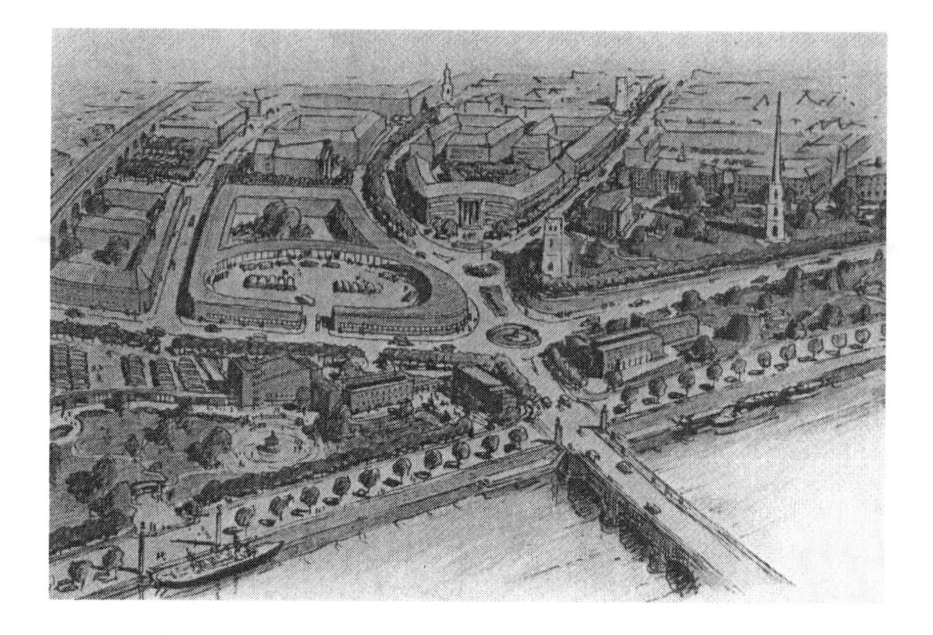

Figure 8.12 Worcester: 1946 proposed redevelopment of the Eastern bridgehead (reproduced from Minoprio and Spencely 1946)

The plan proposed major changes to the street plan, with an inner ring road along the line of the City walls, and three new roads cutting across the city, through the High Street and its largely mediaeval plot pattern, and converging on a new civic centre on the Severn bridgehead area (Figure 8.12). The prevailing attitude to much of the historical centre is epitomised by the view that 'the construction of the new traffic roads will involve interference with existing properties to the west and we therefore recommend that the replanning of the whole area ... should be undertaken at the same time' (Minoprio and Spencely 1946: 54–5). This contrasts with the recognition, as a principal recommendation of the plan, that 'buildings of historic interest or architectural merit in Worcester should be scheduled as monuments of permanent value to the City, and arrangements should be made with the owners for their preservation and maintenance' (Minoprio and Spencely 1946: 71). But this was a period when interest in Victorian buildings was minimal, there was no recommendation on what buildings should be scheduled, and the urgent requirement to renew considerable areas of largely Victorian slum housing and industry is emphasised by contemporary aerial photographs. The colour renderings of this plan show an interesting mixture of large-scale neo-Georgian, mostly for the civic buildings and following in the inter-war tradition of using this style for public buildings, and the flat-roofed monumental Modern toothpaste style also seen in other post-war planning proposals (for example in Oxford: Sharp 1948).

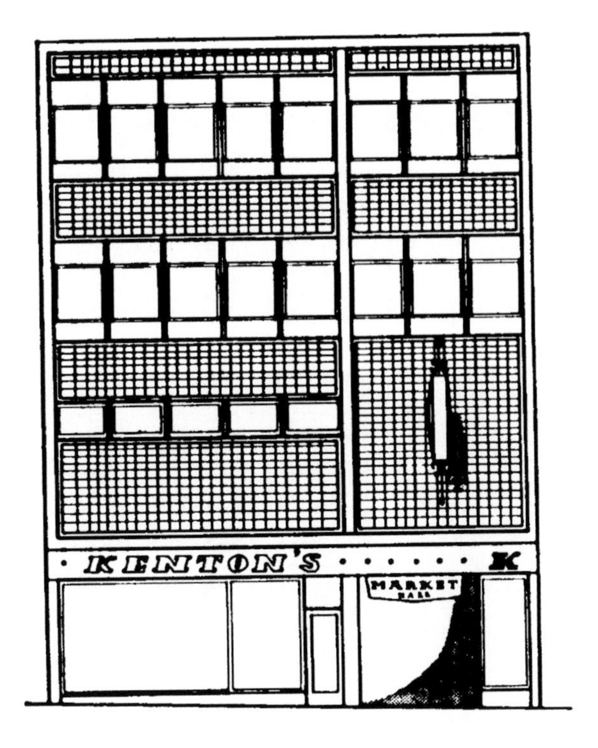

Figure 8.13 City Arcade, Worcester: 1955 original proposal (redrawn from planning files)

Many of the larger-scale schemes proposed in the 1946 plan were not implemented, for a variety of reasons, although clearance of an area of semi-derelict industrial building permitted construction of the City Walls inner ring road in the 1980s, revealing remains of the sandstone wall. Much of the riverside area was redeveloped as a Modern-styled college, and there was a large-scale 1960s shopping centre and multi-storey car-park adjacent to the Severn bridge, and another new retail and hotel development at the Cathedral end of the High Street. Although a number of developments of the 1950s and 1960s did not respect traditional plot frontages (Figure 8.7), a surprisingly large number, including quite significant retail developments such as the Market Hall (City Arcade), did so, albeit using Modern styles and materials (Figure 8.13). The dates, architectural styles and key façade materials used are illustrated in Figure 8.14 (see also Vilagrasa and Larkham 1995). The similarity in the transition of styles between Worcester and the five smaller towns used to construct Figure 8.10 is evident.

By the mid-1980s it was realised that post-war redevelopment, spurred in part by the immediate post-war surveys and plans, had not always respected the special character of the city, despite the inclusion of the major part of the

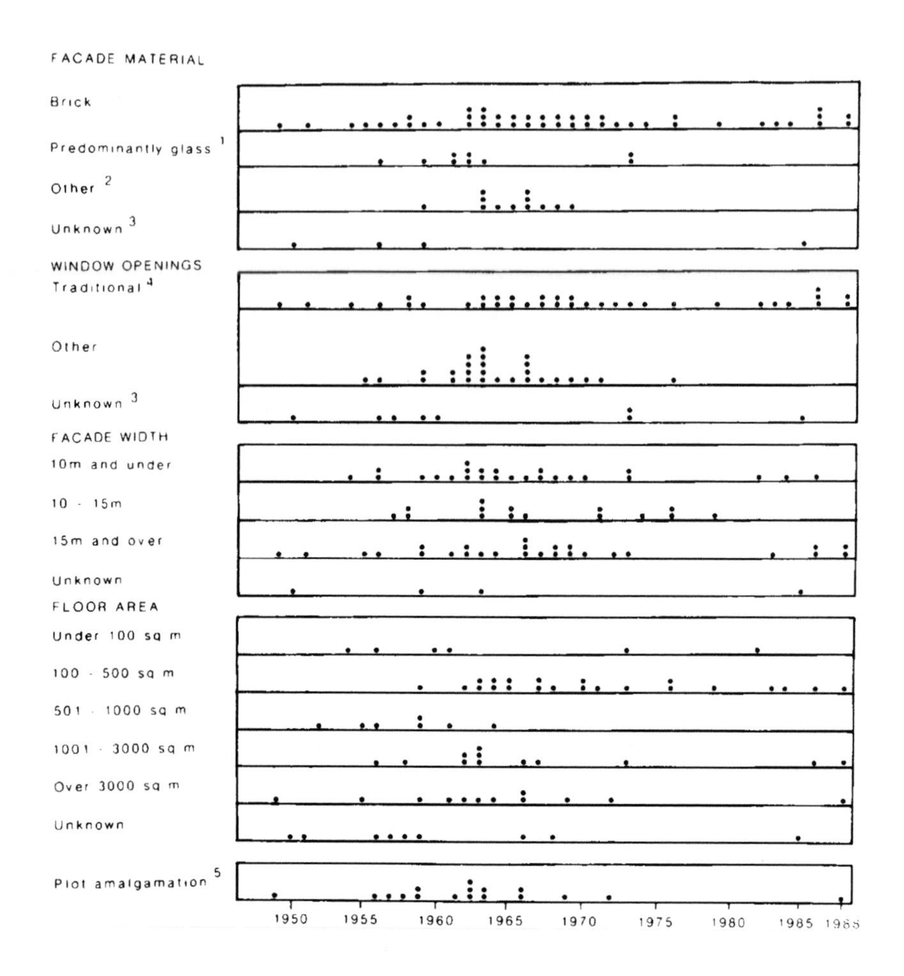

Figure 8.14 Architectural styles and key façade materials of post-war new buildings in Worcester

walled core in a conservation area. This development in conservation planning came after some of the larger-scale, more obtrusive developments. The city centre then suffered from a number of drawbacks that impaired its function as a major sub-regional centre. These included a shortage, poor distribution and low standard of car-parks, a conflict between vehicular and pedestrian traffic in parts of the centre, and a lack of suitably sized premises to meet the needs of modern retailing. This last point was seen as especially significant, since many of the existing older buildings could not be adapted to modern requirements without some loss to their architectural character (Worcester City Council 1985).

One large-scale development proposal of the mid-1980s sought to rectify some of these identified drawbacks of the existing city centre. A developer,

Centrovincial Estates PLC, proposed the redevelopment of two sites, totalling 3.46ha, on either side of Broad Street, on the western side of the city centre. One site included the unsightly 1960s Blackfriars Square shopping precinct and multi-storey car-park (Figure 8.15A). The other site was a temporary surface car-park, but adjoined the listed Countess of Huntingdon's Chapel (Figure 8.15B). In total, some 22,585 square metres of retail floorspace, 1,134 square metres of office floorspace, 728 car-parking spaces and a bus station were envisaged in this proposal. It was considered by a DoE Inspector after a Planning Inquiry that servicing, public transport and highway considerations would all benefit from this scale of development, which would also assist the city centre to fight competition from a number of out-of-town retail sites. In terms of conservation, it was felt that the scale of development, although large, was acceptable, being better than the existing shopping centre on part of the site which was outmoded in retail and stylistic terms. The details of the new street façades were reserved for later approval. Impact on the listed Chapel seems hardly to have been considered, from the evidence of the Inspector's report. The former Sunday School, also on this site, was

> a building of some importance but it is in a very poor condition and would ... require complete reconstruction ... a light glazed structure would be more appropriate in this instance, and the juxtaposition of modern and traditional styles of architecture of high quality would enhance [the development].
>
> (DoE 1988a: 30–1)

Although the development permitted an archaeological investigation on part of the site, it is evident from the plans that a considerable proportion of the mediaeval core, albeit of already much-altered plots, has been lost. Some very substantial buildings have been permitted on backland sites, which are, in this case, very visible from the river. However, this scheme has been the subject of a lengthy negotiation process between the original developer, the LPA, English Heritage, the Royal Fine Art Commission, the public, and other bodies since the initial decision to improve the 1960s Blackfriars Square shopping centre was taken in December 1979 (Frederick Gibberd Coombes & Partners 1988). Interestingly, it is suggested that both the developer and the City Council were originally unwilling to negotiate changes to the proposals for amenity reasons, and that the plans were eventually amended, by agreement, owing to the developer's appointment of a new conservation consultant, the improving financial climate in the late 1980s, and the securing of a prestigious national retailer to occupy the main, 'anchor', retail unit (English Heritage 1988a).

A modified version of this scheme was completed in the early 1990s after the site had been acquired by other interests, including the Crown Estate. The outdated Blackfriars Square was updated – in stylistic terms it was 'de-modernised' – and a new centre, the Crowngate Centre, built to slightly revised designs around the listed Chapel. The LPA's perspective

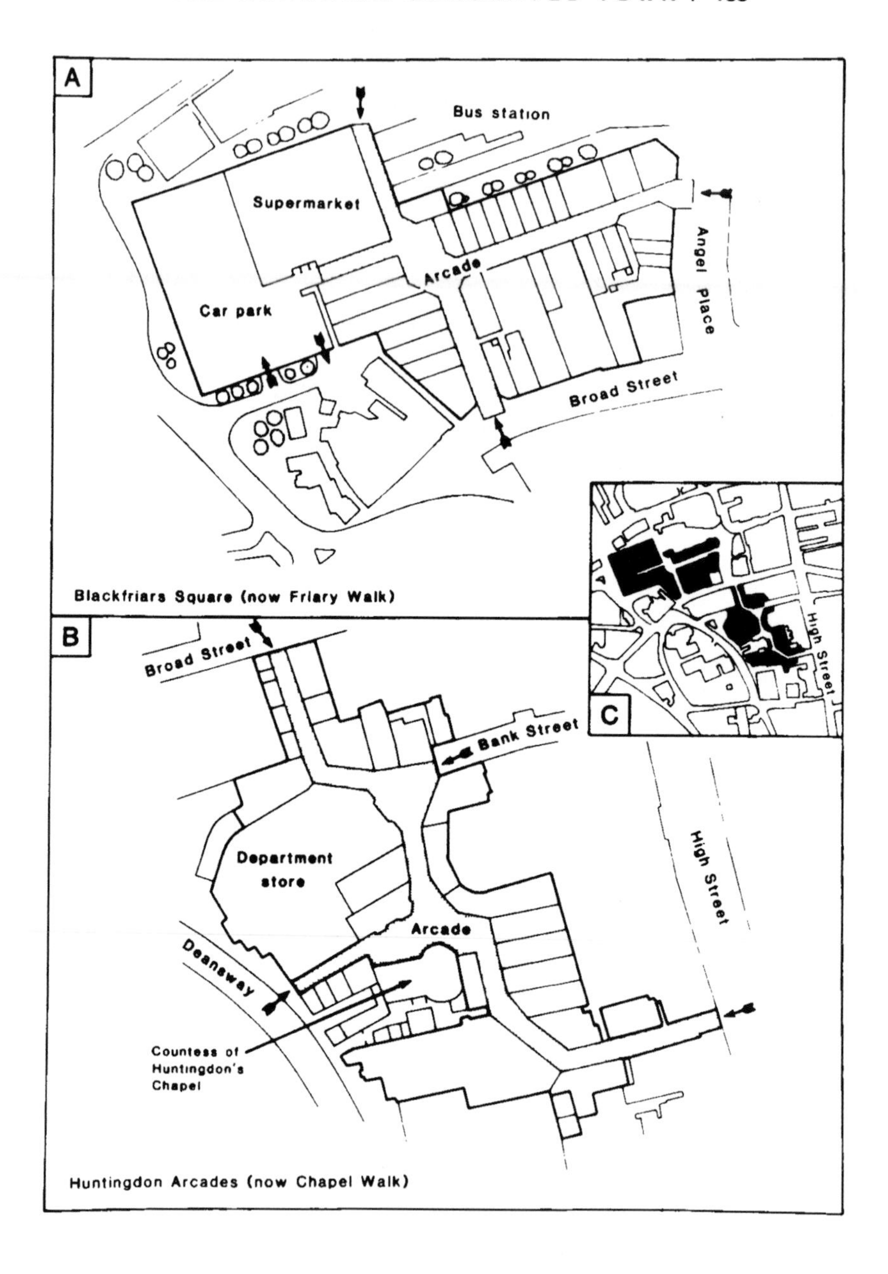

Figure 8.15 Major shopping centre redevelopments, Worcester city centre, late 1980s (redrawn from Centrovincial Estates publicity material)

of this development and its evolution is given in Appendix 2.

This development raises the problem of what happens to historically and architecturally important buildings fronting main streets when major developments take place behind them, and when access to these developments is proposed to cut through these buildings. Façadism is not always the appropriate solution. Part of the building itself may be used, or a new access developed immediately adjacent; both occurred in the Worcester redevelopment. The problem of the impact on historic buildings was considered by the Public Inquiry and by English Heritage, and found to be acceptable. However, as construction progressed in the early 1990s, with a new development company, it transpired that the carefully negotiated design solution agreed by all parties had to be considerably amended. The fire authorities insisted that the entrances to the shopping malls must be six metres wide, in order to cope with the expected volume of shoppers in any emergency. This involved significant changes to the design (Figure 8.16), and occasioned comments from one local councillor that 'what were to have been narrow intimate shopping alleys are now to be broad malls', while another admitted that the new designs were 'disappointing' but that it was necessary 'to make the best of a bad job' (*Worcester Evening News* 1991).

Henley-on-Thames

Similar considerations arise in the case of a proposed shopping mall scheme in Henley-on-Thames. This is another small, planned mediaeval town, with considerable survival of burgages into the post-war period. Many plot tails were acquired by the local authority through the 1950s and 1960s, and were amalgamated to form two car-parks. During the 1980s, several proposals were put forwards to create a large shopping arcade and superstore on one of these car-parks, behind Bell Street (Figure 8.17). The major considerations here, in terms of the impact on existing buildings, included (i) the style and bulk of the new proposal; (ii) the effect on trading patterns (and thus profitability) of existing buildings; (iii) the consequences for neighbouring listed buildings and others of local historical or architectural interest; and (iv) the archaeological consequences.

In terms of style, one of these proposals was strongly criticised by the LPA's own architect.

> Pseudo-Palladian portico gateways, balustrading from Bournemouth municipal gardens and 3-ball light fittings from Burlington Arcade is a recipe for outstanding success of which the developer/architect is no doubt well aware. It offers we British the continuing advantage of living in the past while at the same time providing the illusion that, in some strange fashion, one is 'up-to-date'.
>
> (planning file, South Oxfordshire DC)

But what is being described here is the eclectic historicism of the post-Modern style, in the period when it had decisively supplanted the Modern

A
WOOLWICH
Mall entrance
Existing shopfronts

B

Figure 8.16 Crowngate shopping centre, Worcester: amendment for large new entrance. A: entrance as originally proposed (redrawn from planning files); B: as built, 1992 (author's photograph)

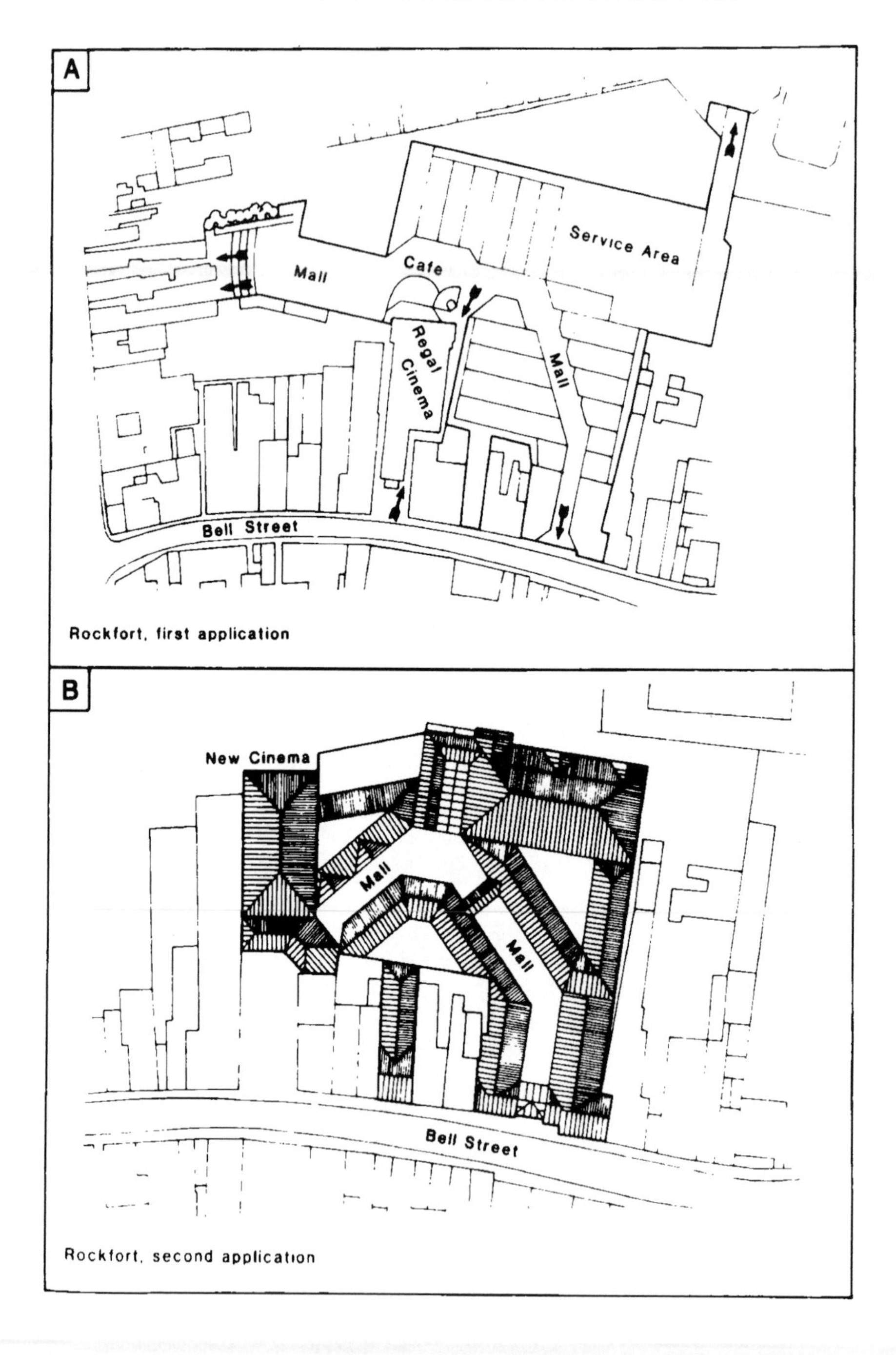

Figure 8.17 Competing shopping centre proposals, Henley-on-Thames (redrawn from planning files)

style in all five town centres studied in detail (Figure 8.10). It was then commercially a very successful style, the richness of decoration and materials proving attractive to retail customers, as the experience of many town centre, edge- and out-of-town retail developments shows throughout the 1980s. This style *per se* was not necessarily in conflict with the existing townscape in Henley, instead it is likely to be much better than the earlier generation of Modern-styled developments. Conflicts may, however, occur with respect to the size, scale and massing of these proposals, both in their location on previously open ground in the historic urban core (formerly plot tails, latterly a car-park) and their proximity to existing historic buildings. The local topography is often important in this respect. In the Henley case, the site itself is relatively flat and, from the main streets, the shopping mall will be concealed by the existing two- and three-storey frontage buildings; it would be visually intrusive only from some of the later development on the surrounding hillsides. In several cases in Shrewsbury, in contrast, new shopping centres are obviously out of scale with their surroundings, and this is made more evident by the steep slopes from the town centre to the river.

The second problem is economic. The effect of a development of this size on established trading patterns is likely to be significant. As customers are attracted to the novelty of new shops and new surroundings, they tend to desert older established concerns, who suffer significant economic recession. Likewise, retail tenants themselves often relocate into such new, purpose-built accommodation in order to take advantage of such trade and, often, more desirable facilities. Such relocations thus frequently leave vacancies in older, often converted, town-centre buildings. This has significant implications for building neglect since, even if new tenants can be found, they are frequently more marginal traders, able to afford only lower rents and sometimes for only short-term lets. Such traders usually only use the ground floor; all such factors lead to the disrepair of upper floors and advancing structural decay. This problem was noticed in central Shrewsbury following completion of several new shopping schemes in the mid-1980s (Martin and Buckler 1988).

Thirdly, there are direct consequences for neighbouring listed buildings and others of more local interest. In particular, the manner in which the proposed mall would be accessed by shoppers on foot from the High Street and Bell Street is significant. Here – and in the Worcester example – a combination of re-using existing historic buildings suitably adapted to form 'gateways', albeit with some loss of their own intrinsic architectural integrity, and new construction was proposed. Figure 8.17B, the more radical proposal, shows not only the scale of such new frontage development, occupying four original plots, but also the proposed redevelopment of the 1930s Art Deco Regal Cinema. Here, pressure from a significant local and national group convinced the LPA to support appeals to the DoE to spot-list the cinema. When this failed, and despite a 30-page legal opinion commissioned by the group, the LPA had a change of heart over the intrinsic architectural value of the cinema, and accepted the arguments for redevelopment. A local landmark of considerable interest, particularly for its Art Deco interior, would thus have been swept away to suit the developer – who proposed a replacement cinema, re-using the cinema organ of the original!

The last significant consequences of such developments are archaeological. In archaeological terms, very little is known of such 'backland' sites: they are the rears of plots, 'plot tails', and little is known of the uses to which these were put. Most excavation has been carried out, often hastily, on the site of the main building on a plot, usually fronting the main street, or, when such buildings are not demolished, on small sites immediately adjacent to their back walls. Yet, where excavation is allowed in advance of such development, significant information on the local patterns of urban development and land use is often found, as was the case in Worcester and in Shrewsbury (Baker 1988). However, developers often have a jaundiced view of archaeology, owing to the costly delays which a successful excavation can impose, as was the case with the Rose Theatre site in London.

Also within the Henley conservation area, there is an interesting case where planning policies related to issues other than conservation or aesthetics had a significant visual impact upon a proposed development. In the case of the former Methodist Church, Duke Street, Henley-on-Thames, it is interesting to note that a series of proposals for a new office block to replace the redundant church were amended partly for aesthetic reasons, and there was a disagreement between the District Council's architect and conservation officer over the suitability and sensitivity of the proposal. More significantly, however, the floorspace, and therefore height and impact of the building, were reduced largely because of Structure Plan restrictions on office development being allowed only for local users, there being a wish to halt an influx of new office users into the town centre (Figure 8.18A–B). Interestingly, this policy was not upheld at appeal for other developments in the area, but the Duke Street scheme did not go to appeal. Eventually, the architects substantially changed the design of the lowered building (Figure 8.18C), eliciting the response from the LPA that, given the circumstances and history of this application, it was an infinitely superior proposal to anything that had thus far been seen (planning files, South Oxfordshire DC). A land-use policy thus played an unintended part in influencing the volume, and thus form, of a new building. Figure 8.18 illustrates the rear of the building, visible from a public car-park, which most clearly exemplifies the visual importance of the lower height and changes in massing. The front elevation was amended following a meeting with the conservation officer, in which a sketch retaining much of the massing but using local vernacular detailing was produced (Figure 8.19). In this case, the conservation officer played little part in reducing the building's bulk and visual impact, but had considerable influence over the detail by which the street frontage would blend in with its surroundings. The officer concerned held a doctorate and had written books on architectural history, which might have influenced his decision to press for a contextual design.

The same policies in the Oxfordshire Structure Plan, against new office

Figure 8.18 Proposals for rear elevation, Methodist Church site, Henley-on-Thames (redrawn from planning files)

development which would draw additional population into the county, and in the Henley Local Plan, presuming against the creation of new offices in the town centre unless required by an existing local firm, were also tested when

Figure 8.19 Alternative proposals by developer and Conservation Officer for front elevation of offices, Methodist Church site, Henley-on-Thames (redrawn from planning files)

a local brewery attempted to develop its malthouse site. The town's river crossing position had encouraged substantial development of coaching inns in the eighteenth century, following which brewing became a major industry. Breweries and associated specialist buildings became significant features in the townscape; one brewery, Brakspears, surviving to the present. Its main brewery premises and malthouse lay on opposite sides of New Street, Henley. Linked applications to convert the disused and listed malthouses, together with significant changes to improve working practices at the brewery, were submitted in the mid-1980s. On the malthouse site, the proposals were to convert the building to office use, together with 'the demolition, alteration and/or erection of buildings and structures to provide four residential units, car parking and garaging with associated landscaping and works to boundary walls' (planning files, South Oxfordshire DC). The listed malthouse was to receive substantial internal alterations, large new window openings, dormers and rooflights (Figure 8.20). Brakspears argued that the considerable conversion costs incurred in returning the building to any viable use led to the requirement for office use. Nevertheless, all consultees and letters received from members of the public on these applications were in favour of refusal. The Henley Society registered strong objection both in principle to the change of use and in detail to the conversion, while the Campaign for the Protection of Rural England suggested that policies seeking to prevent speculative office development within the conservation area should not be breached and that the malthouse should be retained for storage purposes. Planning permissions were refused largely on the grounds of the Structure and Local Plan policies already mentioned, and listed building consent on the principal ground that 'the development would result in excessive and inappropriate alterations to the structure and external appearance of the existing Malthouse building, detriments to the character and appearance of the building which is listed as a Grade II building' (planning refusal notice, South Oxfordshire DC).

The Structure and Local Plan policies had clauses providing that, despite the presumption against such office development, favourable consideration might be given to the change of use of redundant buildings where retention is desirable for architectural, historic or aesthetic reasons, and these were cited by Brakspears. Nevertheless the LPA continued to cite the office control policies. Perhaps it was felt that the unsatisfactory nature of the proposed conversion of the listed building itself did not merit overturning them, and other alternative uses had not been explored in great depth. Nevertheless, the problem of a large, under-used and decaying listed building, prominent in the townscape, remained.

Ludlow

The small rural market town of Ludlow, a multi-phase planned mediaeval town much studied by morphologists and historians (Conzen 1966, 1988; Slater 1990a), is typical of those small rural towns which were largely

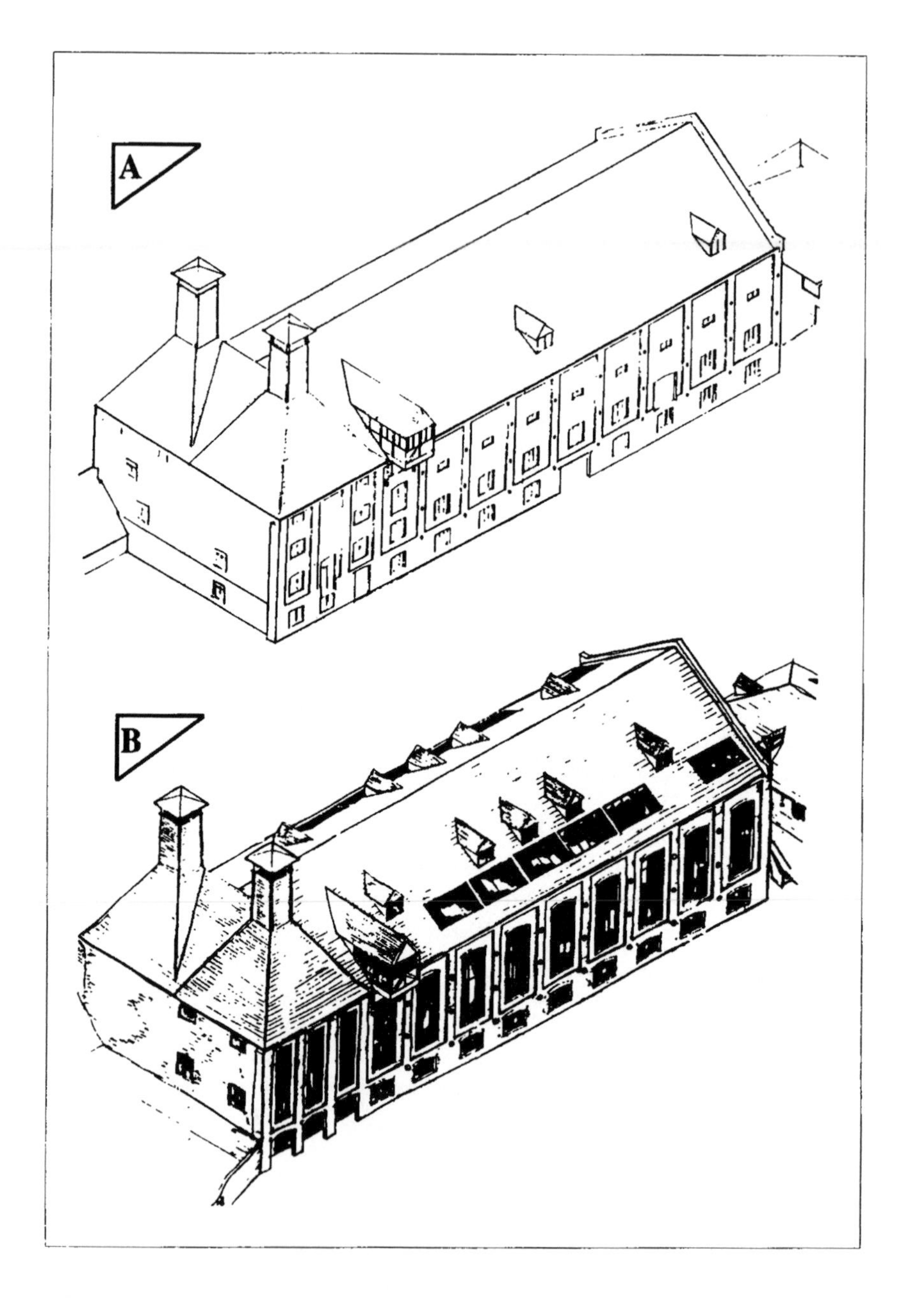

Figure 8.20 The malthouse, Brakspear's brewery, Henley-on-Thames. A: the building in 1985; B: proposed alterations (isometric drawings amended from planning files)

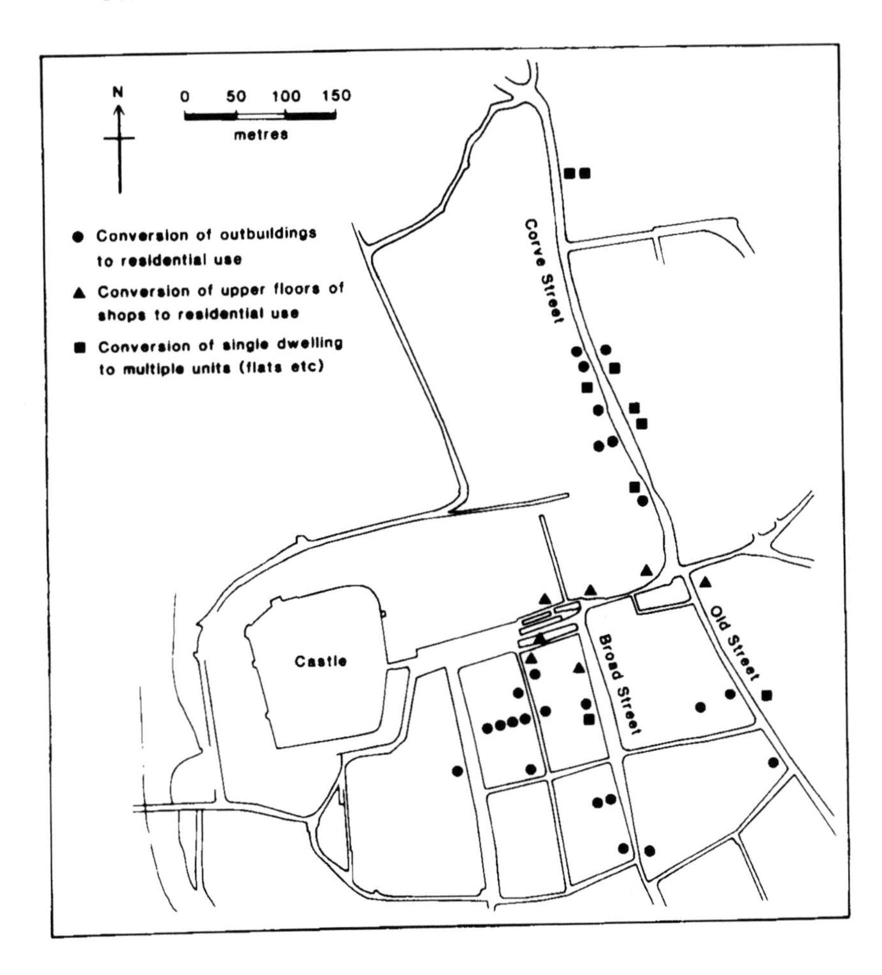

Figure 8.21 Location of applications for intensification of use, Ludlow, 1970–86

bypassed by the Industrial Revolution: Victorian and post-war redevelopments are in a minority, while the mediaeval street and plot patterns remain largely intact. Until relatively recently, residential land-uses have been important in the centres of most towns, particularly these. Increased land prices, together with a more flexible approach to planning by LPAs (encouraged by the DoE: DoE 1987a), have tended to result in pressures to re-use buildings and parts of buildings that had become functionally obsolete. Typical of this pressure is the demand in Ludlow for the conversion of outbuildings, often stables, into dwellings. Ludlow has a relatively large stock of such outbuildings owing to the number of fashionable Georgian town houses on deep burgages, and the easy rear access for horse traffic afforded by the pattern of rear alleys. The pressure for this type of conversion is shown in Figure 8.21. The dwellings thus created have been very small; some include only a living room/kitchen downstairs with a bedroom/bathroom above. Some have no bathroom, instead a 'shower room' is provided. Indeed, some such developments were refused as they 'would constitute a form of residential redevelopment which is so severely restricted in space and outlook as to result in substandard living conditions' (planning files, South Shropshire DC).

Nevertheless, interviews with local developers and estate agents suggest that there has been a demand through the 1980s for such smaller property in Ludlow, mainly from young single- and two-person households of low income, often of local origin, who cannot afford the high prices commanded in that period for traditional homes even of 'starter' sizes.

A more common problem is the under-use of upper floors of buildings in commercial or retail use (DoE 1978). Legislative changes introducing new grants and assured shorthold tenancies are beginning to encourage the conversion of such under-used storeys into residential units (Petherick 1990). However, although the potential for such conversions is clearly present in Henley and Ludlow, the number of actual developments is surprisingly small (Figure 8.21).

CONCLUSIONS: CHANGE IN THE CITY CENTRE

General points: the key questions

The town-centre examples cited have provided valuable information towards answering the key 'what' and 'how' questions posed in Chapter 1. In all cases, the overwhelming motivation for change has been the prospect of economic gain, which might be considerable in a development of the size of the Crowngate Centre.

The plotting of rates and types of change shows the rates of change to be high, varying over time as they relate to national and regional building cycles (cf. Lewis 1965). In numerical terms, the numbers of substantial changes (particularly new buildings) are low, yet individual changes may be immense, affecting numerous historic structures and plots, as in the cases of the shopping centres in Worcester and Henley. Minor changes, particularly those attributable to commercial activity such as signs, shopfronts and even blinds (counted here in the 'miscellaneous minor' category) can be numerically high. It is the cumulative impact of such individually relatively minor changes on the conserved townscape which is simultaneously not assessed by the LPAs and a matter of rising concern.

Worcester's Crowngate Centre illustrates a degree of unanimity by decision-makers that a large-scale development was acceptable and that one listed structure, the Sunday School, should nevertheless be removed. There was less unanimity over some details of the scheme, particularly the changes imposed at a late stage by 'third parties', principally the Fire Officer's requirement for changes to the entranceway within a historic frontage building.

In Henley-on-Thames, there was again general agreement over the scale of development on the backland car-park sites, which had long ago been acquired and cleared by the LPA. There was disagreement in detail both within the LPA and amongst the public over detailed design issues and over the scheme's impact on surrounding buildings, particularly the Regal Cinema. This is an area with a vociferous, well-educated and wealthy public; a pressure group was formed, a QC's opinion sought and the DoE lobbied to list the building. At first the LPA was in support of retention. However, when the DoE refused to list the cinema, the LPA changed its opinion and supported removal, albeit with a new cinema incorporated into new proposals. This illustrates the key fact that agents may change tack over proposals, particularly if key circumstances change (e.g. political make-up of a council or committee) or if the process is prolonged.

The Methodist Church site illustrates how conservation objectives may be indirectly served by non-conservation-specific policies. Here, the intensity of development and, therefore, the bulk of the building was limited by employment, not conservation, policy. Nevertheless, the role of the LPA conservation officer was significant in negotiating changes to the façade. This emphasises the importance of a scholarly, experienced and effective negotiator, and, in examining the contextual design suggested, confirms this book's emphasis on the importance of building façades in conservation areas.

The brewery site develops further examples of conflict: between the developer and public groups, and between different LPA policies. Here, specific conservation policies on the changes of use of redundant historic buildings conflicted with presumptions against office development. Conversion would have secured the retention of the building, albeit in modified form. The office control policy would leave the building to decay further and the owner to seek a more acceptable alternative use. Which policy carries more weight in this circumstance is a purely political decision.

In Ludlow, the key issue is of changing land-uses and the pressure for backland development. There was local pressure for small dwellings, even of the size considered by the LPA to be 'substandard'. Central government has encouraged local flexibility in policy and decision-making, particularly with respect to historic buildings, but how far can this flexibility be taken?

The issue of 'backland development'

A significant issue in many of these historic town centres remains that of 'backland development'. Widely held to be a sensitive issue, it usually involves the creation of large development sites through the amalgamation of numerous tails of often historic plots dating to the mediaeval period. It is a key tactic of prospective developers in seeking development sites in closely built-up areas, particularly those with acknowledged historical, architectural and archaeological value. It is argued that this form of development has minimal impact upon the historic built fabric.

As in Henley, it is often seen to be the easiest way to insert large modern retail developments into sensitive conservation areas, particularly when land assembly is facilitated by the LPA, as occurred in Henley. The historical and archaeological value of these plot tails is often little heeded, neither is the potential use in psychological and urban design terms of such parcels of open space, ranging from private to semi-public and wholly public, within the built-up urban core. One such redevelopment in Ludlow, for example, proposed new development on individual burgages, in a street of obvious burgages, and yet fierce public reaction alleged that it constituted a 'violation of the historic pattern of building and space in the inner town', and that 'such crowding in such a small site . . . opens up the possibility of site infilling along the length of the old gardens of Old Street, either by separate or group[s of] developers – thus changing forever the character of the centre of old Ludlow'; yet the infilling of burgages is an integral part of their development and has a long history (see the discussion of the burgage cycle in Conzen 1962 and Larkham 1989). There are obviously confused views on the actual historic pattern of plot development in this town, and the values placed upon plots by both the LPA and those affected by development schemes vary widely (Larkham 1990a). Despite the problems of dealing with the history of plots and structures, in morphological terms backland development has a long history and there is no intrinsic objection to such forms of development.

In more general terms, much obvious history has already been lost, and archaeological remains have been damaged or destroyed in many of these historic urban centres, prior to the recent pressures for intensive development on amalgamations of, often mediaeval, plots. The morphological frame of streets and plots, which Conzen suggests should have a constraining effect on new development (Conzen 1975; see also Larkham 1995b), has little evident influence in current planning proposals and planning decision-making. These developments have a large number of potential impacts upon the urban centres, not all of which are explicitly morphological: many are economic, with later morphological consequences in the neglect and decay of older buildings. Changed pedestrian traffic patterns may shift the centre of gravity of the business area (e.g. Buswell 1984a, 1984b) with yet further morphological impacts. The demands for such extensive and 'improved' retail facilities are rarely proved on either economic or functional grounds, and economic fluctuations in recent years are such that rents in similar large developments in Birmingham, for example, have dropped dramatically as trade is far below projected figures (*Chartered Surveyor Weekly* 1988). Yet the mid-1980s was 'the most dynamic phase in town-centre redevelopment since the era of comprehensive development' (Debenham, Tewson and Chinnocks 1987).

Historic town centres depend to a large extent on their characteristics of form for their tourist appeal, as is clearly evident from their attempts at 'place marketing' (Ashworth and Voogd 1990) and, indeed, their attempts at recreating at least the superficial ambience of such historic quarters, for example in Germany (cf. Soane 1994). Yet, in the UK, the reactions to individual planning applications and the development of related policies suggests that such historic towns have not yet successfully balanced user demand for retail and office space with developer and investor demand for development with those very characteristics important to the historical ambience. It should now be self-evident that very large-scale retail developments are inappropriate to the centres of historic towns. It is virtually impossible to accommodate modern large-scale developments in historic streets without losing their scale and character. Local planning authorities should look to the longer term, and at the social costs, when assessing planning applications whose apparent purpose is short-term speculative gain (English Heritage 1988b).

9

CHANGES IN RESIDENTIAL CONSERVATION AREAS

The surroundings householders crave are glorified autobiographies ghost-written by willing architects and interior designers who, like their clients, want to show off.

(T.H. Robsjohn-Gibbings, 1961)

INTRODUCTION

The centres of historic towns are undoubtedly of considerable significance, being places where change is most readily evident to most people, and many have been designated as conservation areas relatively early (Jones and Larkham 1993). Yet a similar importance should also be accorded to the considerable areas of well-established residential areas. To date, rather less attention has been paid to residential conservation areas than to urban centres. This may be caused by the relatively low number of areas designated specifically as residential conservation areas: Pearce *et al.* (1990: xxiii) note only 618 conservation areas occurring within urban areas but outside the town centre (about one-tenth of the then total number of designated areas), and two out of three of these comprise only small portions of the town. Admittedly, some of the core designations extend beyond the commercial centre to include adjoining residential areas. This suggests that the residential areas that have officially been designated vary considerably in size, character and location.

The designation of conservation areas consisting wholly of residential areas has recently been increasing: these now form probably the fastest-growing category of conservation area (Pearce *et al.* 1990). Not only are Georgian and Victorian suburbs being designated, but some early twentieth-century municipal housing schemes, rare areas of Modern Movement houses (Figure 9.1) and even standard speculative inter-war suburbs (Figure 9.2) have now been designated. Remembering that, in UK law, conservation areas must demonstrate 'special' interest, it is sometimes difficult to justify some of these designations, and this has led to considerable debate over whether the coinage of the conservation area has become debased (Suddards and Morton 1991; Morton 1991).

Some areas are of relatively homogenous character. The Chapelfields area

Figure 9.1 Modern Movement houses, Amersham-on-the-Hill (author's photograph)

Figure 9.2 Semi-detached houses in School Road conservation area, Birmingham (author's photograph)

in Coventry, for example, is a conservation area designated in 1976, comprising several streets of terraced houses some 2.4km from the city centre (Figure 9.3). The development dates from 1845, and the houses were designed to combine the functions of dwellings and workshops for watchmakers. The prosperity of the master watchmakers led to the development of larger houses with front gardens and architectural detailing, whilst the houses of the journeymen are more plain and have a rear workshop at first floor level with a characteristic long window (Coventry City Council 1979). In sharp contrast in history and style, but similarly homogenous, is the conservation area of Ashleigh Road in Solihull, designated in 1985. It is a single short street of 44 houses, developed between 1902 and 1913, and the houses are large and almost entirely in exuberant Edwardian styles (Figure 9.4) (Metropolitan Borough of Solihull 1985). These two conservation areas are both residential, but very different in character. Chapelfields was designated partly for its historical importance in the growth of the local watchmaking trade, while Ashleigh Road was designated as a good and relatively unchanged example of mature low-density early twentieth-century suburban development, which was under some threat of plot subdivision and infill development.

Examined in detail, however, even the boundaries of some villa suburbs, widely acceptable for conservation owing to their distinct character of mature landscaping and large villas of high quality and individual architectural character set in large plots, may be doubted. Recent studies have shown the high rate of infill development in such estates in the post-war period: the trends to 'building in the back garden' or 'densification' are quite evident (e.g. Jones 1991; Whitehand, Larkham and Jones 1992). Many designations include areas of such modern infill, which may comprise a significant proportion of the conservation area and which may make character appraisal more difficult: Figure 9.5 shows one such designation of a Victorian villa suburb, which has suffered considerable recent infill, in Blaby district. It must be noted, however, that the strong view of one district is that some recent developments have their own character and may, in the future, be worthy of preservation: their 'accidental' inclusion within a present-day conservation area should not be a cause for concern (Roy Vallis, Stroud District Council, pers. comm.).

It is particularly these areas of originally low-density, 'first-cycle' residential development that surround the centres of the majority of English towns that are of considerable interest in terms of the management of the conserved urban landscape. It is here, rather than in the high-density workers' housing of areas such as Chapelfields, that increasing efforts are being made to increase residential densities, often to the detriment of the character and appearance of the area (Whitehand and Larkham 1991a; Whitehand, Larkham and Jones 1992). These areas have, until relatively recently, been little explored in morphological terms. Such areas can now be termed 'mature', as they have existed for decades, some indeed for well over a century. The garden planting, hedges and trees, important visual elements in many such urban landscapes, are certainly mature. These also tend to be

Figure 9.3 Chapelfields conservation area, Coventry
(redrawn from City Development Directorate publicity material)

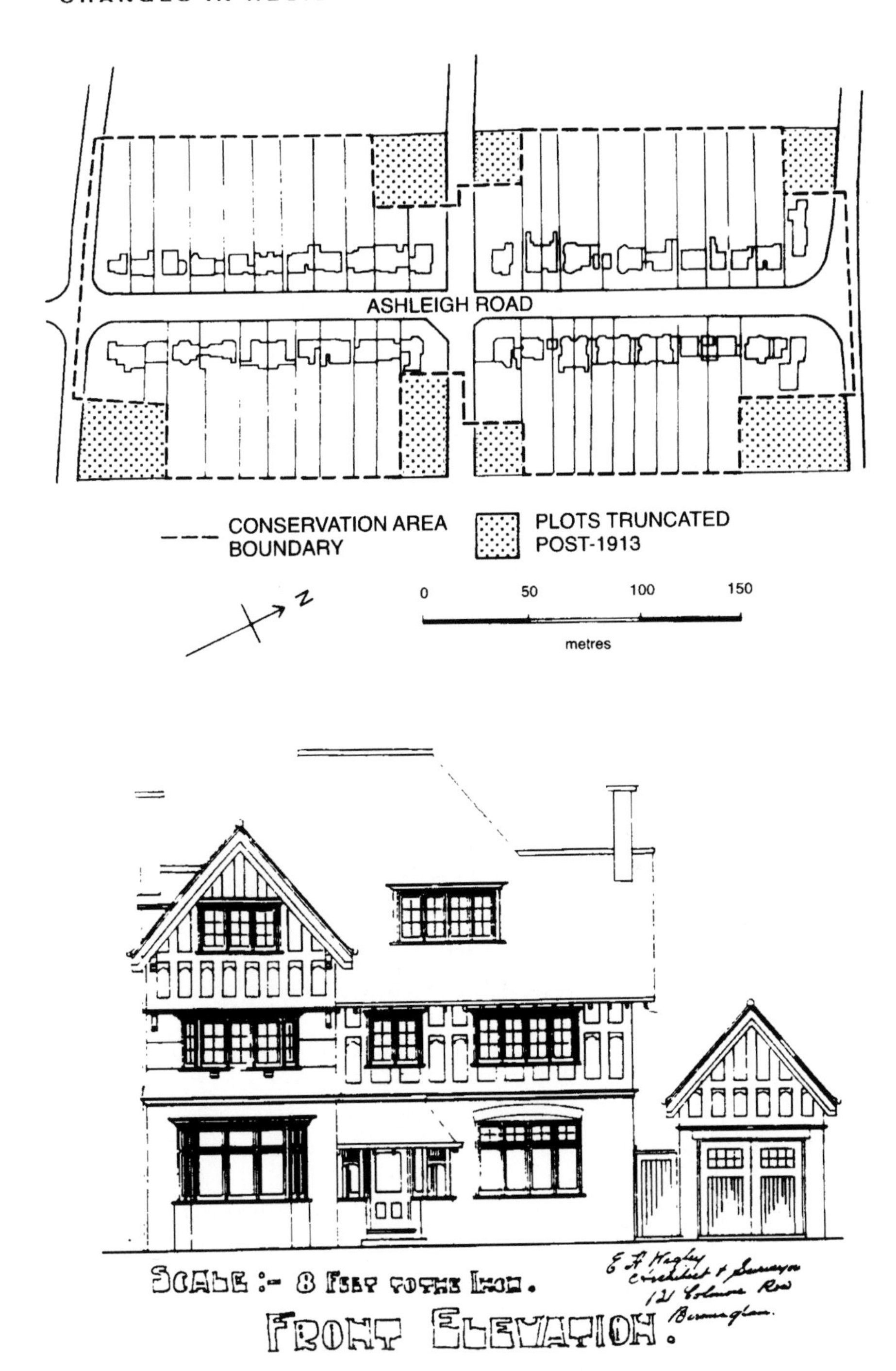

Figure 9.4 Ashleigh Road conservation area, Solihull: area plan and typical elevation (redrawn from building surveyor's plan)

Figure 9.5 Kirby Fields conservation area, Blaby, near Leicester. Conservation area boundary shown in bold dotted line; heavy unbroken lines outline recent residential cul-de-sac developments (amended from LPA designation map)

townscapes that embody a large cumulative investment in terms of human endeavour and in this sense, although not necessarily in a strictly economic sense, they could be regarded as being of great value.

These residential areas were usually created when the needs of society were rather different from current needs. Many were located towards the then urban fringe, away from the noise, pollution, and undesirable social conditions of the central areas. The availability and price of land in these locations frequently led to suburban development having characteristically extensive plots and substantial detached houses, some of which are of notable architectural value (as is the case, for example, in Ashleigh Road). Indeed, these are the areas of the nineteenth-century ornamental villa and its successors. The archetypal ornamental villa stood in its 'own pleasure-grounds, with an approach road, or sweep, up to the house, a walled kitchen garden and stable offices, the extent of the gardens being from two to ten acres' (Loudon 1850: 43). 'The houses themselves were normally of two storeys, with extensive attic accommodation for the servants. Architecturally, they were amongst the earliest buildings to be designed in the full range of nineteenth-century "revival" styles' (Slater 1978: 130; see also chapters in Simpson and Lloyd 1977). The exact characteristics of any one area of this form of first-cycle development vary, depending upon (i) position of the settlement within the urban hierarchy; (ii) location of the area within the settlement; (iii) exact socio-economic class catered for; and (iv) date of design. The villa of the turn of the nineteenth century was very different from that of a century later, as this description and house type descended rather rapidly through the social hierarchy (Slater 1978). By the later years of this period, semi-detached 'villas' were not unknown (Barrett and Phillips 1987). Nevertheless, large buildings and large plots remain characteristic of these mature residential areas. There are many studies of this type of area, mostly from the socio-economic and historical perspectives, and most have some discussion of morphological characteristics (see, *inter alia*, Summerson 1945 (general, on Georgian London); Olsen 1976 (general, Victorian London); Cannadine 1980 (specific, Edgbaston and Eastbourne) and Spiers 1976 (specific, Victoria Park, Manchester)). The experience of other countries with this form of suburban development is rather different, but some parallels do exist (see, for example, Holdsworth 1986).

Social and economic conditions have changed, often quite markedly, since many of these areas were originally developed. The expansion of urban areas has engulfed most of these mature suburbs. Changes in accessibility to the town centre, together with changes in bid-rents and use values, have led to land-use and often built-fabric changes. The inevitable and inexorable structural decline and obsolescence of buildings (cf. Cowan 1963; and Chapter 3) is paralleled by the change from exclusive single-family occupancy through multiple occupation, business/commercial use and, in extreme cases, redevelopment. Investment in repairs and restoration may, in some cases, temporarily halt what has been seen as a cycle (Slater 1978: 142–3; Whitehand 1987a: Chapter 6); a residential parallel to the burgage cycle familiar in historical town centres. What is widely seen as an increasing demand for all types of development during the late 1980s has prompted detailed examination of the incidence and impact of infill development and redevelopment, more succinctly known as 'second-cycle development'. This is of considerable importance in these mature residential areas since their extensive plots, fragmented landownership and the vulnerability of their owners to changes in the family life-cycle render them particularly susceptible to pressure for change during the economic situation prevalent during the late twentieth century. The density and social characteristics of such second-cycle development differ greatly from the original, and its impact could markedly change the original character of the areas involved.

A pilot survey undertaken in south-east England (Whitehand 1988) showed the incidence of changes to original plot patterns to be particularly high, and to vary with the position of the town in the urban hierarchy and in its relationship with London. For convenience, this pilot study examined development pressure in a 500m × 500m area close to each town centre, and another similar area near each town fringe. The square was based on the National Grid. The artificiality of imposing such a 500m × 500m grid square upon the townscape proved problematic in some cases where, for example, it has meant the inclusion of townscape elements other than the desired mature residential areas into the area selected for study. Nevertheless, this method was felt to be superior to that of delineating the appropriate townscape region, which could expose the study to criticisms of subjectivity. The problems of the definition of conservation area boundaries are most relevant in this context. Experience in other studies using large volumes of data acquired from development control records (for example, those discussed in Chapter 8) suggested that such a 500m × 500m region would contain sufficient relevant cases for detailed study, and that examination of larger areas would be unduly time-consuming for little actual return.

Four such study areas in comparable mature residential areas were selected for detailed study. Two are in towns in the south-east region, Epsom and Amersham, and two in the Midlands region, Tettenhall (near Wolverhampton) and Gibbet Hill (near Coventry). These study areas form part of the Tettenhall Greens and Kenilworth Road conservation areas studied in Chapter 7. Results from the south-east study areas, and selected comparisons of the characteristics of this form of suburban development in both regions have already been published (Whitehand 1989a, 1989b, 1990a, 1990b; Whitehand and Larkham 1991a, 1991b). This section deals with detailed examples derived from the two Midlands areas. Previous work in this region has already provided some knowledge of the likely numbers and types of development pressures in the region (Larkham 1988a, 1988b).

RESIDENTIAL CONSERVATION PLANNING: MIDLANDS EXAMPLES

Tettenhall

Tettenhall is now a high-class residential suburb of Wolverhampton but was originally two hamlets, probably of Anglo-Saxon origin. During the mediaeval period, the church became a Royal Peculiar (Denton 1970; for the early history see also Greenslade *et al.* 1984). The area was selected for study as a region of large Victorian plots is situated along the crest of the north-east/south-west running sandstone ridge that separates the original two settlements (Figure 9.6A). Later development extended to the west, back down the dip slope into farmland, where further development is now constrained by the Green Belt, registered public open space (Upper Green) and a golf course. The parish remained rural until the nineteenth century. In 1776 Arthur Young, having travelled through Wolverhampton from

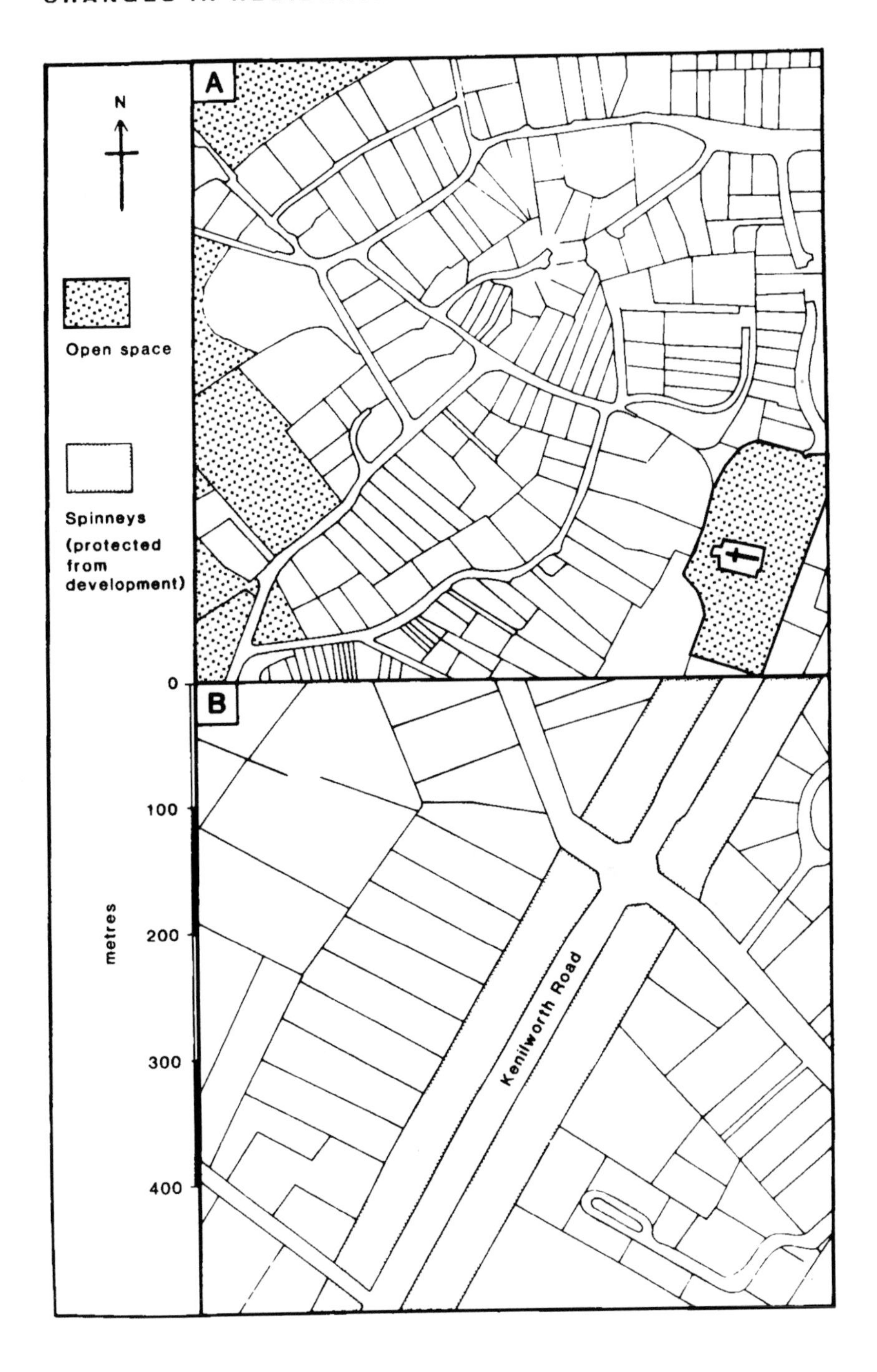

Figure 9.6 Midlands residential study areas. A: Tettenhall; B: Coventry

Birmingham, marvelled at the contrast in Tettenhall, which he considered 'as retired as the Ohio' (Young 1932 edition). In 1812, the 'delightful view of Wolverhampton and the adjoining country' was used as a selling-point for a house on the ridge (*Wolverhampton Chronicle*, 23 December 1812). By 1846 Tettenhall was described as

> daily more intruded upon by clumsy proprietors of scraps of ground, by imported shopkeepers, and by those architectural affectations, termed 'country houses', in which the hardware gentility of Wolverhampton carries on its evening and Sabbath masquerades, at a convenient distance of a mile and three quarters from the locks, nails, and frying-pans of the productive emporium.
>
> (Palmer and Crowquill *c.* 1846: 33–4)

In 1894 the village formed the centre of Tettenhall Urban District. By 1928 the Tettenhall district, together with Clifton (adjacent to Bristol), were referred to as the two finest suburbs possessed by any large manufacturing town (Morton 1928: 198). In 1966 most of the Urban District, including the study area, became part of Wolverhampton, thus ending the administrative (if by then hardly physical) exclusivity which its residents had sought to preserve.

Gibbet Hill

The second residential study area is a part of the Kenilworth Road in Coventry, close to the present city boundary (Figure 9.6B). This road has been described as the finest approach road to any city in the British Isles, and it owes much of its character to the spinneys bordering the road. These belts of mature trees are the three-row-deep avenue of oaks planted in the eighteenth century, which replaced 'the grove of elms at Stivichall', apparently a feature of the area in the mid-seventeenth century. Originally part of the holding of Stoneleigh Abbey, founded in 1154, much of the area was owned by the Gregory family from *c.* 1587, and they appear to have been responsible for the oak avenue.

This whole area remained free from the encroachment of suburban housing, and was rural in character until the early twentieth century. At this point the estate was broken up and sold, most of it to the Corporation (Watts 1967). The Corporation sold off blocks of land to private developers, and much of the resulting development was of large detached houses in extensive grounds, lying behind the spinneys and thus hidden from the road itself. Some residential estates were developed prior to 1965, since when there has been continual pressure for plot subdivision. The LPA has attempted to resist this pressure, on the grounds that this is the last sizeable area in the city in which numerous large plots have remained intact. This approach has not been received sympathetically on appeal to the Secretary of State. A development control policy for new housing in this part of Kenilworth

Road, adopted in 1976 and amended in 1980, has been a more effective means of control.

The fabric changes

In examining changes to the built fabric of both residential study areas, figures showing the principal elements of applications for fabric change have been made in a manner and to a scale identical to that used for the town centres (compare with Figure 8.2). These two residential areas give results in marked contrast to those found for the town centres (Chapter 8), to the residential areas in the south east (Whitehand 1989b, 1990b), and to each other.

Overall, there appears to be little pressure for development shown by this index for the period 1960–86. Some categories of fabric change are altogether absent, for example façade alterations and refurbishment from Coventry. Others are present in minimal numbers, and for still others there are lengthy gaps between the small numbers of changes. Nevertheless, if all proposed change for both areas is added together, some broad trends appear evident. These are of a low level of development pressure in the early to mid-1960s, rising to a peak *c.* 1972, thereafter subsiding into the early 1980s. Both areas show this trend, although Tettenhall also shows a resurgence for 1985–6 while Gibbet Hill shows only a small peak for 1985 alone. This resurgence in the mid-1980s is also demonstrable, strongly for Henley-on-Thames, Malton and Solihull, and also to some extent in Ludlow (Larkham 1992). In all of the town centres and residential areas examined in detail using local authority data sources, the minor changes to the built fabric significantly outnumber the major changes. This is frequently asserted but, until these studies, not quantified. As an illustration, Table 9.1 compares ratios of building extensions – additions

Table 9.1 Comparison of major and minor fabric changes in town centres and residential areas

Area	*Ratio of additions to major rebuilding and new building*
Wembley central area*	5.6 :1
Solihull High Street†	5.0 :1
Knowle village centre†	4.75:1
Watford central area**	4.3 :1
Northampton central area**	3.7 :1
Aylesbury central area*	3.4 :1
Tettenhall Greens†	2.8 :1
Kenilworth Road, Coventry†	2.6 :1

Sources: *Freeman (1986: 71); areas include parts of conservation areas
†Larkham (1986: 126); areas are complete conservation areas
**Whitehand and Whitehand (1983: 489); areas include parts of conservation areas

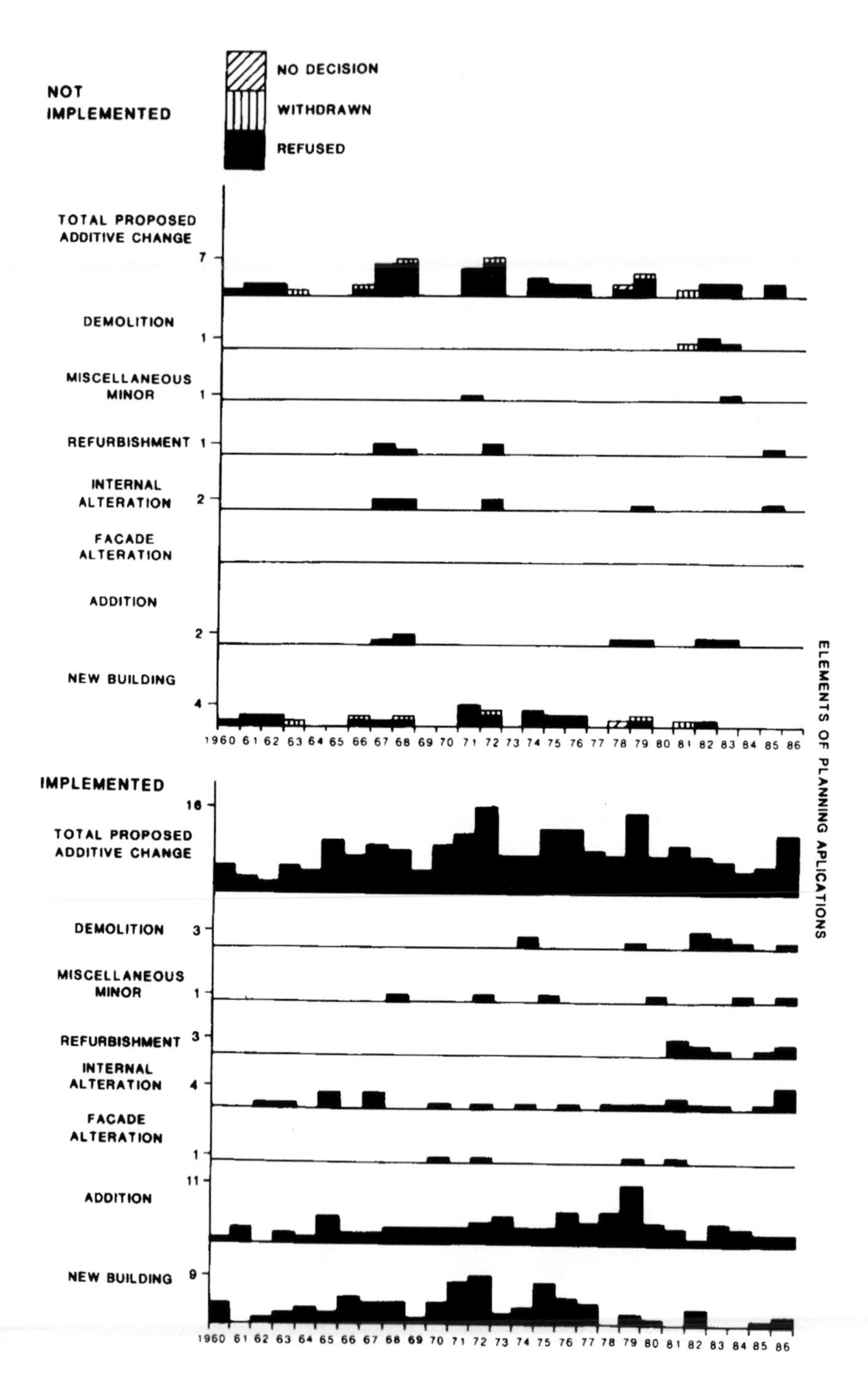

Figure 9.7 Analysis of built fabric change, Tettenhall

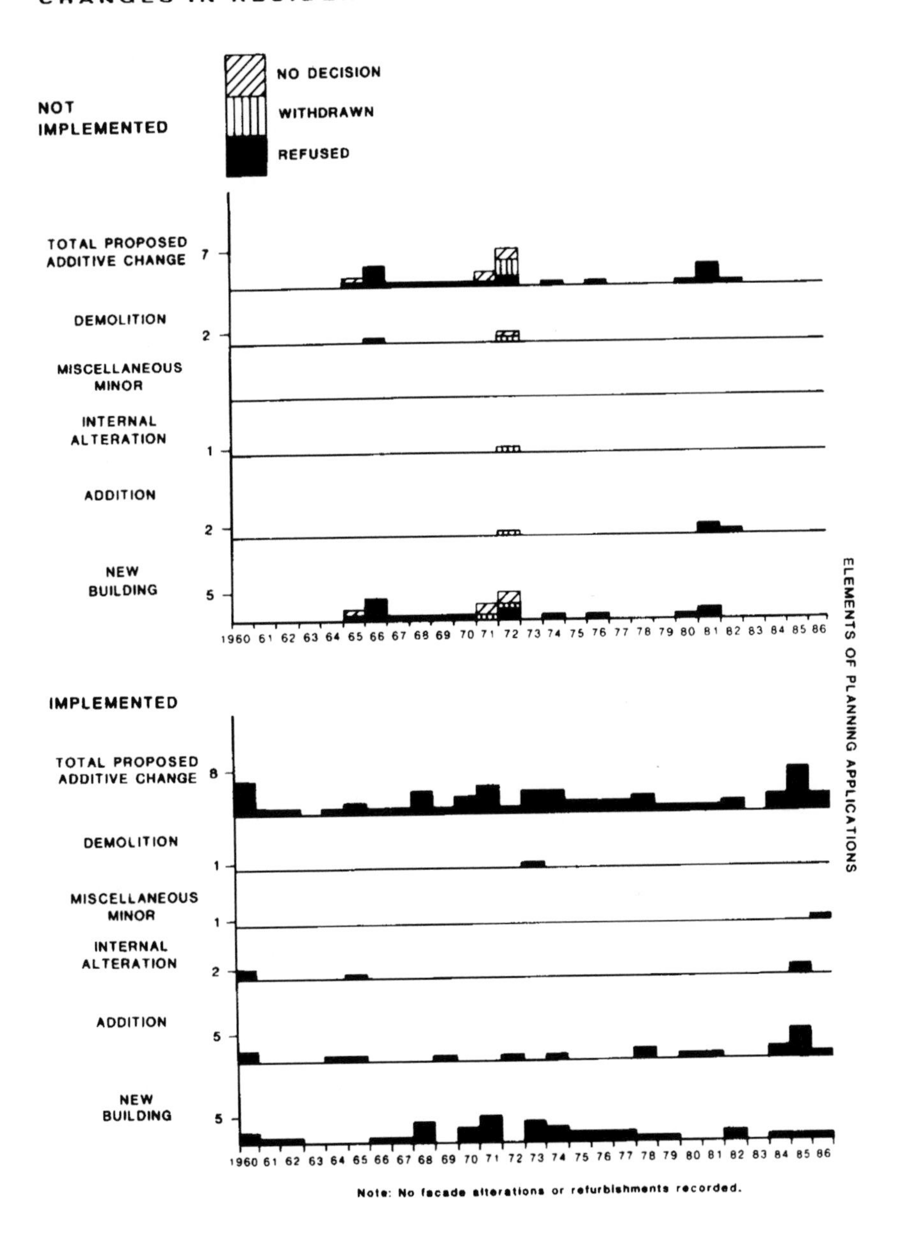

Figure 9.8 Analysis of built fabric change, Gibbet Hill, Coventry

of floorspace – to new building in eight areas. Town centres show many more extensions than new buildings, while the residential areas have rather lower ratios, with more new construction being proposed on subdivided or vacant plots.

Gibbet Hill has altogether lower development pressure than Tettenhall, although the numbers of new building applications are higher relative to the other types of change (Figures 9.7 and 9.8). The success rate for new building applications is lower in Gibbet Hill, with the ratio of approvals to other decisions (refusals, withdrawals, etc.) being 38:23 for Gibbet Hill and 78:30 for Tettenhall. This high rate in Tettenhall owes much to the large number of applications for single, 'infill', new dwellings, often on subdivided gardens. Many such applications were made early in the study period, most were granted, and the decision times were short. Tettenhall has higher pressure for new building and additions, with the former peaking in the early 1970s and the latter in the late 1970s.

Discussion

The general trend in development pressure as recorded by this index may be compared with the theory of building cycles (Lewis 1965; see also Whitehand 1987a). A basic development of Lewis's explanation into the 1980s, using several of his indices (Larkham 1986: Figures 3.9–3.11) suggested that the available data showed the crest and trough of one cycle since 1969, and that the cycle was on the upswing by the mid-1980s. The extrapolated wavelength of such a wave is some 20 years. The increasing development pressure evident in 1985–6 might conform with this trend. However, the 1986 study suggested a rise in activity in the 1970s, anomalous since at that time the national trend was sharply downwards, interest rates were high and the country was suffering from the social and economic aftermath of the oil crisis and the property crash. The present work, wider in scope in terms of space and types of areas studied, appears to suggest a peak in about 1972, with 1973 being a poor year in terms of pressure for development. This would appear to be more in line with national data.

Although both study areas were selected for their characteristic extensive plots and sizeable dwellings, they do differ greatly in their development history, and this is reflected in their different plans (compare Figures 9.6A and B). The extensive plots in question in Tettenhall are late Victorian, while those in Gibbet Hill are inter-war, developed at a time when first-cycle development in Tettenhall was virtually complete. The first cycle in Gibbet Hill is only just complete, with a large estate on formerly agricultural land being completed after the study period, in the late 1980s. However, in both areas a second cycle of development was actively contemplated, if not actually begun, by the mid-1960s. One might, therefore, expect more modifications to existing, older buildings in Tettenhall than to more recent buildings in Gibbet Hill. The position of the Tettenhall area, close to a village nucleus and two miles from Wolverhampton, may account for the greater development pressure than that in the Gibbet Hill area, which is a quiet residential backwater. The smaller plots in Tettenhall may suit different types of second-cycle development, compared to the very extensive first-cycle plots in Gibbet Hill. In the latter case, all development has been extensive, with detached houses and substantial bungalows in their own grounds. Actual and proposed development in Tettenhall has been more intensive, certainly at particular times during the study period, and this point deserves further attention.

INTENSIVE AND EXTENSIVE RESIDENTIAL REDEVELOPMENT

Intensive development in Tettenhall

The second-cycle development of new dwellings proposed in Tettenhall may be divided into four types: detached houses, flats, bungalows and a category of 'residential development', unspecified in extent and nature. The latter category is exclusively found on outline planning applications. No semi-detached houses, terrace/town houses, shops or other commercial developments were proposed in this area. Table 9.2 shows the number of units of the first three types, and the number of developments of the fourth, throughout the study period.

There appears to be a low, but reasonably steady, stream of bungalow applications, trailing off in the 1980s. The period 1965 – *c.* 1977 is the peak period for detached houses. There are two boom periods for flats, from the start of the study period to 1962, and 1971–4. The latter is in the middle of the housebuilding boom, while virtually all of the earlier 'boom' (if it can be so called) is the sole result of applications made by an elderly lady invalid, owner of 46 Clifton Road but herself resident elsewhere. All proposals were refused on the grounds of overdevelopment, detriment to the amenities of the area and the neighbours, and similar reasons. Figure 9.9, redrawn from a planning application, is an impression of one of these proposals, refused *inter alia* because

> the building of a six storey block of flats on this site which is situated between two storey dwellings would be prejudicial to the amenities of those existing dwellings by reason of appearance, noise and additional traffic to the premises and would be incongruous with its setting, creating a strident feature in the street scene.
>
> (application STU/1662, refusal letter dated 10 January 1962)

In the mid-1970s, this site was redeveloped with two neo-Georgian detached houses. The history of this particular site, which does raise interesting points, consists of seven applications between the period January 1960 to December 1975. Initially, prior to any conservation concerns, the non-resident owner wished to capitalise on the development value of her property. The early applications for flats were submitted following an architect's advice that the site would be suitable for a block of 20 flats; following refusals she accused the LPA of having 'prevented me from developing my property from motives of partiality for local agents and building contractors who are interested in my land but will not pay me a fair price for it' (planning files,

Table 9.2 Development types in a 500m × 500m sample area within Tettenhall Greens conservation area

Type of building	*Numbers proposed*
Detached house (including bungalow)	163
Semi-detached house	2
Terrace/town house	—
Flat/maisonette	131*
Residential, unspecified type and number of dwellings	9 developments

Note: *There are also five flat developments where the application did not state the number of dwellings

Figure 9.9 Proposed redevelopment of flats, Clifton Road, Tettenhall, 1962 (redrawn from sketch in planning files)

Wolverhampton MBC). Yet the surviving evidence suggests that three applications were refused on the grounds of insufficient information being submitted, while the fourth (for 12 flats) was refused on the grounds of excessive density and prejudice to the amenities of neighbours and the area itself. By 1971 the owner had died, and an outline application for two blocks of two flats was submitted by an executor of the estate, possibly wishing to market the land at an increased price with the benefit of a planning permission. In 1975 a detailed application was submitted, following the precedent of the previous outline. By this time the site was within the

Figure 9.10 House developed at 46 Clifton Road, Tettenhall, in the mid-1970s (author's photograph)

designated conservation area, and the blocks of flats were seen as 'gross overdevelopment'. LPA officials were clearly concerned at the possible consequences of the still valid 1971 outline permission, and were relieved when the applicant decided not to pursue an appeal but to sell the site. In late 1975 a new application proposed two detached houses, which speedily received permission and were constructed (Figure 9.10). Nevertheless, an internal memo reveals some concern:

> it is worrying to think that a rash of houses of this identical design may be built on infill sites throughout the Tettenhall Greens conservation area, but on the other hand there are no design details which we could claim did not conform to our conservation policy.
>
> (planning files, Wolverhampton MBC)

In a similar manner, the development history of the Eynsham House site shows a concerted attempt over several years to develop a prime site (Figure 9.11). Such repeated applications do distort the figures of Table 9.2, but do emphasise the pressure for development. Here, the executors of the deceased owner wished to increase the value of the property for sale purposes, and applied for the first permission (not shown on Figure 9.11). This was withdrawn, apparently when a sale was agreed. The new owner was evidently conscious of the potential of the site (Figure 9.11B). Further proposals all retained the original house while a developer, having acquired the garden,

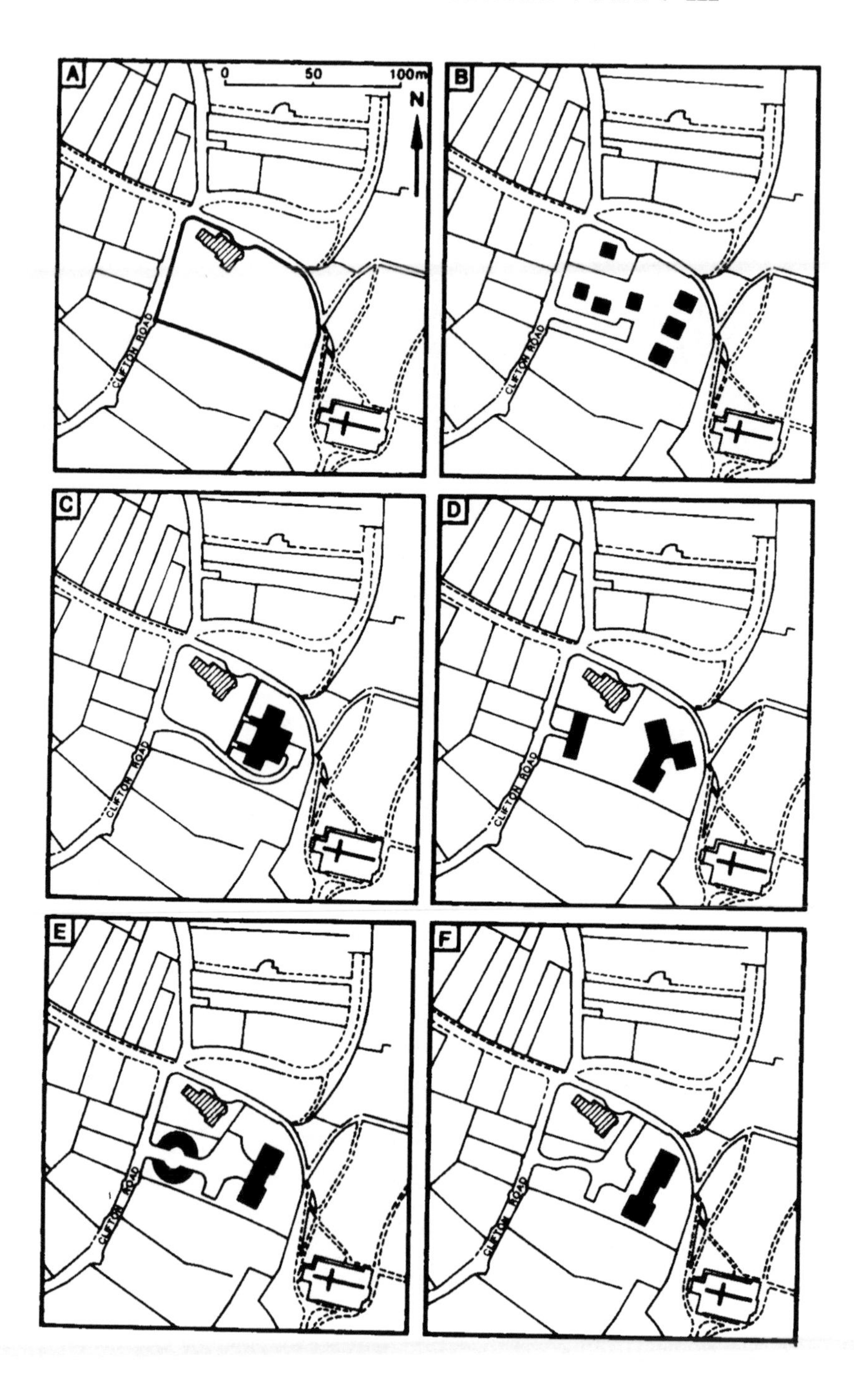
A
0
50
100m
N
CLIFTON ROAD
B
CLIFTON ROAD
C
CLIFTON ROAD
D
CLIFTON ROAD
E
CLIFTON ROAD
F
CLIFTON ROAD

submitted various proposals during negotiations with the local planning authority. This eventually resulted in a successful development (at least from the developer's view). In the early 1980s, the main house became vacant, and was eventually purchased by a charity for use as an old people's home, and extensively refurbished.

Another significant type of intensification is the conversion of existing ancillary buildings into dwellings. This process is also found in the cores of historic towns such as Ludlow (Chapter 8) where retailing and other urban business functions have in the past displaced residential uses, which are now seeking to re-colonise vacant buildings. In residential areas, once again, little-used outbuildings such as stables and coach-houses belonging to the large Georgian, Victorian and Edwardian villas are particularly vulnerable. Some applications propose 'granny-flats', where the new unit is ancillary to the occupation of the main dwelling; others propose the creation of entirely separate dwellings. The latter have more important consequences for the urban landscape, as they involve plot subdivision with often substantial walls and fences, and the creation of new driveways. In several cases in Tettenhall, the original fabric of a coach-house proved too dilapidated for conversion, was demolished, and an entirely new dwelling constructed in a mock-vernacular style. In the case of Eynsham House, discussed above, on conversion of the main house the coach-house was separated and converted to form a new dwelling, and a new plot was formed from the original garden area. Some years later, this house was further extended and outbuildings constructed, considerably reducing the amount of open space on this already-constricted plot, and further altering the domestic outbuilding character of the house itself.

One significant reason for the continued intensification of development is economic. Agents representing landowners make it clear that it is their professional role to maximise the return for their clients, and 'maximisation

Figure 9.11 Series of proposals for redevelopment of Eynsham House site, Tettenhall (redrawn from planning files)

Key: A Eynsham House site as shown on Ordnance Survey 1:1250 sheet of 1973.

B Outline application 1801 of 1968 for 7 dwellings and demolition of original house. Withdrawn: no documentation microfiched.

C Outline application 780 of 1971 for 12 flats. Refused: 'height and design prejudicial to amenities of area'.

D Outline application 1181 of 1971 for 'flat development' (including garage block). Refused: 'height and design prejudicial to amenities; overdevelopment; overlooking'.

E Full application 1755 of 1971 for 9 flats and circular garage block. Submitted 'following recent discussions for a revised scheme for the site'. Granted.

F 1972. Alteration of 1755/71 to give 8 flats without separate garage block. Granted.

Not shown: Outline application 446 of 1966 for 2 houses in the garden of Eynsham House. Withdrawn; no documentation microfiched.

of site value is still generally construed as "getting as many as possible on the site"' (Bishop and Davison 1989: 18). Further, some of the most intensive proposals are put forward by owners wishing to gain outline planning permission, thus increasing the site's value before a sale. This often occurs following a major change in family circumstances, such as death of the owner (for a case study see Whitehand 1989a).

Extensive development in Gibbet Hill

In contrast to Tettenhall, there are no applications in the Gibbet Hill study area for flat development. All second-cycle proposals are for detached houses, bungalows, or unspecified residential development (Table 9.3). Development in this study area is far more extensive than that in Tettenhall. The character of the area, and the stated policy of the local planning authority to retain the characteristic large plots, are significant in this respect, since most householders and prospective developers must have been made aware of the LPA's attitude and policy.

Yet proposals for intensive development have been made, both before and after conservation area designation. In particular, in 1966 a series of outline applications for two fields, parts of which now lie within the conservation area, proposed an unspecified amount of residential development on 11.25 acres: these were refused on the grounds of inclusion within the Green Belt and traffic generation. In the same part of the conservation area, distant from the city centre, a speculative residential development was proposed in 1965 on undeveloped land. Ironically, this was after the Ministry of Transport had directed refusal of a previous proposal for two dwellings on the same site owing to problems of traffic generation. Detailed discussions of the plan layout and house type took place between the LPA and the developer's agent. A 'high class patio development', giving houses close to the road with south-facing patios, was agreed:

> this appeared to be an excellent architectural approach, but I asked what safeguards would be taken against the conventional chalet bungalow etc being requested by the purchasers of any plot. [The developer would] probably build the first three dwellings speculatively, and this would set the pattern.... I queried the size and status of the development being appropriate to the Kenilworth Road area, and the answer [from the developer's agent] was that the proposal represents approximately 6 dwellings with a sale price of approximately £10,000 ... this appears to satisfy this point.
>
> (case notes, Coventry City Council)

It should be remembered that, at the time, £10,000 was over three times the price of a standard semi-detached house. However, the application submitted in December 1965 actually proposed a cul-de-sac of 20 dwellings (Figure 9.12). No formal decision was taken on this application owing to problems of access and drainage; the applicant rapidly submitted a

Table 9.3 Development types in a 500m × 500m sample area within Kenilworth Road conservation area, Coventry

Type of building	*Number proposed*
Detached house (including bungalow)	111
Semi-detached house	—
Terrace/town house	—
Flat/maisonette	—
Residential, unspecified type and number of dwellings	2 developments

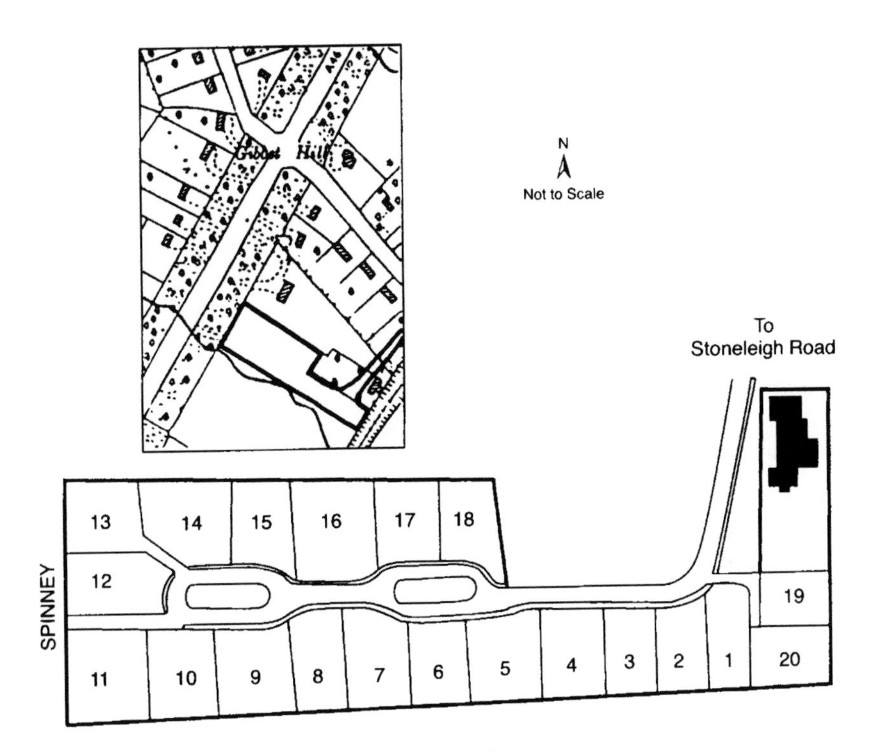

Figure 9.12 Original site layout for Beverley Drive, Coventry (redrawn from planning file: not to scale). Note the difference in size between original plots in this area (on the location map) and the 20 proposed in this application

revised application for 10 dwellings. Considerable disquiet was expressed by local residents, particularly over the question of density.

> The whole of the development in the area has to date been broadly on the basis of 1 house per acre and I submit that to grant a development

> on the lines of the application . . . would be a complete negation of good planning and would destroy completely the character of the area.
>
> (objection letter, Coventry City planning files)

The Planning Committee agreed to a certain extent and requested a revision of the application to five dwellings, which was granted. Thereafter, until the early 1970s, there was a series of individual applications as these plots were sold, and bespoke houses designed. The issue of area character was clearly significant even before conservation area designation. There has also been a series of applications for individual plots along the Kenilworth Road, proposing various forms of plot subdivision and construction of sizeable new dwellings, albeit on relatively small and constricted sites. In all cases, the original building was to be retained.

The LPA introduced a development control policy in 1976 as a response to the perceived increasing demand for further residential development within the entire Kenilworth Road conservation area, within which the Gibbet Hill study area is located. The policy stated

> (1) That new development or redevelopment entailing the intensification of the land use along Kenilworth Road be not generally permitted.
> (2) That in those very exceptional cases where development or redevelopment is considered to be acceptable the following criteria be taken into account:
>   (i) that such development or redevelopment shall not be so located as to necessitate the removal of or cause damage to trees or associated ground cover;
>   (ii) that the height of any development or redevelopment be restricted to two storeys (excluding accommodation within the roof space);
>   (iii) that existing drives be not utilised to provide access to more than one additional dwelling;
>   (iv) that no new vehicular access be provided on to Kenilworth Road;
>   (v) that adequate foul and surface water drainage facilities must exist.
>
> (Coventry City Council 1976)

The 1981 amendment to the Plan was introduced to take into account the changing circumstances of the preceding five-year period, and particularly the continuing pressure of applications proposing new development in the area. The revised Control Plan's objectives included the following statement:

> To conserve the appearance [of the Kenilworth Road frontage] it is necessary to restrict any proposal which would in any way detract from the spacious, natural, wooded character of the area. Buildings readily apparent from Kenilworth Road, new accesses onto Kenilworth Road, and the removal of trees, undergrowth or grass verge all fall into this

> category. Extensions or new development which fill the spaces between buildings, so detracting from the spacious character, should also be restricted. A complementary objective is the need to maintain a stock of high quality individual houses within the city. The redevelopment of Kenilworth Road houses for smaller houses or their use for multiple occupation would conflict with this objective in addition to probably detracting from the visual appearance of the area.
>
> (Coventry City Council 1981)

One of the applications prompting this policy was submitted in 1975, and was for the erection of one detached house on a plot of 1,042 square yards adjoining 139 Kenilworth Road. The comments of the LPA planning officers were that the area was of limited size given the nature of the proposal. In discussion at a planning committee meeting, it was noted that 'the site was extremely narrow with the result that the proposed house was totally out of character and inappropriate' (memo on meeting, 21 January 1976). A refusal on these grounds was upheld at appeal.

Despite the operation of the control policy from 1976, applications for plot subdivision and new building continued. An appeal against refusal of one such proposal, adjoining 138 Kenilworth Road, was upheld.

> The appeal site lies behind the woodland on the north-west side of the road ... and comprises just over 3/4 of an acre of the approximately 2 acre curtilage of your client's detached house.... It would, of course, alter the appearance of this low density residential area ... [reducing] the spaciousness apparent in the immediate vicinity.... My opinion is that the proposed development would be more in keeping with the existing pattern in the immediate vicinity, although the appeal site would be marginally smaller than the nearby plots, rather than constituting an interruption to the rhythm and pattern of buildings as the council contend.
>
> (DoE 1981c)

These examples graphically demonstrate the size of some of the original plots, the sizes and awkward shapes of the proposed subdivisions, and the varied success of the control policy. One noteworthy point is the emphasis continually placed in the control policy, in LPA files and in Inspectors' Appeal decision letters, on the importance of aspects of the mature planting of the area. The Spinneys are obviously the most important single characteristic of this area. This concern for mature landscaping in these residential areas is shared in other areas studied, including Tettenhall, Edgbaston, Barnt Green and Northwood (Jones 1991). But, because of the Spinneys, conservation of the natural (as opposed to built) form is more successful in Kenilworth Road than in these other areas.

The example of the two phases of the Marshfield Drive development (Figure 9.13) draws together many of the arguments and issues arising from

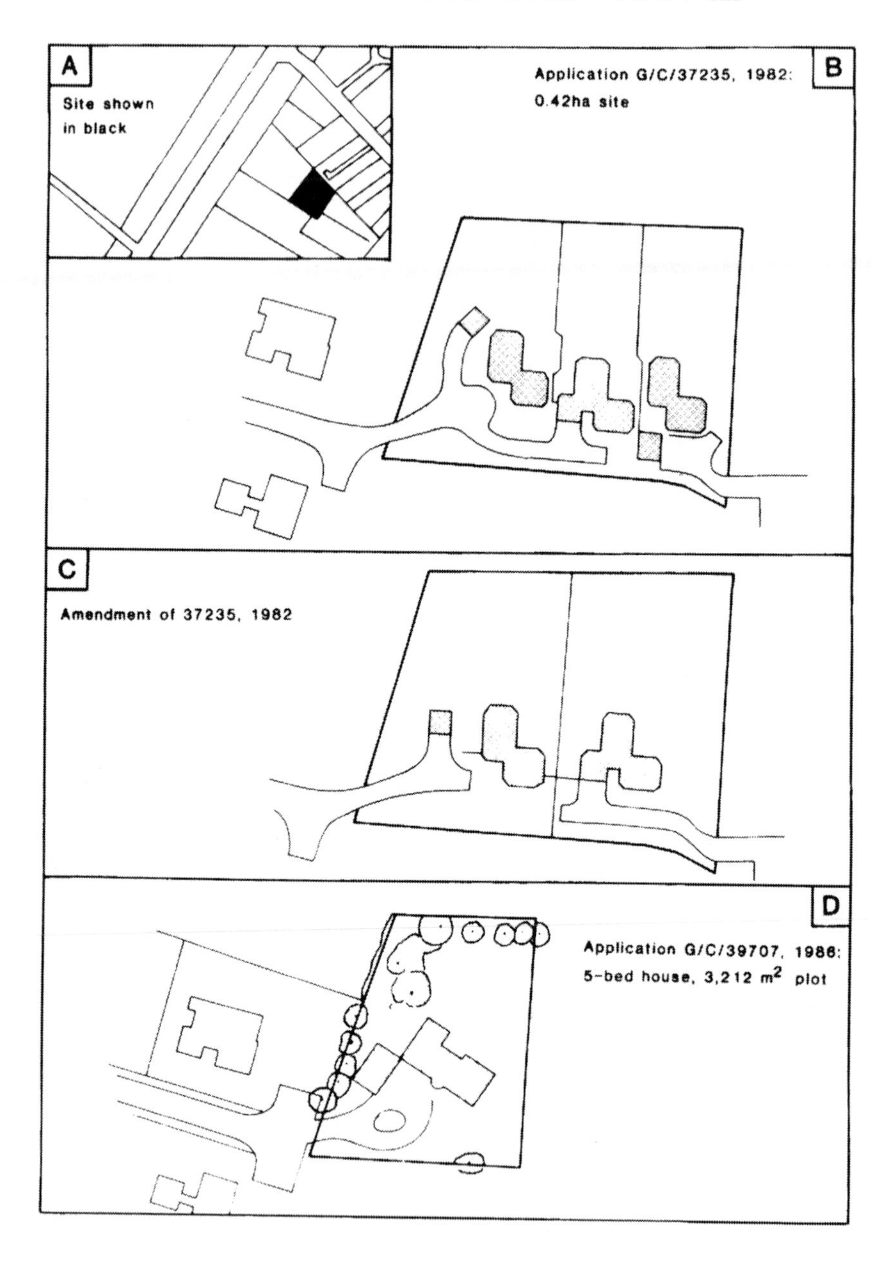

Figure 9.13 Site history, Marshfield Drive, Coventry (redrawn from planning files: not to scale)

these discussions of development in historical areas. The first phase concerned 2 and 4 Stoneleigh Road, with a series of proposals for both plot subdivision and the demolition and redevelopment of both original houses. The combined area of the two original plots was approximately 1.5 acres. Most applications were permitted. Although contacted, no residents objected to the specific proposal in 1973 for 3 houses and 1 bungalow on this site. The City Engineer was concerned only about the lack of mains drainage in this area, and the consequent requirement for cesspits. Permission was granted for the two houses forming Marshfield Drive in 1975; the drive layout was amended in 1980, when it was noted that the plots were 900m^2 and 1,100m^2 in area.

From 1982, a further series of applications proposed development to the rear of Marshfield Drive. The site was owned by the previous owner of Playfels Grange, a large house facing on to Kenilworth Road, who retained 'substantial ownership of land in this area'. Three plots, each of 1,300m^2, were proposed. The City's Conservation Officer reacted favourably: approving well-designed houses worthy of this high-class location. However, there were many objections from neighbours. A typical reaction was that the 'density is inappropriate to the area ... it constitutes "backland development" increasing the density of the area above an acceptable level' (consultation responses, Coventry City planning files). The substance of protests from the new owner of Playfels Grange was that this scale of development was not planned when he purchased the property, and that he would not have made the purchase had he known of this scheme. Other protests were that the plots were too small, and that substantial two-storey houses were inappropriate on a site adjacent to an estate of substantial detached bungalows. After consultation, the proposal was amended to two dwellings, and approved in this form. Detailed plans for the dwellings were submitted between 1982 and 1986, when there was some concern that the original 'split-level' dwelling type was changed first to a bungalow, then to a substantial five-bedroom two-storey house.

DISCUSSION: RESIDENTIAL DEVELOPMENT IN CONSERVATION AREAS

These sagas of development raise a number of significant issues. First, the nature of development. 'Backland development' is a term used by both planning authorities and numerous vociferous objectors to refer to proposals such as this. Yet in an area of some pressure for new development (however measured), and where access is relatively easily available as in this case via an existing cul-de-sac, there are few sustainable counter-arguments. Appeal decisions for comparable cases involving proposed development on the tails of long residential plots were quite common in the boom period of the mid- to late 1980s; with one in Radlett, Hertfordshire, showing the then-current DoE Planning Inspectorate thought on this issue.

> In principle there can be no valid planning objection to the residential redevelopment of the site situated as it is in the heart of a residential area, and this use of presently under-utilised land would be wholly in accordance with current Government policy to make maximum use of urban land and to supplement the supply of private sector housing.... I find no proper justification in this instance for supporting the view that the main backland part of the site should remain as undeveloped garden land, in order to preserve for the sake of the existing character of the locality the pattern of frontage development with very long rear garden plots ... inherited from the inter-War period.... With regard to the fear of establishing a precedent for further *cul-de-sac* development at the expense of the special character of the area ... I consider that the depth of this area of backland makes it unique in the context of that area, and that this should not in principle frustrate its re-development for residential purposes.
>
> (DoE 1988b)

This is supported by numerous cases of backland development. In most, use of the land is not in dispute; what causes problems are issues such as access, roads, sewers and drainage. Examination of the history of a much larger backland residential site, for 100 dwellings in Shrewsbury, confirms this (Davies *et al.* 1989: 58–9). Piecemeal acquisition of areas of backland by developers in order to assemble a viable development site, in some cases accompanied by piecemeal construction, is a common tactic in these mature residential areas, even in conservation areas (Figure 9.14) (Larkham and Jones 1993). Although some residents profit by selling land to facilitate such schemes, others lose through the loss of amenity value caused by the processes of subdivision, construction and intensification.

Secondly, perceptions about areas change over time, owing in part to continuing and increasing development pressure, and in part to the increasing awareness of environmental issues and conservation. In these cases in Gibbet Hill, public reaction was very different to the similar applications of 1973 and 1982. There were also differences of opinion on the part of neighbours, the developer, potential purchasers and the professional LPA officers over what would be an acceptable density of development. Many of the original dwellings in the area have plots of between one and two acres. No objections were raised to the Marshfield Drive development (plots of 900 and 1,100m^2), but strong objections were expressed to the proposal for three houses on plots of 1,300m^2 each: one neighbour stating that only one house on this entire site (of 3,900m^2) would be acceptable.

Thirdly, there is the question of the impact of this type of development, if continued, on the character of the area. This is part of a designated conservation area. Any such infill development of necessity changes the relationship of buildings to plots, and this is one feature of which the LPA is aware. However, attempts to limit infill by reference to the social and environmental need to retain Kenilworth Road and Gibbet Hill as the last

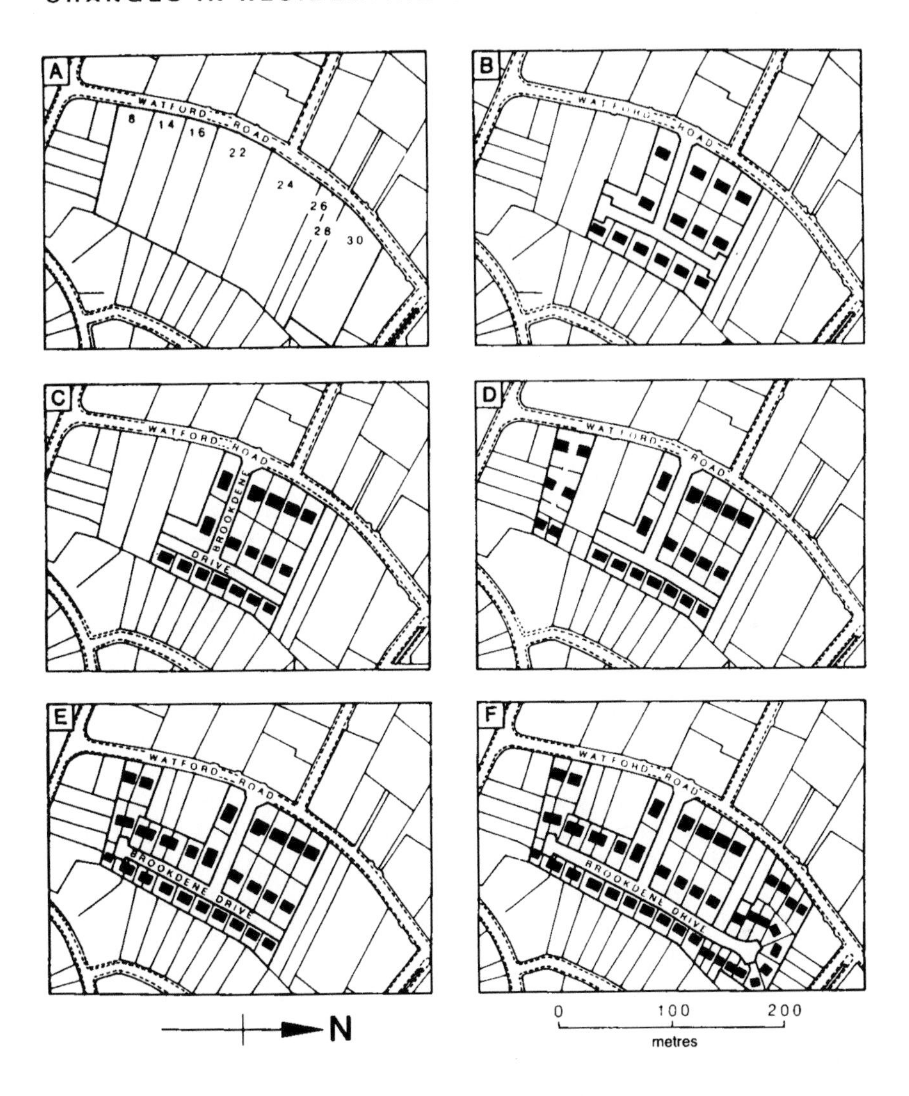

Figure 9.14 Gradual acquisition and development of a backland site on the London fringes (from Larkham and Jones 1993)

area of large dwellings on extensive plots in Coventry (a standard refusal reason for such applications: Larkham 1986: 96) has not been successful at appeal. This brings into question the entire problem of attempts by LPAs to 'manage' these valuable historic environments, using such tactics as the designation of conservation areas, introduction of Development Control Policies, and refusal of individual planning applications. Not all LPAs monitor development proposals in this type of area; not all have consistent policies; and the reactions of residents, developers and the DoE to such attempts at management vary widely.

Figure 9.15 Regency Drive, Coventry (author's photograph)

Fourthly, it is clear in the areas studied that the architectural style of new residential development is, relatively speaking, a minor issue. Styles have broadly followed the progression evident in town centres and nationally (Chapter 8 and Jencks 1985) from Modern to post-Modern, the latter encompassing vernacular and historicist styles. In Clifton Road, Tettenhall, therefore, vernacular forms albeit with standard 'Georgian' small-paned windows, and standard speculative Georgian with fibreglass pillars and detailing, stand adjacent to substantial listed Georgian and Victorian villas. In the Kenilworth Road conservation area there is a complete speculative cul-de-sac named Regency Drive comprising housing clearly borrowing design idioms from the Georgian/Regency period but re-using them in an ahistorical manner – the intrusive but now ubiquitous double garage door detracts considerably from any styling applied to the house itself (Figure 9.15). These developments owe far more to the then-current fashions in speculative housebuilding and buying than to anything in the character or appearance of the surrounding area. Similarly, little overt concern was expressed at the insertion of similar culs-de-sac of standard Bryant houses into the largely Victorian villa suburb of Edgbaston conservation area, yet these developments have significantly changed the area's character and appearance in terms of street layouts, plot patterns and sizes, building forms and massing and architectural styles (Jones 1991). The same adverse impacts are visible in the Kirby Fields conservation area (Figure 9.5). Only once in the examples studied, in Tettenhall, did an LPA officer note some concern over the proliferation of a standard speculative style seeming to bear little relation to the area's character, and this key conservation issue was not further addressed by the LPA.

Issues of density are commonly raised in these areas; most clearly so in the Kenilworth Road cases cited. It has been observed that much development in this type of mature residential area consists of intensification, building on undeveloped plots or subdividing existing plots and building in garden spaces; moreover, it is common that modifications to proposals by developers after planning permission has been granted frequently lead to further increases in density (Whitehand and Larkham 1991a, 1991b). Density policies, where they exist, are commonly calculated for entire districts, and are not sufficiently fine-tuned to act as suitable controlling mechanisms in these residential areas. The actions of the DoE, in overturning applications refused by Coventry City Council on the grounds of subdividing large plots characteristic of the area and thus detrimental to its character, clearly suggest that conservation is a low priority.

CONCLUSIONS

Again, the examples studied should be related to the key questions posed in Chapter 1. Again, financial profit is a major motive for development proposals; most obvious, perhaps, in the applications for new dwellings submitted by speculative developers. Nevertheless, even individual owners have sought profit by, for example, subdividing large gardens. With regard to more minor alterations, the changes in family life-cycles are an important motive, where alterations are intended to increase the use, rather than sale, value. It is an estate agents' commonplace that many residential alterations will never recoup their investment when the property is sold. Again, as with town-centre areas, there is a constant pressure for change of various types, fluctuating somewhat in line with national and regional trends.

Some areas differ in their character; in particular, whether original extensive plots are rendered vulnerable to large-scale infill development on individual plots or parts of plots. The assembly of development sites of economically viable size is easier in such areas. Areas of an originally smaller-scale development form limit the amount of infill development possible without the acquisition and amalgamation of several plots. In such areas, development pressure remains high, but development type is more restricted to individual alterations to individual buildings.

These areas also show fashions in both development type and architectural style. Most noticeable is the trend away from blocks of flats in Tettenhall, at about the time when conservation was becoming a higher planning priority nationally and locally. Speculative infill development also clearly shows the prevalence of neo-Georgian styles in the 1970s and the rise of neo-Tudor in the 1980s. Throughout, however, there was a form of pseudo-vernacular which, however, did not relate closely to true vernacular in either area. This raises the interesting question of how the design of these new buildings does fit into the character and appearance of the areas. In Tettenhall, for example, the original character of large Victorian villas in extensive plots has been considerably eroded by infill developments of quite different architectural styles and scales. The new developments do relate better to each other, in part owing to the predominance of designs by one architect from a neighbouring village. But, without doubt, the character of this conservation area is being eroded.

There is a problem when an LPA amends development control policy in the light of continued development pressure, as Coventry did for that part of Kenilworth Road studied. It is evident that owners may see precedents of earlier development, and wish to undertake similar work. They are aggrieved when this is not possible. This emphasises the self-interest of the owner-occupier, and the very different (and wider scale) value-set of the LPA.

Further conflicts arose when residents objected to developments. In Tettenhall, objections were largely on the grounds of the physical characteristics of the proposed buildings. In Tettenhall, with its closer development pattern, the buildings would be more visible. In low-density Gibbet Hill, however, objections were almost exclusively limited to comments about building densities. Owners wished to retain the extensive, low-density development pattern. The form and style of individual dwellings was not a key issue. Again, the self-interest of the owner-occupier is paramount, rather than any real concern for area character or appearance.

10

THE IMPACT OF CHANGE ON THE CONSERVED TOWNSCAPE

> Large buildings in London and elsewhere today are too often designed in the lift going down to lunch.
>
> (Sir William Holford, 1960)

INTRODUCTION

Previous chapters have examined the nature and extent of, and processes shaping, changes in a variety of conserved areas, and attention can now be turned to a closer study of the impacts of change in historical townscapes. Several key questions arise. First, how is change accommodated? Secondly, what architectural and urban design strategies are employed by architects, developers and planners? Thirdly, is such change 'managed'? Given the phrasing of UK law and policy, it would be expected that a principal aim of any policy of 'management' in historical areas should be to ensure that new buildings do not look 'out of place'. This expectation draws upon the evident importance placed upon the façade, above virtually any other aspect of historic structure or plan, in the UK 'townscape' tradition of planning in the post-war period, as has been evident in some of the examples already cited. This chapter will thus concentrate on some of the design trends which have been developed in response to growing concerns for conservation and contextualism, some (but not all) of which have been raised in examples discussed in the previous two chapters.

However, developments that were unsatisfactory in this respect of contextual design abounded in the UK during the 1950s and the 1960s, associated with the widespread use of the Modern style of architecture. In size, scale, bulk, massing, materials and style, many such post-war buildings stand in stark contrast to their neighbours (Figure 10.1). The reasons for this ignoring of local context are difficult to elucidate, but are bound up in the prevailing architectural and planning ethos. Context was, at this time, relatively unimportant. It is not inconceivable that, in the Aylesbury example shown in Figure 10.1, the original intention was for piecemeal redevelopment to occur along the entire street frontage. Any such intention was evidently not carried out, and the street remains as a palimpsest of both actions and intentions. The prevailing fashion among both planners and developers was

Figure 10.1 Intrusive Modern building, Aylesbury (photograph by M. Freeman)

for comprehensive redevelopment, sweeping away the old in favour of the new (cf. Esher 1981). But, although such terms are in common use particularly among pressure groups resisting development, we should not really talk about the 'evil planner' or 'evil developer'; in reality these agents are acting or reacting to changes in fashion and taste; their actions are constrained by what the sources of funding will pay for, or consider to be a safe investment.

This fashion can be seen particularly in the plans put forward for the rebuilding of devastated areas following the Second World War. Buildings of this post-war period used new materials, predominantly plate glass and concrete; they had few, if any, decorative features, and may be typified in the

Figure 10.2 Warsaw Old Market Square: post-war reconstruction, now a World Heritage Site (author's photograph)

'toothpaste architecture' sketch proposals in key post-war planning documents such as the plans for Exeter and Oxford (Sharp 1946, 1948). There is a significant difference between this attitude, prevalent in much of Britain throughout the early post-war period, and the reconstruction in many of the war-damaged cities of Germany and Poland. In these towns, the reconstruction of the central areas was either facsimile of such accuracy that, for example, Warsaw old market square is now a World Heritage Site (Figure 10.2), or a simplification of historical precedent or local vernacular which has been so successful in, for example, Nuremberg (Soane 1994). The widespread reaction in the UK during the late 1960s against such large-scale and insensitive developments and the destruction of historical townscapes that they caused is evident in the rise of the conservation movement (Cherry 1975) and the parallel emergence of a new architectural style, often and misleadingly called 'post-Modern' (chronicled in Jencks 1985, 1991). There are various aspects to this style, but most are characterised by the use in one way or another of historicist architectural elements, particularly the columns and arches characterising the 'post-Modern Classical' (Jencks 1991).

By the mid-1970s, the number of buildings being built in the Modern style or its debased variants had fallen dramatically in many towns in Britain, and the number of post-Modern buildings, in several variants of that style, was increasing (Figure 8.10). The larger parts of each of the five centres studied to compile Figure 8.10 included designated conservation areas. This broad, pro-conservation movement produced a number of contrasting approaches to the problem of fitting new buildings into old and valued townscapes (cf. Brolin 1980). Some have already been discussed briefly in previous chapters. The most significant of these approaches appear to be (a) deliberate contrast; (b) the use of local architectural idiom; (c) disguise; and (d) the use of historicist architectural styles. Each of these approaches can be demonstrated with the use of examples ranging from individual houses and shops to large superstores.

DELIBERATE CONTRAST

Despite the criticisms of Vallis and others (Chapter 7) that enhance and contrast are mutually exclusive, the concept of 'deliberate contrast' is frequently met. It is argued that there is a distinction between buildings that contrast with their surroundings but are specifically designed to do so, and buildings that simply ignore their surroundings. While the latter group merely ignore the *genius loci* for a variety of reasons, the former seem to attempt to add something wholly new to the *genius loci*. The new structure often deliberately seeks not to distract attention from pre-existing structures through repetition or pastiche, but to add a new, well-designed structure which could of itself be an attraction, and potentially listable in the future. Being particularly extensive in their land-use requirements for large single-storey structures with large car-parks, retail store buildings often provide useful examples of well-designed contrast.

The superstore built for J. Sainsbury in York during the mid-1980s stands on a problem site, formerly derelict, just outside and in full view of the city walls (Larkham 1988c: 52–4) (Figure 10.3). It was designed from the outset to contrast with its surroundings. It adjoins a Georgian County Hospital (a Grade II listed building, now converted to offices as part of this scheme). Although having nearly 54,000ft^2 of gross floor area, this building is very low in relation to the tall hospital. It is clad in glass panelling giving a light appearance, and a feature is made of the structural steel frame (Figure 10.4). The City of York Planning Department itself requested that the design of this store should be adventurous, avoiding what it regarded as the clichés of the more usual brick-clad and mansard-roofed designs common in the early 1980s. Considering the size of the store and the operator's requirements for a single-storey building, it was felt that the eventual design was better than a brick structure in which cosmetic contrivances were used to relieve the visual impact. The Planning Department played a substantial role in this scheme, being responsible for initiating the reversal of a previous planning policy opposing retail uses on this site. After some negotiation, the final development included a large car-park, housing, and the conversion of the hospital to offices. Despite the considerable visual contrast between the store and its surroundings, the scheme appears to be very successful in the views of the developer,

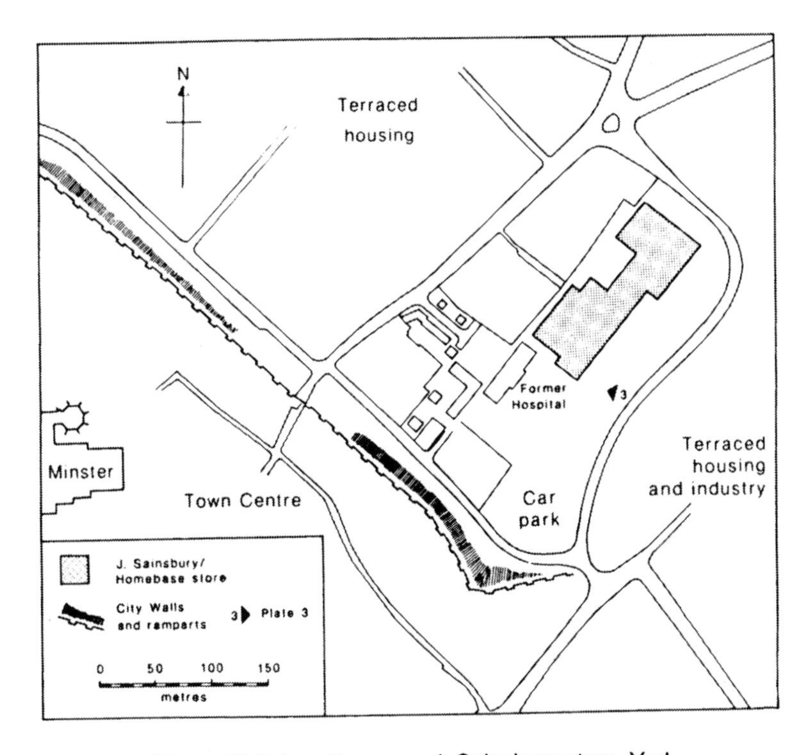

Figure 10.3 Location map, J. Sainsbury store, York

Figure 10.4 J. Sainsbury store, York and adjacent Georgian hospital building (author's photograph)

Figure 10.5 J. Sainsbury store, Wolverhampton, abutting the Classical-styled church which now forms the main store entrance (author's photograph)

the planning department and the public using the store.

Significant contrasts are also evident in Sainsbury's stores in the key historic city of Bath and the industrial town centre of Wolverhampton (Larkham 1988c: 54–6). Here, major new store buildings are integrated with existing disused listed buildings (a railway station at Bath and a church at Wolverhampton). At Bath, the design emerged from discussions between Roy Worskett (then City Architect and Planning Officer, and a noted writer on townscape and conservation issues) and Sainsbury's architects. The store was designed from the outset to be 'modern' rather than a pastiche in Victorian style, and its scale was to complement the refurbished station building. To this end, the height of the eaves, height and pitch of roof, and wall cladding materials were designed to match those of the station. At Wolverhampton, the building has a flat roof and is clad in GRP panels and concrete blocks. It was considered that other materials, including brick, and other styles would be inappropriate for a building directly abutting onto the Classical-styled sandstone church, which has been refurbished as part of the scheme (Figure 10.5). Some of the final details, including canopies and entrances, have been amended following the Planning Department's consultations with local amenity groups and further discussions with the developer. Although some have seen such a re-use of a redundant church as incongruous, the only alternative proposals had been the LPA's suggestion to level the walls at a height of one metre, leaving the remains and the steeple standing in a public open space. The Sainsbury's scheme has retained the entire church, a notable landmark adjacent to the town's ring road, although the original interior, already badly damaged by vandalism and neglect, has been destroyed. Nevertheless, this scheme was awarded a Civic Trust Commendation and was supported by the town's Civic Society.

LOCAL ARCHITECTURAL IDIOM

The increasing movement against what has been seen as the blandness of Modern architecture, combined with the rising tide of conservation, has led to a new style of architecture that attempts to use local architectural idiom in terms of style and materials in order to blend new development with old. Unfortunately this so-called 'pseudo-vernacular' style soon became almost as undesirable as its predecessors, because little or no attempt was made to incorporate true vernacular, that is, local, characteristics. In many cases, only token efforts were made to study and understand the *genius loci* before designing a development proposal. Instead, a virtually nationwide bland 'vernacular' developed, characterised by false mansard roofs, so-called 'hand-made' (but actually mass-produced) bricks, and some little detailing such as arched window-heads or curved brick lintels. This trend clearly spurred the requests for other designs from planning departments such as those at York and Bath.

However, remaining for the moment with the example of Sainsbury's large stores (Larkham 1988c), a number have used vernacular features reasonably well, for stores in both residential areas (albeit mostly new), where pitched roofs rather than flat roofs may echo those of the surrounding housing, such as the store in Dunstable. In existing townscapes, where traditional materials are used, even these large stores may blend quite well, as has occurred in Walsall town centre. The use of vernacular brick detailing at Selly Oak in Birmingham owes much to constraints of the site, principally levels and access, and the developer's internal layout requirements (Larkham 1988c: 50–2). These factors determined that the store should face onto the car-park, rather than towards the main road and existing shopping area, and that windows on the main road façade would be impracticable owing to a drop

Figure 10.6 J. Sainsbury store, Selly Oak, Birmingham: vernacular façade (author's photograph)

in levels across the site of between one and a half and three metres. However, the planning officers thought it important that the existing shopping area should not be faced by a blank wall, and the architects were required to submit a 'highly modelled and detailed' façade that would overcome the monotony of a windowless blank wall, and would reflect in scale, features and materials the character of the shops opposite. The resulting detailed design has a pitched, tile roof with gables, a red brick wall with blind windows and arches, and an entrance leading around to the main store entrance facing the rear car-park (Figure 10.6). But it is no more than a skin on a steel box structure: in the words of the American architect Venturi, clearly it is a 'decorated box' (Venturi 1966). Given the site and operator constraints, this treatment appears reasonably satisfactory, yet in scale and detail it is clearly not ideal. It is the result of a lengthy process of negotiation, and this compromise is found acceptable by the planning officers, the developer, local residents and traders.

DISGUISE

A third method of blending the new with the old is to disguise it. The disguising of new buildings commonly takes two forms. The first is an attempt to make a wholly new, large development appear less intrusive by creating the impression that it is subdivided. The plot sizes common in many old-established town centres in Britain relate to mediaeval burgage dimensions, or to subdivisions of those plots. The standard English burgage width is some 2 perches (approximately 33 feet or 10 metres), giving adequate street frontage for a timber-framed town house of two structural bays. Many new developments of the 1950s and 1960s greatly exceeded this traditional plot width (Figure 8.6). Developments of the 1980s and 1990s have been similarly extensive, but long façades are now commonly subdivided so that they give the outward appearance of plots of traditional widths. This was the case, for example, in the Grosvenor Centre, a major retail redevelopment scheme in Northampton (Figure 8.8). It should be noted, however, that the mere use of this technique alone does not guarantee success. Many such attempts show the plot subdivision, but the style and scale of the actual building are wholly out of character.

The second tactic of disguise is commonly known as 'façadism'. At first sight, this is one type of building activity that might seem to reconcile LPA conservation policies with commercial pressure to change buildings, and public pressure to retain the existing appearance of buildings. In such developments a virtually new building, built to modern standards with up-to-date facilities, is constructed behind the façade of an existing building, the remainder of which is demolished to make way for the new development. In this way what are essentially new buildings may be slotted into an existing townscape with minimal visual disturbance, retaining the existing and familiar façades and rooflines. Obviously, the historical and architectural significance of the original is lost but, if commercial pressure (for example the need to obtain maximum revenue from a site) would have dictated redevelopment in any case, it may be felt that the retention of at least the external visual appearance may be preferable to a total loss of the building. Several applications involving façadism have been made in Solihull High Street (Table 10.1). Solihull as a planning authority feels strongly that façadism is a tactic of last resort, being contrary to its basic premise of retaining the original building (Chief Planning Officer, Solihull, pers. comm.). Yet six of

Table 10.1 Planning applications involving façadism, Solihull conservation area, 1968–84

Application	*Site*	*Decision*	*Date*
910/70	148–152 High Street	Granted	24/08/70
1730/70	76–90 High Street (retain façades of 82–84)	Granted	13/01/71
830/72	116/120 High Street	Granted	28/11/72
2253/73	35 High Street	Withdrawn	
561/75	48–66 High Street	Granted	22/04/76
2732/79	48–52 High Street	Granted	09/02/80
64/80LBC	48–52 High Street	Granted	03/03/80

Source: Planning files, Solihull MBC

Figure 10.7 Façadism: redevelopment on Colmore Row, Birmingham, showing new mansard roofs providing additional floorspace and visually linking the diverse lower elevations (author's photograph)

the seven proposals received planning permission: the seventh was the subject of debate between the LPA and the developer, who wished to demolish the entire building and create a replica façade, while the LPA wished the original façade to be retained. The developer eventually withdrew from the scheme owing to this disagreement.

The number of these conversions already existing is surprisingly large, and many of them cannot easily be identified from the street. In the Colmore Row and Environs conservation area, covering much of Birmingham city centre, for example, façadism and related schemes have been proposed for a total of 68 plots (Barrett and Larkham 1994: Figure 7). The impact of such schemes is clearly shown by the Barclays Bank proposal at 55–73 Colmore Row. This scheme involved the demolition of all but the façades of the Victorian Italianate palazzo-styled buildings owned by the Bank along Colmore Row, Church Street and part of Barwick Street. The LPA wanted greater retention of the buildings, many of which were listed. However, the application was not determined within the statutory period, and was taken to appeal following a breakdown in communications between the developer and LPA. The scheme was granted on appeal, with the Inspector indicating that, in his view, the scheme satisfied conservation aims, concluding that the façades alone were of real importance in satisfying conservation aims, and that the proposed buildings provided an economic re-use of the site. This decision was also based upon the presumption in favour of development at the national legislative level, and the relatively low emphasis placed upon Victorian architecture in the national context. It illustrates the differences in opinion concerning the worth of Victorian architecture between the LPA and DoE. The LPA did manage to secure retention of the Banking Hall in the new development, but failed to win the desired greater retention of internal features. The façades, felt to be most important by the Inspector, are now visibly linked internally by continuous open-plan offices; the new floor levels do not always respect the original window openings, and significant new floorspace and servicing has been provided in multi-level mansards at roof level. These latter are particularly noticeable from the adjoining cathedral churchyard, creating a significant visual intrusion into the townscape (Figure 10.7).

Façadism is often a useful technique to save otherwise derelict buildings that may have lost many of their internal features through neglect, but in some cases, as at Colmore Row, the buildings were not so poor that they could not have been restored. In this case at least, the economics of providing far more lettable floorspace clearly prevailed over conservation. Again, there can be bad examples of this technique, and unscrupulous developers have been known to report the 'accidental' demolition or falling of a vulnerably propped-up front wall.

'REVIVAL' STYLES

There is a last tactic that should be considered. This is the use of 'revival' styles, and this has become a very popular tactic in recent years. The idea is related in some respects to the use of vernacular, local, details mentioned earlier, but here often involves the re-creation of historic building styles, often with some accuracy. One example in Northampton, for instance, is a reasonably accurate copy of its Victorian neighbours, although some external detailing has been simplified (Figure 10.8). It fills a gap in a terrace where originally there was a gateway and drive to a large house; on redevelopment of the house site the driveway became redundant and itself open to redevelopment. A near replica was a deliberate design choice to attempt to 'repair' a gap in the terrace, although that gap itself was original! (information from Northampton planning files via J.W.R. Whitehand).

Revival styles of various types are currently very popular in residential conservation areas, although in effect they are really only slightly better examples of the styles used by speculative housebuilders throughout the country. These are not true replicas, but rather re-interpretations, using some key design elements to recall the original style. 'Neo-Georgian' was probably the most popular residential style in the 1970s, for example, but rarely done well or accurately. In virtually every case, the anomalous double garage completely spoils the effect (cf. Regency Drive, Coventry: Figure 9.15). The style is still very popular in the high-class, expensive residential estates now being built in the USA, more for reasons of fashion than conservation (Knox

Figure 10.8 Revival styles: infill in a Victorian terrace in Northampton (author's photograph)

Figure 10.9 Revival styles: speculative neo-Tudor in Kenilworth Road conservation area, Coventry (author's photograph)

1992). Since it was re-introduced in south-east England in about 1978, neo-Tudor has also become popular, but again often simplified and debased (Figure 10.9) (Larkham 1986).

A BELGIAN EXAMPLE

These issues and design solutions are not peculiar to UK conservation areas, but it is interesting to see that one of the most thorough and thoughtful recent approaches to these problems has been in the historic city core of Brussels, albeit a part where the townscape has been much mutilated. The site was partly occupied by the 1960s office tower headquarters of the AG 1824 company, which wished to redevelop part of the site for its offices and part, fronting the Rue de Laeken, for shops and housing. In conjunction with a local cultural association, the *Fondation pour l'Architecture*, an architectural competition was held with the dual aims of attracting young European Community architects and 'to demonstrate the aesthetic, functional and economic value of the traditional lot [i.e. plot] in the construction of quality housing in a historic city centre' (AG 1824 (1995): 93).

A jury selected 24 architects who each submitted proposals for two plots on the site. The jury then combined these proposals in various permutations to create an 'ideal street' (Figure 10.10) and, in addition, recommended the restoration of two historic houses which had been identified by one participating architect. The varied experiences and national backgrounds of the architects resulted in a wide range of designs and details, all of which were contained within traditional-sized plots. This gave the 'ideal' street an interesting visual variety and richness, and suggests both that these are individual buildings (which is true), and that they were not necessarily the product of a single building 'campaign' (which is false). The company's view of the design approach selected is interesting.

> There are two possible approaches when reconstructing a section of a street.
>
> The first, which is generally adopted for a development on this scale, consists of designing a single building with, around each vertical circulation core, a series of usually standardized apartments. This option, built over four or five floors, would have produced from 12,000 to 13,000 square metres – making a hundred apartments – or a surface area equivalent to that of the same building owner at the same time on the same block [i.e. the new corporate headquarters].
>
> The second option, and the option chosen, consists of building a series of small juxtaposed constructions and of reducing the overall size of the development by one half, with a total of 6,000 square metres rather than the possible 12,000. This reduced scale, which may at first appear incompatible with the need for the site to be financially feasible, had a positive effect on the project. The size of the buildings is perfectly in keeping with the traditional building techniques which were able to

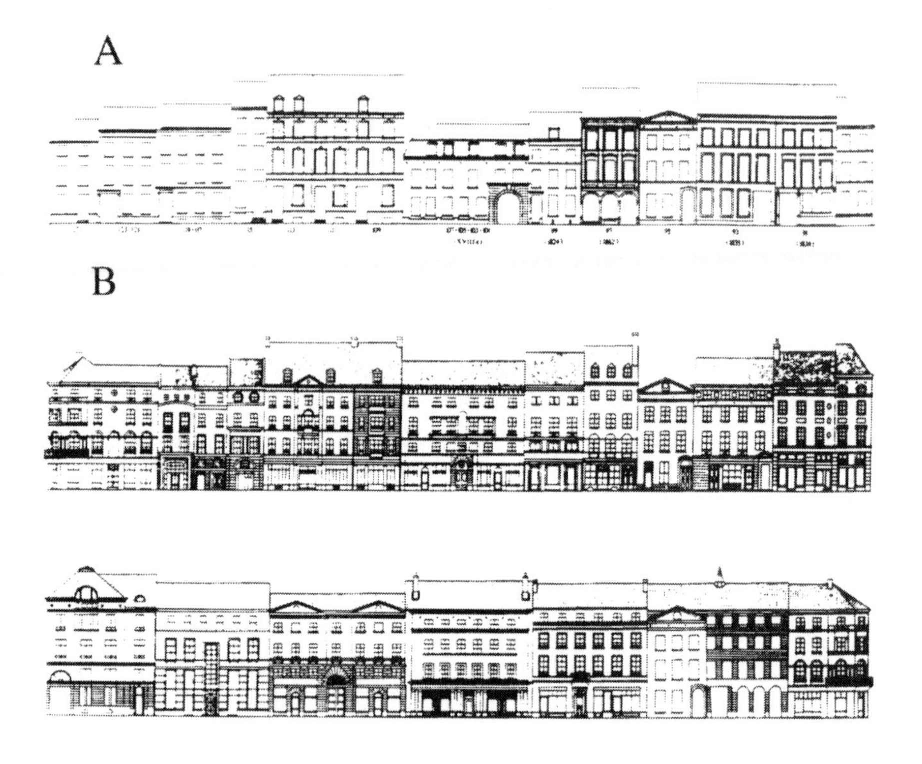

Figure 10.10 The reconstruction of Rue de Laeken, Brussels. A: the street *c.* 1865 reconstructed from building permits and photographs; B: two of the permutations of projects to form 'ideal streets' (drawings by A.A.M. 1990; reproduced by permission of AG 1824)

> prove competitive both in terms of the quality of the result and the cost involved … the construction of small autonomous units dispenses with the need for many common areas – corridors for example – and lifts.
>
> These potential savings are accompanied by another factor which is perhaps the most important of all. By building fewer apartments you are able to build better apartments – and without this, the new apartments could well be left standing empty like so many others on the Brussels property market.
>
> (AG 1824 (1995): 22)

Two problems occurred during construction which had implications for the overall design. First, the company was not able to acquire all sites on the street block. The design for the new buildings thus had inbuilt flexibility for extension to replace these buildings, should they become available at a later date. Secondly, the jury recommended retention of two original buildings, which had not at first been envisaged. That at no. 95 has been 'rebuilt

Figure 10.11 View of the Laeken street block project (photograph reproduced by permission of AG 1824)

"identical" to its original construction', repairing the seventeenth-century roof structure but retaining the later pedimented façade; while the much-altered rear elevation was reconstructed with the aid of some surviving details. Again, in retaining the early nineteenth-century house at no. 117, major rebuilding was necessary: 'the stability of the existing house was soon revealed as too fragile and it was quickly decided to opt for an "identical" reconstruction' (AG 1824 (1995): 39). Thus, although the original designs were extensively reworked to cope with these retentions, what exists today is largely replication rather than original.

The entire project required an investment of some 500 million Belgian Francs. It provides 41 apartments and houses, 13 shops, 87 parking spaces, 600m^2 of private courtyards and gardens, and required 140 cubic metres of stone, 1,600m of ceiling mouldings and 3,800m^2 of wood flooring. The project took three years to construct; the architectural competition and design work took a further two years. The office headquarters is bland but reasonably sympathetic, but it is the variety and attention to detail which marks this rebuilding of a street façade (Figure 10.11). In particular, the use of individual architects for individual buildings differentiates this from the monolithic office block at Richmond Riverside, hidden by a variety of Classical-detailed façades. The fact that the site owner, AG 1824, was willing to take this financial risk, and perhaps to recoup less profit than from a more intensive development, is a key factor in this example.

RECENT ATTITUDES IN BRITAIN: QUESTIONS OF CONSERVATION

Having seen some of what is happening in the UK under the general banner of 'conservation', supported by the current upsurge in pro-conservation sentiment, some key questions arise. The examples and design principles discussed in this chapter have revealed some of the problems of coping with change in historical townscapes, and some of the ways in which these problems have been dealt with, particularly during the past two decades. The different approach used in the Belgian example usefully contrasts the British methods, as does their seemingly rather more relaxed attitude towards replication.

A first question is 'who is important in the process?' The importance of planning departments in recommending possible solutions, particularly in the case of major developments such as the Sainsbury's superstores, is sometimes apparent. This is clearly shown by the growth in various forms of design guidance (Chapman and Larkham 1992) and, increasingly, as Supplementary Planning Guidance to Local Plans (Punter *et al.* 1994). Because of the process of public consultation undertaken in Local Plan preparation, such guidance is afforded considerable weight in the quasi-judicial aspects of planning decision-making (Jones and Larkham 1993). In most cases, however, the planning department is an indirect agent in the process of change. Planners are merely able to react to proposals put forward by those who are here termed the 'direct' agents of change – the developer, owner, and architect. These also have the ultimate sanction of withdrawing from a scheme if the constraints imposed by a planning authority are felt to be too onerous.

There is a great variety of potential design solutions to the problem of fitting new buildings into historical townscapes. The majority of the examples cited here have been thought to be generally successful by planners, developers and building users. Nevertheless, there are many poor examples of all four design approaches, and all four have been heavily criticised. Sainsbury's stores have been described as 'gestures at traditional effects', 'gestures at the high-tech shed that has become so fashionable', and 'horrible brick and metal lumps' (Gardiner 1985). One proposal to have 'a bogus façade above ground floor level to increase the apparent height of the building' which, superficially at least, would give the building a similar façade to its neighbours, has been described as 'a dreadful device' (ibid.). One view of façadism is that it 'makes a complete nonsense of the concept of conservation. It is ridiculous to have a street made up of historic front walls' (Bearman 1982). Vernacular styling has been described as 'the architectural equivalent of reproduction Chippendale' (cited by Aldous 1978) and, in the view of Matson, discussing the application of an historicist façade to disguise a multi-storey car-park at St Mary's, Colchester, it is 'a curious idiosyncrasy of design insecurity' (Matson 1982). Wise, an experienced conservation architect and director of the influential Institute of Advanced Architectural Studies in York, told the Historic Houses Association in 1990 that

> for the first time in the history of urban development, the present generations have introduced the sham on a large scale – and I think we shall regret it, because the net effect is not to improve quality but to erode it.
>
> (Wise 1990: 20)

In short, as Max Hutchinson, then President of the Royal Institute of British Architects, said in response to the Prince of Wales's influential book and television documentary in 1989,

> the nostalgic copying of the past forms and the mindless conservation of all that was built before this century will destroy the rich potential that lies in the future, and cities will become museums.
>
> (Hutchinson 1989)

It must be admitted that Hutchinson was responding to a particularly eloquent and well-illustrated view, taken by many architects to be an attack on their profession. In leaping to the defence of architectural creativity Hutchinson and other architects perhaps over-state their case and fail to recognise the plurality of positions and design solutions which could be acceptable even within a conserved townscape.

There is no consensus, either amongst planners, architect/designers nor the public, on how change should be incorporated in historical townscapes. The nature and scale of the problem, possible solutions and their successes or failures are all debated. However, this debate remains mostly at the level of vague generalisations or specific case studies. What is conspicuously absent is any idea or theory of townscape management, any ethic of conservation planning. Worskett has identified four needs. First, a redefinition of our ways of perceiving the historic town: instead of an emotional reaction against the twentieth century, a more positive and considered approach on a sound philosophical basis is needed. Secondly, the development of that philosophical argument into a more widely accepted conservation ethic, dealing with the constraints on and opportunities for coping with change in historic areas. Thirdly, it is necessary to demonstrate the practical application of that ethic. Finally, there is a need to stimulate public awareness and debate (Worskett 1982). These requirements should be considered when assessing the potential contribution of urban morphology to townscape management.

Part 3

IMPLICATIONS OF CONSERVATION: ETHICS, THEORY AND PRACTICE

11

TOWNSCAPE CHANGE: IDEAS AND PRACTICE

> We must beware of contempt for old buildings just because, like old people, they can be frail, muddled and squalid. That contempt can easily become a sort of architectural fascism. Not all our slums are slums. Piecemeal renewal, each piece in scale with the place, is not necessarily a wrong answer just because it is an old one.
>
> (Viscount Esher, 1964)

INTRODUCTION

Thus far, this book has examined aspects of the pressures which have supported conservation, the history of conservation, the UK's legal-judicial conservation planning system, and planning policies, practices and developments in designated conservation areas. Thus far, the resulting picture is one of confusion of ideals and practices, and conflict throughout the procedures. In this last part, we need, therefore, to return to the book's starting-point and to review some of the key ideas and questions of conservation in the light of the examples already used, and other relevant cases.

Although the adaptation and renewal of an ageing townscape is one of the most pressing problems facing mature settlements, it has received relatively little attention until recently. Visions of the future have predominantly concentrated upon creating anew. In town planning, ideas about new towns and villages, and extensions to existing towns, have generally taken precedence over the seemingly more mundane task of reshaping existing townscapes. Theoretical bases for tackling this problem in a manner sensitive to the existing townscape and its role in society remain ill-developed, despite the current high priority afforded to conservation. There is no accepted theory of conservation. The little relevant theoretical work that does exist is almost unknown to practising town planners, even those with special responsibilities for designated conservation areas.

Much of our attitudes towards the aesthetics of built forms and their potential conservation are coloured by culture: society and experience. In particular, the majority of the industrialised West is dominated by the aesthetic of the Renaissance, with its emphases on regularity, repetition and symmetry. In terms of the urban plan, the tenets of the *beaux-arts* aesthetic again dominate, with the concepts of regularity, planned vistas, and formal geometric layouts creating both familiarity and surprise. The history of this aesthetic approach is well documented, for example by Rasmussen (1951), Kostof (1991) and Sutcliffe (1981b). This aesthetic superseded that dominating the mediaeval and earlier periods, which was based on the local availability of materials and building technology: hence the rise of the concept of the local vernacular. Irregularity of architecture and form dominated, even in some cases where there were ideal regular plans (cf. Slater 1987). Some developments in twentieth-century urban aesthetics have moved away from *beaux-arts* aesthetic ideals, principally in the rise of the Modern Movement, favouring clean lines and some types of regularity (Frampton 1980) and in the burgeoning of suburbia with its typically arcadian winding irregular layouts and wide varieties of architectural style (Edwards 1981).

The conservation aesthetic, as developed in practice in the urbanised Western societies and as demonstrated in operation in this book, is rather different. Driven by both the highly educated professional and by a rising tide of public opinion, the dominant aesthetic in late twentieth-century conservationism is that of 'blending in' or, in the increasingly fashionable architectural term, 'contextualism' (Groat 1988). A plethora of planning guidelines, design advice and policies emphasise this. As some of the examples in Chapters 8 and 9 have suggested, planning practice is moving towards the creation of more uniform conserved urban environments, and architects and developers are helping through deliberate use of tactics such as pastiche, replication, façadism and use of vernacular (Chapter 10). In this way, it is argued (using the English legal terms) that area character and appearance are preserved and enhanced. It is agreed that some developments are 'out of keeping' with their conserved surroundings, and it is these examples which are often analysed by critics, to the detriment of any overview of how entire areas are being changed. It is this wider spatial and temporal view which this book seeks to develop. However, in some cases, and particularly where analyses of character and appearance are poor or absent, what is being created is a new aesthetic of regularity: a blandness, or 'conservation-area-architecture' (Rock 1974). The types of design ideas prevailing, and policy and guidance exercised, are tending to 'create areas of uniformity through mundane historic replication' (Tugnutt 1991). The resistance to new forms and styles of development on the part of decision-makers and public groups is a very negative view; the reactions to the concept of 'enhance by contrast', for example, have already been discussed. Yet there is a strong counter-argument that some degree of change is inevitable, even within conserved areas. UK planning legislation clearly recognises this. Conservation should, it is argued, be more receptive to development that will 'be recognised as being of positive value to the area's appearance once completed' (Biddle 1980: 11).

ETHICS IN OTHER FIELDS

One significant problem underlies any further detailed consideration of heritage planning. This is the general lack of any accepted theory, or ethic, of conservation (Worskett 1982). The few attempts to promulgate such an ethic have been published in low-visibility sources or have not met with great success among planning practitioners (e.g. Briggs 1975; Faulkner 1978; cf. Larkham 1990a). Yet, in fields other than the built environment, there are considerable parallels in heritage and conservation problems but far greater advances in conceptualising. In art, for example, there is a general consensus over the delimitation of fake, restoration and replication. The *intent* to deceive is here viewed as a major consideration, more significant even than actual deception, which may be more a factor of the perceiver rather than the perpetrator. There is also a widespread acceptance that repairs and restorations should be reversible, even if not readily discernible to the naked eye. These arguments are well explored in Jones (1990). Indeed Lowenthal suggests that

> many fabrications are essentially mental rather than material; the fake inscription or manuscript is simply an adjunct to an intended historical deception. Yet their supposed veracity, sanctitude and uniqueness makes fraudulent physical objects seem essentially repugnant.... Although it is now evident that artefacts are as easily altered as chronicles, public faith in their veracity endures: what can be seen and touched cannot lie. Material objects attest to the pasts from which they came because they are tangible and presumably durable.
>
> (Lowenthal 1990: 21)

To exemplify the problems inherent in art history, and the approaches which have been developed, Jones (1990: 28–49) cites the cases of replication, the 'collector's copy', artists' copies, the persistence of tradition, and deception. All are approaches to the elusive concepts of 'original' and 'fake', and it must be said that different cultures and historical periods have placed different values upon these concepts. To illustrate outright fakery, Jones uses a terracotta Etruscan sarcophagus, purchased by the British Museum as a piece of the sixth century BC; within a year of purchase, inscriptions were shown to be copies from a brooch in the Louvre, and the poses and clothing of the figures were unlike any precedent from antiquity. Here, there was clearly an intent to deceive.

There is no such attempt where the same culture may continue manufacturing items for a long period, or may (even after the passage of centuries) revive traditions of manufacture and use. Likewise, items copied by artists as part of their training, particularly in the studio of a master, were not intended to deceive; indeed were accepted as 'genuine', and often marketed by the master as such; an accepted practice in the Renaissance period. Copies of original artworks, produced by craftsmen for collectors, became popular from the Classical revival brought about by the aristocratic Grand Tour. Such works were often regarded as works of art in their own right, and only when collections were dispersed were some passed off by traders as genuine. Since the same time there has been a demand for mass-produced replicas as souvenirs, but many such are not direct copies, being rather historicist re-interpretations, or are clearly cast and moulded using new materials including resins and silicone rubber.

In the majority of such cases there is no original intent to deceive. Deception, when it occurred, was a later stage in the life-cycle of these products; often after several resales when the original provenance of the item was lost or conveniently forgotten. As Jones (1990) shows, the number of such deceptions practised by middlemen is enormous. Yet, in their original state, with full provenance, the same items were widely accepted, sought after and manufactured, and today, when research has uncovered historical details of a 'fake', some items are now re-valued as key examples of craftsmanship and revival styles.

Yet there are many controversial cases in the art world, as a court case dealing with an alleged Schiele painting showed (Alberge 1995). The painting was sold as a Schiele and was initialled 'E.S.'. The court held that it has been heavily overpainted by a restorer, and that the restorer had sought to conceal the extent of the overpainting. Christie's, the vendors, argued unsuccessfully that no amount of overpainting would make this a forgery if the overpainter had followed the design of the original artist. There appears in this case to be a genuine painting by Schiele, where the nature and extent of 'restoration' renders it a 'forgery'. The reported court verdict thus seems confusing: when did the genuine article cease to become genuine? How much overpainting would have been legally permissible?

The movement over the past two decades or so towards authentic re-creation of musical performances offers a further insight to changing attitudes. Use of original instruments, or supposedly faithful copies, together with strict adherence to original scores is held to produce a faithful version of a piece as the composer originally intended. Yet many factors, such as the acoustics of modern concert halls, the lack of *castrati* and the interpretations of modern performers inevitably deviate to an unknowable amount from this ideal. Further, the same movement also castigates the adaptations and interpretations of earlier music so common from the nineteenth century to today as being unoriginal or fraudulent: Lowenthal (1990: 20) draws the parallel with the attitudes of Ruskin, Morris and others towards the cathedral restorations of the nineteenth century: they castigated Giles Gilbert Scott for his lack of fidelity to the original. Certainly, a full concert orchestral performance of, for example, Bach's Brandenburg Concerti by the Berlin Philharmonic Orchestra recorded in 1965 (von Karajan's performance on the *Deutsche Gramophon* label) contrasts markedly with the 'authentic' sound of the same pieces from The English Concert, conducted by Trevor Pinnock and recorded in 1982 (Archiv recording). But to what extent is one interpretation better, or more valid, than the other? Karajan's orchestral version is in the nineteenth-century tradition of adaptation and re-interpretation; Pinnock's in the late twentieth-century tradition of originality. Lowenthal suggests that such differences, and the widely different values placed upon them by recent critics, show that there is a break in musical history and tradition: instead of adapting Bach's music to retain it in the popular repertoire, we are told to keep to the original. 'To restore a true past the nineteenth century consciously altered it; today we likewise alter the past, but habitually blind ourselves to our own impact on it' (Lowenthal 1990: 20).

In dealing with mechanical devices, however, it appears that constant use means constant replacement of damaged or worn pieces, and yet it is widely accepted that despite such intervention, the 'identity' of a vehicle may remain unchanged: this was found to be so in the High Court after the purchaser of the racing Bentley known as 'Old No. 1' challenged its provenance in 1990. Its chassis had been replaced after race damage; racing engines were frequently replaced; and the body was a reconstruction. Nevertheless, it was held that, irrespective of the date of its constituent parts, in this case the vehicle itself had always retained a coherent and distinct identity. Indeed, there are many cases of car restorations where a vehicle, accepted as authentic, has been constructed from various parts. Hawtin (1995) writes of a 1924 Aston Martin where the body had vanished following the collapse of the garage in which the car was stored. No data existed on the original tourer body, and a wholly new body was constructed 'based on an original line drawing of the 1924 Motor Show car, which is known to have been similar' (Hawtin 1995: 38). Yet the review of this car is wholly favourable: 'the arrival of [this] superbly-restored example on the vintage scene will, I am sure, be welcomed by all' (ibid.: 40). And acceptable restorations have been carried out from even less: an Edwardian Austin is being restored from 'most of the major components', which seem to consist of engine, gearbox, axles, wheels, petrol tank and steering column, together with enough chassis and springs to allow fabrication of new parts. 'Nevertheless, what is there forms a good basis for restoration' (Worthington-Williams 1995).

Still more unusual is the general acceptance of new vehicles or aircraft, manufactured to original drawings and standards by the original company years after production originally ceased, as being 'original' rather than 'replica'. In this respect, Dron (1991) discusses the example of the 'new' Aston Martin DB4GT Zagato, manufactured by the original company and using four chassis serial numbers originally allocated to this model in 1960 but never used. Although the four new cars have some modifications to overcome performance problems with the originals, they also contain some original parts from Aston Martin Lagonda's spares stocks. Accepting these well-documented changes, Dron argues that these four cars should not be seen as replicas, but rather a second production run.

> It is a shame that the word replica has been abused in recent years: we have tried to point out time and time again that a replica can only be made by the original manufacturer and that it must be an exact copy of

> the original car. It seems funny that everyone knows what reproduction furniture is, while reproduction cars which bear no mechanical resemblance to the classic machinery which they resemble in appearance are called replicas.
>
> (Dron 1991: 29)

Conserved aircraft, however, have different ethical 'rules', depending upon whether the machine is to be restored for flying or for static display, and whether the airframe has intrinsic historical merit or is merely an example of a marque. A Spitfire XIV recently restored to flying condition has been given South-East Asia Command camouflage and roundels since the operators were 'seeking interesting colours and something other than the usual "temperate European"' (*Aeroplane Monthly* 1995: 5). The safety implications of flying often require remanufacture of structural components, insertion of new instruments, brakes or engines, as was the case with a restored F4F-3 Wildcat. Here, 95 per cent of the original fuselage and 25 per cent of the wing were re-used, the remainder was either manufactured anew, rebuilt or restored. A later, improved engine was installed (*Aeroplane Monthly* 1995: 8). Authenticity is much less important than safety. Undistinguished aircraft are often repainted to represent historic aircraft, or those associated with particular individuals.

It is also interesting that concepts from built environment conservation might be applied to the conservation of machinery. The Department of National Heritage has commissioned a pilot study on horse-drawn carriages in an attempt to decide how to list objects of national importance.

> I would personally like to see the same 'listing' process applied to commercial vehicles and then, hopefully, to cars. There are so few totally original or else correctly restored commercials, that the task would be relatively simple. The problem might come, as with historic buildings, when the owner of a listed vehicle wanted to do something unsuitable like painting it with dayglo stripes. However, I feel that most owners would be proud to have their fine specimens listed and would do all they could to keep them in original form. . . .The same should apply to some extremely nice examples of private car, which so often lose their lovely old upholstery or hoods just when they have acquired the patina of a lifetime. Surely a bit of fading, damaged stitching or cigarette burns are far more important to a vehicle's history than a gleaming new interior, quite often in inappropriate materials?
>
> (Baldwin 1995)

PARALLELS IN THE BUILT ENVIRONMENT

Returning to the historic built environment, do these parallels from other fields assist in making conservation-related decisions? Do these parallels support, for example, the practices of façadism, replication, or the use of historicist styles? Probably not much since, although there is some measure of agreement over fakery and the intent to deceive, there is confusion over whether a building should be considered as a work of art – and thus have parallels in museum curatorship – or as a machine, requiring regular maintenance, modification, repair and replacement of parts. Historic buildings and townscapes seem to produce higher levels of debate, particularly among the general public as well as the informed élite, and issues of fakery, façadism, replication and originality are central to the argument of much Westernised conservation (Larkham 1995a). This dominating Western ethic – or aesthetic – is quite contrary to the aesthetic of some other countries. Japan is a notable case where the regular demolition and rebuilding of key national monuments, particularly Shinto shrines, is a widely accepted historical and religious tradition. Some shrines are demolished and reconstructed, on their original sites and in identical forms with identical materials. Obviously, however, the physical degradation of years or decades of wear and weathering is lost; the structure is as new, but retains all of its religious and historical associations. The Ise Shinto temple, which is rebuilt every twenty years on a new site, is preserved not as a physical continuity but as a record of the techniques of the original craftsmen. The Japanese aesthetic does not have the concerns over authenticity versus replication so prevalent in much of the West; nor is the 'patina of age' (cf. Pevsner 1976) so revered.

This might be contrasted with Lowenthal's entertaining example of 1930s neo-Georgian houses in Hampstead Garden Suburb. Some had open shutters, bolted to the walls, as a decorative element of the original design. During the Second World War, when the threat of bombing raids was great even in these London suburbs, the occupants of one house undid the original bolts, added hinges and closed the shutters as a defence against blast damage. 'After the war, the local preservation society insisted the hinges be removed and the shutters bolted back: they were authentically *fake* shutters' (Lowenthal 1990: 16).

Issues of originality have also been raised in a Court of Appeal ruling over the reinstatement after extensive fire damage of a Grade II listed building, converted from several 300-year-old cottages (*Beaumont and Another* v. *Humberts and Others*, 11 July 1990; Burns 1990). Although this case revolved around the extent of financial liability, the interpretation of 'reinstatement' was central. If this was held to mean an exact copy, the rebuilt structure would clearly be in breach of modern building regulations in several counts, and would be deemed unfit for human habitation. The owners wished the house to be reinstated substantially as it was before the fire: 'an exact copy in so far as was practicable', and had relied on a surveyor's 'value for insurance reinstatement purposes' to be sufficient to undertake this. A dissenting judgement of the Court noted that the surveyor's valuation would bring about a pleasant house, but did not constitute reinstatement; by a majority, the Court decided that, had an estimate for an exact copy been desired, it should have been asked for in such words, rather than relying on the standard 'insurance reinstatement purposes' wording. The Court did not address the issue of whether a reinstated building, to whatever standard,

Figure 11.1 How far should originality go in restoration or repair? (reproduced by permission of David Austin)

would still retain its listing, but the question of the amount and degree of 'originality' in repairs or restorations is clearly significant (Figure 11.1), and has been highlighted by the fires at Windsor Castle and Hampton Court. Indeed, English Heritage has reportedly suggested that, if more than half of a building is destroyed, then reinstatement may be such a sham that it might be best to clear the site, put up a plaque and sell it on (*Chartered Surveyor Weekly* 1991: 19).

The issue of authenticity has also been raised over the restoration of the Queen's House, Greenwich (Wilcock 1990). Unlike many, this was a deliberate 'museumisation' with the decision being taken to restore this building to its 1661 state. Yet some short-cuts were taken, apparently for financial reasons, including the retention of the 1708 fenestration and the 1690s paint scheme in the hallway (grey-green rather than the original white and gold). More controversially, much of the house has been furnished with modern replica furniture and fabrics.

> An even worse insult to the purists is the attempt to recreate the original ceiling in the hall. The nine painted ceiling panels by Gentileschi were removed to Marlborough House in the eighteenth

> century. Using the Scanachrome process, transparencies of the ceiling paintings were scanned by laser to generate a computer image. Missing sections were regenerated on computer and the final images were then produced at the correct size by spraying paint on to canvas.
>
> (Wilcock 1990: 74)

The extent to which this showpiece is thus restoration or fake is debatable. By the standards of the court adjudging the Schiele case, the ceiling is undoubtedly a forgery!

Similar issues arose in the restoration of Barley Hall, a fourteenth-century house and fifteenth-century hall-house in York. It has been researched, dismantled, and 47 original and over 470 new timbers used in the reconstruction. The research led to the listing status being upgraded from II to II*. Yet the Society for the Protection of Ancient Buildings (SPAB) feel that this is 'reproduction heritage: meticulously researched and beautifully executed fakery, but fakery nonetheless' (SPAB 1992). SPAB feels that York City Council failed in its duty to keep the Society (a statutory consultee) updated on the planning decisions relating to the building, and that the restorers, the York Archaeological Trust, have destroyed a genuine mediaeval structure. Indeed, following this debate, the DoE again reviewed listing status, and downgraded it to II.

> The York Archaeological Trust . . . has a strong record in excavation and presenting the past to the public. It has pioneered challenging and imaginative approaches in this latter area. Never before has it come up against an adversary so well qualified to comment as is the SPAB. . . . The work has been done and is available for all to see. It is perhaps one of the best examples of the philosophical clash that can occur between the conservation lobby and the heritage lobby.
>
> (Society for Medieval Archaeology 1993)

A last philosophically contentious case is that of Dee House, Chester. This is a fine Georgian house of *c.* 1730, which became a Catholic convent school in 1854. Two wings, of 1867 and *c.* 1900, now flank the original house. But the house is constructed over part of the remnants of the Roman amphitheatre, which was discovered in 1929. That part outside Dee House's grounds has been excavated and laid out for public display. In 1986 a private developer suggested that he would complete the excavation if he was also given permission to construct a visitor centre and reconstruct one quadrant of the amphitheatre. This, obviously, entailed demolition of Dee House. This demolition, and the 'expendability' of the house, was the main feature of the Public Inquiry in 1987. The Planning Inspector was unconvinced that the retention of Dee House and the pursuit of archaeological knowledge were mutually exclusive and recommended refusal of the scheme. However, the then Secretary of State, Nicholas Ridley, disagreed, and over-ruled the Inspector's recommendation.

> Listed building consent has now been given and Chester will soon be blessed with the curious sight of twenty-five per cent of a reconstructed amphitheatre, with extras promised in Roman costume and 'authentic' spectacles in the arena itself.
>
> (Ancient Monuments Society 1989)

The scheme has not yet been developed. Nevertheless, this did raise the interesting question of the relative weighting given to conserved structures. A Georgian house, with some fine original features surviving despite later extensions, was felt by the Secretary of State – the highest judge in these matters, although with no specialist training – to be expendable. The amphitheatre, one of nine known and three possible in the UK, gained priority. Yet the decision was not made on purely conservation grounds: politics were clearly involved. It must be remembered that Ridley and the then Conservative government were demonstrably pro-business in their planning decisions. And, moreover, the tide of archaeological thinking has since changed, with English Heritage being reluctant to sanction excavation of monuments unless there is a demonstrable threat to their continued existence.

Thus the ethical concepts of factors such as repair, originality, replication, forgery and usability, derived from other aspects of conservation, clearly have something to offer the field of built environment conservation. Yet there are problems, not least because of the confusion over whether individual buildings – let alone an area of a town – are works of art or machines. Nevertheless, these parallels are worth pursuing. Recently, the heroic restoration of the National Trust's Uppark House, recreating items using original materials and craft labour, and re-using as many salvaged pieces as possible, seems to be a good parallel with the Aston Martin case cited earlier, and to have received as much favourable publicity.

PRACTICAL APPROACHES

> Engulfed by so much residual evidence of history, a real problem does exist for the British of reconciliation of past with future but the challenge of this conflict gives direction and real architectural opportunity. There are differing views of the past and different weighting given to what remains, but the only genuinely uncreative interpretation is that which argues that the collective memory needs to be erased in order to progress.
>
> (Terry Farrell, 1984)

This section discusses some practical approaches to the management of conserved urban landscapes, given the demonstrated dichotomy of the 'high Art' approach to conservation – the view of the architectural and historical purist – and the contrasting, possibly more pragmatic, view of the conserver society, the general public. The two cannot be reconciled in every example. Façadism, for example, retains only a front wall; although the impact of

Figure 11.2 Caricature of the 'evil planner' approach to conservation (reproduced by permission of *Private Eye*)

redevelopment on the townscape is thus minimised, it creates the problem of streets of preserved historic front walls: a sham, screen-set form of townscape. Worskett's complaint over the lack of a conservation ethic is, once again, clearly relevant (Worskett 1982).

The perspective of this volume here returns to the people, corporations and institutions vital to this perspective of urban change and conservation – the agents of change. What is vital is not the polarising of attitudes caricatured, for example, in the NIMBY approach. Development proposals should instead be preceded by more open discussion and negotiation than at present. In the past few decades, the development process has been characterised by (a) finance-led design solutions and a hasty approach, requiring fast decisions for a fast return on investment; (b) increasing bureaucracy and delay, as planning authorities attempt to react to the growing flood of development proposals; and (c) public feelings of helplessness in the face of the 'evil developer' and 'evil planner' (Figure 11.2). The whole process of change in historic and conserved towns needs to become more considered, more open, more democratic: even at the expense of some delay and some lower financial return. There are grave consequences here for the urban heritage of the developed world in general if the prevalent approaches and administrative systems are not changed.

There have been very few attempts to turn the slippery concepts of conservation into a generally acceptable and practicable set of policies. Two of the most significant are discussed below, and are reproduced in Appendix 3.

> Imbued with a message from the past, the historic monuments of generations of people remain to the present day as living witnesses of their age-old traditions. People are becoming more and more conscious of the unity of human values and regard ancient monuments as a common heritage....It is essential that the principles guiding the preservation and restoration of ancient buildings should be agreed and be laid down on an international basis, with each country being responsible for applying the plan within the framework of its own culture and traditions.
> (International Congress of Architects and Technicians of Historic Monuments 1964)

Such principles were defined for the first time in the Athens Charter of 1931. This was a significant contribution towards the development of an extensive international movement, particularly in the post-war period when the need to restore bomb-damaged monuments and urban landscapes was a high priority. By 1964 it was felt that problems were becoming increasingly complex and varied, and that there had been sufficient experience and research to enable principles of conservation and restoration to be re-formulated. This was done in the Venice Charter of 1964 by the Second International Congress of Architects and Technicians of Historic Monuments.

The Venice Charter remained probably the most significant document dealing with principles of conservation for over two decades, until the ICOMOS International Charter for the Protection of Historic Towns was adopted in 1987. Although the Venice Charter dealt specifically with 'historic monuments', it contained much of relevance to historic urban landscapes. The ICOMOS document is much more detailed, and deals specifically with the problem of historic towns. Being international in context its approach has to be very general, and this may be a failing when it is applied at a national, or even local, level. It is quite comprehensive in the physical features that are listed for preservation (Article 9). It highlights some important points, particularly the fact that the protection of districts has implications for tourism, and thus increases pressure on the preserved area (see Ashworth and Tunbridge 1990 for further discussion of this), and also that protection must be a social, as well as physical, process – the requirements and aspirations of the town itself are important. But whose heritage is to be preserved? This question is particularly difficult to answer in colonial towns and towns that have undergone phases of conquest and immigration (Ashworth and Tunbridge 1990; Newsom 1971; Tunbridge 1984). The recognition that urban functions change over time, and that their location within the urban area also changes, is significant when assessing impacts on urban landscapes (Buswell 1984a, 1984b; Tunbridge 1986). In short, this Charter is a very important document, having significant implications for the practice of historical townscape management.

In practice, however, and as the preface to the Venice Charter cited earlier suggests, different countries may require their own interpretation of, or extensions to, these principles. The English Heritage policy statement is one such. During the mid-1980s, a tremendous pressure was evident in England for the construction of extensive retail developments inside, on the edge of, and outside towns (see, for example, Larkham and Pompa 1989). This affected historical towns both directly, by the destruction of sizeable areas of their centres, and indirectly, by the diversion of trade away from the centre (Martin and Buckler 1988). English Heritage thus attempts to address this problem of retailing in historical town centres. This is again a significant document, on a national scale. It is perhaps more realistic than the ICOMOS Charter, since it is able to state the morphological characteristics of English historical town centres and to suggest that as many as possible of these be preserved, while the ICOMOS Charter must generalize (Article 9). This provides guidance that would be of greater and more immediate practical use than ICOMOS could hope to give. Nevertheless, it is not a set of rules: it again generalises ('the external detailing and materials of the structures should respect the existing character of their surroundings'), and leaves the details to the individual local planning authorities, the decision-making bodies that in England manage historical urban landscapes on a day-to-day basis.

It is thus evident that some form of hierarchy of policy documents is required to translate high ideals and theories into everyday practice. Appendix 3 gives one significant example of the highest level in that hierarchy – the ICOMOS International Charter – and one from the middle rank – the English Heritage document. Local policies are so specific to diverse local conditions that it is inappropriate to illustrate them by example here: they are dealt with where appropriate in the main text of this book.

One last practical approach, albeit extensively dependent upon a set of values which remains unstated, is the series of interlocking questions posed by Wise (1990). He suggested that for any new development in a conserved area, whether it involves rehabilitation or demolition and replacement, the following questions must be answered.

- Is the existing building *fundamental* to the visual quality and *unity* of the environment?
- Would the quality of the environment suffer *significantly* if the building was to be demolished and replaced with another, but different, building; even of high quality?
- Does the answer 'yes' to either of these questions over-ride all other considerations?

Wise suggested that the number of times that the answer to any or all of these questions was 'yes' would be rare. If 'no', then further questions arise.

- Is the building of such importance *individually* as to justify its rehabilitation and conservation?

- Is that importance due to its intrinsic quality or because it is a good example of its style and type?

If both are answered negatively, then one final question emerges.

- If the building has no particular importance, either for the total visual environment or its own intrinsic qualities, does its conservation and rehabilitation make sense financially; aesthetically; morally; in terms of our responsibility to the continuum of urban development; or on purely sentimental grounds?

This, therefore, returns us to some of the key conservation issues outlined in Chapter 1. Although Wise's questions are framed for individual buildings, they can be readily adapted for wider areas. Nevertheless, they do make some assumptions over value-sets and aspirations regarding the historic built environment, and, more importantly, one must question exactly who is answering these questions, and whether their values represent those of a wider society. Again, examining Wise's challenging questions suggests that there is a lack of a widely accepted theoretical position to underpin conservation and the management of historic urban landscapes.

A THEORETICAL APPROACH TO MANAGEMENT

Although planning controls influence many developments to some degree, their bases generally remain unstated. In so far as any rationale for change may be recognised, it seldom seems to be connected directly with the townscape. Townscape change is much more likely to reflect the working-out of the objectives of particular organisations and individuals or, indeed, simply result from their changing circumstances, than any notion of townscape management in a planning sense. Even at an academic level, the notion of townscape management is seldom considered, and theoretical bases for such an activity have seldom been suggested, let alone coherently articulated. It is appropriate, therefore, to conclude by reconsidering M.R.G. Conzen's rare attempt to provide an essentially historic-geographical rationale for townscape management (Conzen 1966, 1975; see also Larkham 1990a).

Conzen's theoretical basis for townscape management has its foundation in his view of the historical development of the townscape (discussed in Whitehand 1987c). Fundamental to his perspective is the idea of the townscape as the objectivation of the spirit of a society. This philosophical concept can be traced back to studies on the philosophy of culture by German philosophers such as Freyer (1934) and Spranger (1936). In geography, it appears first in the work of the German geographer Schwind (1951). It is rooted in the fact that landscapes embody not only the efforts and aspirations of the people occupying them at present, but also those of their predecessors. In this way, the townscape may be seen as embodying the spirit of society in the context of its own historical development in a particular place. This objectivation is individualised in the physical arrangement of the townscape. It becomes the spirit of the place, the *genius loci*. In Conzen's view, it is an important environmental experience for the individual, even when it is received unconsciously. It enables individuals and groups to take root in an area. They acquire a sense of the historical dimension of human experience. This stimulates comparison and encourages a less time-bound and more integrated approach to contemporary problems. Landscapes with a high degree of expressiveness of past societies exert a particularly strong educative and regenerative influence. The Conzenian townscape is a stage on which successive societies work out their lives, each society learning from, and working to some extent within, the framework provided by the experiments of its predecessors.

Viewed in this way, townscapes represent accumulated experience, historical townscapes especially so, and are thus a precious asset. This asset, according to Conzen, is threefold. First, it has practical utility at the most basic level in providing orientation: our mental map and therefore the efficiency with which we function spatially is dependent upon our recognition of the identity of localities. Secondly, it has intellectual value by helping both individual and society to orientate in time: through its high density of forms a well-established townscape provides a particularly strong visual experience of the history of an area, helping the individual to place him or herself within a wider evolving society, stimulating historical comparison and thus providing a more informed basis for reasoning. Thirdly, and more contentiously, the combination of forms created by the piecemeal adaptation, modification and replacement of elements in old-established townscapes has aesthetic value: for example, in the maintenance of human scale, in the visual impact of, and orientation provided by, dominant features in the townscape, such as churches and castles, and in the stimulus to the imagination and the visual surprises provided by variations in street width and direction. Clearly, all three assets are inter-related with, though not necessarily dependent upon, appreciation of historical and geographical significance. They have been considered individually by other scholars, including Lowenthal (1975) and P.F. Smith (1979), but rarely have the connections between them been made explicit.

It is on the historical expressiveness or historicity of the townscape that Conzen places most emphasis. It is accordingly the nature and intensity of the historicity of the townscape that provides his major basis for devising proposals for future townscape management. The translation of the concept of townscape historicity into practice still awaits completion. Some of the main facets of Conzen's thinking can, however, be understood as an extension of his division of the townscape into three basic form complexes: town plan, building form and land-use. These are regarded as to some extent a hierarchy in which the building forms are contained within the plots or land-use units, which are in turn set in the framework of the town plan. These three 'form complexes', together with the site, combine at the most local level to produce the smallest, morphologically homogenous areas that might be termed 'townscape cells'. These townscape cells are grouped into minor townscape units, which in turn combine at different levels of integration to form a hierarchy of intra-urban regions. Clearly the patterns of the three form complexes frequently differ. The town plan is particularly resistant to change and building forms frequently endure longer than the use for which they were constructed. Particularly in old-established towns, surviving lineaments of a mediaeval plan often provide a framework within which modern development takes place (cf. Larkham 1995b).

In his research in the English market town of Ludlow, Conzen (1975) maps a hierarchy of spatial units for town plan, building forms and land-use individually and for the townscape as a whole, viewed as the composite of these three form complexes. The hierarchy of townscape units is the geographical manifestation of the historical development of the townscape and is a basic component in Conzen's ideas for linking townscape historicity to proposals for townscape management.

No other study comparable to that of Ludlow has so far been attempted. Furthermore, no attempt has been made to develop further Conzen's ideas on townscape management, let alone put them into practice. Even the basic building-blocks of his analysis, the streets, plots and buildings, are rarely dealt with at a micro-scale by LPAs developing policy for conservation areas (Chapter 6); a finding reinforced by Fennell (himself a senior planner and a former student influenced by Conzen's teaching) who noted, in a study of Chichester, that

> perhaps the most disturbing conclusion that emerges from this case study is the failure to appreciate the significance of plan form. Little or no attention has been given to the contribution that street pattern, plot shape and building plan inevitably make to the character of the city.
>
> (Fennell 1982: 7)

The Conzenian morphological approach as discussed here does not provide a 'theory of conservation' *per se*. Yet it is conceptually useful in three ways. First, it emphasises the need for some continuity in built form, pressing planners and developers to understand the significance of *genius loci* to all urban users. Secondly, it specifically addresses the concept of 'historicity', although the manner in which this is discussed in the example of Ludlow is too labour-intensive and complex for everyday planning use. Thirdly, it provides an analytical framework – the division of the urban landscape into the hierarchy of streets, plots and buildings – which would be extremely useful in measuring the cumulative processes of change over lengthy periods to inform the decision- and policy-making processes, and to demonstrate precisely how 'character and appearance' are changing.

Conzen himself regards his ideas on management as no more than indicative of a direction in which research and practice might proceed. The divorce between this conception and the day-to-day perceptions and activities of the agents of change discussed earlier would seem to be almost total. Yet the prevailing mood in planning, and in particular the high status now being accorded to conservation, is compatible with Conzen's perspective. Though the agents of change considered earlier are concerned with specific cases, not with theoretical bases, actual townscape changes in the past two decades have been in much greater sympathy with existing townscapes than were developments of the 1950s and 1960s. It would be unrealistic, however, to expect Conzen's theory to be widely understood, let alone adopted in planning practice. Certainly the few planners and urban designers familiar with this body of work have mixed reactions: some lament its thoroughness, with the consequent resource implications, and question its usefulness to design beyond historic urban landscapes (Samuels 1985, 1990; Bandini 1986, 1992). The most that can be hoped for is that the wider promulgation of his ideas will help to promote greater appreciation of the historico-geographical context within which each individual proposal for townscape change should be set.

At this simple level, Conzen's ideas have much in common with ideas of apparently different origin that have gained acceptance among certain groups of architects, especially in Italy and France. Vernez Moudon (1994), surveying these three national approaches, concludes that

> The three schools ... offer an intellectually challenging framework for thinking about the built landscape within the historical context of the city. Italy's provides a theoretical foundation for planning and design within age-old traditions of city building. England's offers a scholarly approach to researching how the built landscape is produced. And France's outlines a new discipline that combines the study of the built landscape with a critical assessment of design theory. Together these schools suggest an order for a formidable agenda of research, planning, and design that takes into account the relationships between space, time, habitat and culture.
>
> (Vernez Moudon 1994)

Because they are more directly grounded in contact with the European tradition of 'urbanism' – a fusion of architecture and planning determining interventions in the built environment – these perspectives are much more likely to have widespread influence in European practice. The direct impact upon UK practice of such approaches requires further investigation beyond the few studies currently available (Kropf 1993; Vernez Moudon 1994). This is a significant direction for future research.

MANAGEMENT AND MISMANAGEMENT?

Strong (1990: 222) suggested that, in general, Britain has a 'chaotic system of piecemeal planning control that has failed to sort out with any clarity what should and should not be conserved'. From the detailed, albeit small-scale, studies reported and cited here, it is clear that, in practice, the operation of the UK's bureaucratic development control system is short-term and short-sighted. Yet it is this system – LPAs acting in their role as indirect agents, reacting to the development proposals of other parties – which far outweighs their direct role in enhancement in terms of what actually occurs within conserved urban landscapes.

It is clear from reviewing hundreds of planning files in different types of conserved areas during the course of this research that there is little that could be described as 'management' of the urban landscape, in the sense envisaged by Conzen. Nor, indeed, are there many developments that could justifiably be described as 'planned' outcomes (Whitehand 1990b), since the variation between original schemes and final development is frequently considerable. Even within designated conservation areas, the effects of Local Plan documents have so far been quite small at the scale of the individual site (Whitehand and Larkham 1991b). Yet these individual sites are the building-blocks of whole landscapes: the large-scale landscape change documented here suggests that policies are failing – where they even exist – because they are insufficiently site-specific.

Indeed, if LPAs are judged by their actions in the residential areas, then highway matters and development densities are far higher priorities than the visual environment. The forms of developments, particularly their architectural styles, have tended to be afterthoughts, when battles over density and access have been concluded. Attempts by applicants to meet or overcome density and highway objections have often been at a cost to the urban landscape in terms of both the relationship between buildings and the survival of the mature trees and hedgerows so important in many residential conservation areas (cf. Whitehand 1989a).

The rather abstract concept of townscape management is clearly replaced by that of profit on the part of developers, who often attempt to obtain different, more profitable, planning permissions once a site has been purchased (Whitehand and Larkham 1991b: 152). On the part of the initiators, little overt thought is given to management, as development is often initiated by a change in the family life-cycle and consequent re-valuing of use values (Whitehand 1990b). Lastly, LPAs clearly value administrative convenience above abstract management, as their reliance on easily quantified density, parking and highway standards shows. Indeed, the dominant impression of the decision-making process leading to development (or non-development) is that it consists of a number of poorly co-ordinated activities in which expediency plays a major part (Whitehand 1990a: 389).

Therefore, although the UK's well-developed system of development control and local planning might be expected to contribute to urban landscape management, the actual landscapes and the records of the bureaucratic system strongly suggest that this is not the case in practice. The detailed morphological research reported in this book and in other recent studies clearly suggests that local development control planning, and the district plan-making process which is having an increasing influence on development control, should place far more weight upon the morphological building-blocks of the urban landscape: street, plot and building. If the urban landscape is accepted as an important factor in physical planning, then the planning system could be much more pro-active in seeking to develop a shared vision of the ways in which the landscape could change, and seek to guide such change through the development of detailed site-specific policies and even design briefs. Many authorities are beginning to do so, particularly where conservation is a key issue, but such tactics could be used much more widely in planning practice.

Although the system is not equipped to consider urban landscape management nor large-scale urban design considerations, it does concentrate to a considerable extent upon small-scale design issues. In some cases, even in these sensitive landscapes, design plays a minor role to easily quantifiable factors such as planning and highways standards, operated seemingly regardless of the outcome in the landscape (Whitehand, Larkham and Jones 1992). Many development proposals are amended by the system to mitigate the worst design proposals, but this tends to result in mediocrity rather than in the encouragement of good design. This may also considerably prolong the development process.

The extensive nature and continual piecemeal operation of the process of change, particularly within historic urban cores and mature residential areas officially protected by conservation area status, is given little heed. Area character is rarely addressed thoroughly, in either historical or morphological terms, either before designation or in any regular review process. The amount of relatively small-scale change is rarely monitored. Without this appreciation of changing character, policy development remains weak and, often, insufficiently robust to withstand challenge at appeal by a determined developer. Only in very recent years have there been any significant attempts to study area history, development, form, character and appearance on a systematic basis, and to use these data in the formulation of robust management policies ranging from area-wide to site-specific. Despite the well-developed bureaucratic development control system, and the conservation area system designed to protect the large number of designated areas, urban landscape management in England over the past two and a half decades can largely be characterised as mismanagement.

CONCLUSIONS

The conservation area as a planning tool

This book's message is that the UK concept of the conservation area, as applied in practice through the quasi-judicial planning process, is a very blunt tool of planning policy. There is much that could be learned from other experiences, for example the French operation of the *secteurs sauvegardés*; while countries adopting the UK model, including Malta and some developing countries, need to learn from the UK experience rather than adopt it unquestioningly.

The undoubted strength of the UK system is its local basis. Conservation areas can be designated very rapidly to counter likely threats; there is little constraint upon what can and cannot be designated, so actual designations can be very flexible in response to local conditions. And, importantly, conservation areas are generally well supported by the local public.

There are drawbacks in the varied justifications advanced for designation. It seems that some areas are preservation areas; some are demolition control areas; some are areas of potential, where designation might encourage confidence, grant aid and inward investment; some are designated to appease local politicians or public groups; and some are designated because local planning officers feel that they do show, as the Act specified, 'special' qualities of character or appearance. This variety, which produces many different types of areas with effectively the same type of protective policy, is where the conservation area concept is blunted and open to both criticism and abuse.

Improvements have been suggested to the designation process to allow for greater public consultation while still affording immediate protection if necessary (Jones and Larkham 1993). The designation should be included in the Local Plan to ensure that appropriate policies – for protection and enhancement – are devised as soon as possible, and given a further opportunity for public comment. The varied aims for each type of conservation area could be made more explicit in policy development. The present, rather vague and general, concept of a protected area could thus be focused much more sharply on the requirements of the individual district.

Conservation and change

> It has been said that a city without old buildings is like a man without a memory. But environment is not just a question of old buildings. Cities must be made environmentally attractive. The essential qualities of all towns require sensitive conservation of the building fabric and relationships which give positive character.
>
> (Rose 1974: 135)

However, a balance needs to be struck between retention and change. Creation of new valued landscapes – urban and rural – has, in the past, been allowed by the destruction of older, also valued, landscapes (Strong 1990: 221). However, how balanced are actions in the UK?

The identification and designation of areas and buildings for conservation purposes does, to some degree, isolate them from the expected cycles of obsolescence and renewal. Conservation presupposes retention in the majority of cases. Yet not only is there an element of choice in this, for not every structure is retained even in a conserved area, but UK conservation practice and policy accept some degree of change and alteration. The case study areas have shown the nature and scale of that change. Although major changes – such as new buildings – are relatively infrequent, more minor changes are extremely frequent. And there are those changes, which in the UK are 'permitted development', for which no record appears in the planning system. All of these changes do fluctuate with the cyclic economic position. The process of change is, therefore, continuous, insidious and – because it is not comprehensively monitored – it is likely to alter the character and appearance of preserved areas over several decades.

It is clear that the planning system exerts much effort in the control of minor changes, and this is often where localised conflict occurs. As such changes, to doors, windows, external decoration, and especially commercial signs, shopfronts, etc., occur relatively often, perhaps a lesson could be learned from the philosophy of art restoration. Where these changes are relatively minor and reversible, and rapid and thus of relatively short duration, perhaps they could be lived with more readily. The scarce resources of planning could be used fully to monitor and understand change, and to fight the larger battles more effectively. With greater public education and guidance, some of the more minor elements in the conserved built environment could be afforded some greater freedom for personalisation.

Conservation as conflict and mismanagement

Bringing together the diverse material contained in this book, returning to the key questions posed in Chapter 1 and seeking to draw conclusions, inevitably suggests the unpopular basic finding that conservation as a planning activity is characterised by confusion, conflict and mismanagement. This is not to belittle the efforts of individual highly qualified, experienced and motivated conservation officers, who may have a considerable influence upon local areas (as was the case in Henley-on-Thames), but it is rather a comment on the UK system as a whole.

In terms of the key 'what' questions, there is conflict between the centralised State-run listing system, with its rigid guidelines and expert assessors, and the locally-based conservation area system. Even listing does not ensure preservation, as later planning officers, and even DoE planning inspectors, may agree that a derelict but listed structure may go. There are, reportedly, conflicts between the expert views of those who list and designate, and the public at large; not to mention other interested parties such as developers and landowners. Conservation area designation, in particular, can be shown to be driven by a variety of pressures. Nevertheless, conservation in principle has widespread public and professional support, perhaps more so at present than at any time in the past.

There is little consistency of view, within LPAs, between LPA and developers, between either LPA or developer and the public, over attitudes to areas and development proposals. Basic principles such as development density are challenged; such challenges do, however, seem to be rooted in self-interest and a desire to retain existing property values. It is far more difficult to rationalise the degree of dissent over aesthetic issues, some of which may be relatively minor.

Local studies can readily catalogue vernacular details, materials, building scales, masses and proportions, which may be used to inform (perhaps rather than 'control') new development. Yet there are still debates over window sizes, pseudo-Palladian light fittings and so on. It is clear that various powerful interests are merely attempting to impose their own value-sets on the built environment. The economic argument remains powerful, and has been particularly supported by the Conservative government during the 1980s.

In the UK there are evident problems in the quasi-judicial nature of the planning process. The intervention of the courts leads to judicial arguments, detailed explorations of the wording and definitions of statute and guidance, sometimes to the detriment of the conserved built environment itself. Judgements can also conflict. One of the principal problems is the issue of precedent. Although each planning application should be decided upon its own merits, court judgements set legal precedents over how applications should be treated, or how they should be seen to be treated.

Some of the confusion perhaps arises through the admitted lack of a conservation ethic or philosophy to guide all parties in a development. Other fields such as art and museum curatorship, where ethics might be expected to be highly developed, also show disarray. Nevertheless, there is merit in the art world's approach to fakery, where the intent to deceive is important. (The legal view, exemplified in the Schiele case, is again less helpful!) The parallels of attitudes towards historic machinery may also be more useful, and could be developed further. Le Corbusier's aphorism of a house as a machine for living may take on new significance if, for conservation purposes, we regard buildings more as machines than works of art. Buildings and areas, like cars, are used, wear out and require replacement parts. In some cases, a skin-deep conservation is acceptable. The built environment parallel is surely façadism. In others, absolute originality is important. But the parallel here is not with Grade I listed buildings, or with the former outstanding conservation area designation.

As our conservation areas change, as they inevitably do, we need to develop a more sensitive appreciation of the dynamics and processes of change, a better philosophical approach to conservation, in which parallels from other fields can help, and a more flexible and locally-based approach to planning.

APPENDICES

APPENDIX 1

REPORT OF NORWICH CHIEF PLANNING OFFICER TO PLANNING COMMITTEE, FEBRUARY 1992, ON PROPOSED EXTENSION TO THE CENTRAL CONSERVATION AREA[1]

BACKGROUND

1 Planning Committee on 21 December 1989 resolved to approve a report on the review of Outer Conservation Areas. Contained within this report was reference to reviewing the Central Conservation Areas and reporting back at a later date.

2 Policy B4 of the City's Local Plan states that the existing City Centre Conservation Areas will be extended.

3 The proposed City Centre Conservation Area, the subject of this report, is generally defined as being the historic area within the City's medieval defensive boundary. This follows the line of the City Walls to the north, south and west and the River Wensum to the east and north-west. In terms of the Local Plan the boundary generally follows this historic boundary to the north, west and south. [In] the east, however, it goes beyond the river to include Thorpe Hamlet wooded ridge, Rosary Road, Riverside and the railway station. There are also other deviations to the north and south to include Jarrolds printing works site (currently the subject of a planning brief) and the former Victoria Station and adjacent land respectively.

4 Approximately three-quarters of this area is already covered by Conservation Areas which are designated as:

City Centre No 1
Magdalen Street/Colegate No 2
St Augustines No 3
King Street No 4
All Saints Green No 5
Bracondale No 6

1. Reproduced with the permission of Norwich City Council

HISTORICAL IMPORTANCE OF THE CITY CENTRE

5 The existing Central Conservation Areas were identified in the Council's 1969 policy document 'Conservation in Norwich'. The boundaries were drawn to define specific areas of the historic centre where there are numbers of surviving listed and historic buildings and other historic features. Other areas within the city walls, although part of the historic core, were excluded because they had largely been redeveloped with modern buildings or later industrial development of no architectural or historic merit.

6 The 1969 policy document was produced as a consequence of the Civic Amenities Act 1967 which empowered, and indeed required, local authorities to designate areas of special architectural and historic interest as 'Conservation Areas'. The report also listed criteria defined by the Council for British Archaeology in assessing the significance of an ancient settlement.

- Town Plan well preserved – Ancient, Georgian or Victorian
- Bridge crossing and approaches – Ancient or Georgian
- Waterfronts
- Town walls, ditch or gates well preserved
- Castle site well preserved
- Major ecclesiastical site or precinct well preserved
- Number of buildings worthy of preservation, particularly dating from medieval, Georgian and Regency or Victorian periods

The City Centre of Norwich, within the ancient defensive boundaries of the walls and the river, fulfils all these criteria as well as being the largest walled city in the country and also relatively unaffected by the Industrial Revolution. The report concluded that protection of these features would be fundamental to the conservation of this important city centre and that therefore much of the central area should be included within designated conservation areas.

7 In considering the historic development and character of the walled city centre it is important to understand that although the area enclosed is large, it was not fully developed with buildings. Until the late 19th or even early 20th Century, large tracts of land remained as open orchards, gardens or grazing land. For example, the Oak Street area and Mountergate were sporadically developed with much later industrial development, although containing isolated listed buildings and other historic structures. By their age and nature many of these later buildings are now obsolescent and subject to redevelopment.

8 In addition, other areas of the centre have been redeveloped both as a consequence of economic activity in the city as a regional centre, and as a result of bombing in the war, particularly in St Stephens Street, All Saints Green and Rampant Horse Street.

PROPOSED NEW BOUNDARIES

9 The policy agreed in 1969 rightly stressed the importance of redevelopment in enhancing the central conservation areas and guidelines set out the design criteria to be applied.

- size
- scale
- proposition
- character
- materials
- skyline
- the design and detail of the space between buildings

The new City of Norwich Plan re-emphasises the same issues.

10 In reviewing the boundaries of the Central Conservation Areas in the light of the criteria set out above, it is now suggested that the approach be to consider the ancient walled city as the basis of a new Central Conservation Area. This would emphasise the importance *as a whole* of:

- the medieval street plan
- the City's ancient defences
- bridge crossings
- waterfronts and quays
- surviving buildings and other features

This would also recognise the importance of views *into* the historic centre from outside, for example, towards Oak Street and Mountergate.

11 This proposed boundary would relate also the boundary line of the archaeologically sensitive area, as defined by the Norfolk Archaeological Unit. However on advice from the Unit, it is not considered to be appropriate to include the area to the north and north-west, beyond the City Walls within this proposal. This is indicated on the Plan in Appendix II [not included].

12 The new enlarged City Centre Conservation Area would allow:

- emphasis to be placed on achieving sympathetic redevelopment in areas such as Mountergate and Oak Street *in the context of the historic centre*
- road improvement to be considered in the context of the historic street pattern
- surviving isolated listed and historic buildings and scheduled Ancient Monuments to be seen in their context as part of the centre
- existing unlisted historic structures to be protected from demolition without consent
- trees to be protected, particularly important in the context of the wooded ridge to the south
- development of a strong policy for redevelopment on the riverside sites related to the historic development of the centre

- appreciation and enhancement of the river valley as an important landscape feature

In conclusion, extending the existing Conservation Areas will result in a more understandable boundary which will enable a more consistent approach in dealing with planning applications.

CONSERVATION AREA EXTENSIONS

13 The areas under consideration for Conservation Area status are as follows and shown on the attached and displayed Plans [not included].

A	North-Western Area	St Martins/Anglia
B	North-Eastern Area	Pockthorpe
C	Southern Area	Rouen
D	South-Western Area	St Stephens

The existing character, historic features and key issues are outlined in Appendix 3 [not included].

14 The existing central Conservation Areas are sub-divided into clearly defined areas. If you agree to the principle of extending the areas to cover the whole of the City within the medieval defences and specific areas beyond, I propose as part of the detail to re-assess these sub-divisions and outlining key policy statements. This would be included within the 1992/93 work programme, together with the procedures leading to designation.

CONSERVATION AREAS, THE LOCAL PLAN AND CIRCULAR 8/87

15 Within the Local Plan for Norwich are policies specifically for aspects such as employment, housing, traffic, built environment and natural environment which aim to enhance the character and appearance of the city centre. Extending the existing Conservation Areas to cover the whole of the historic core of the City and certain areas beyond will be in line with and give further weight to these policies and also restate the City Council's attitude towards the preservation and enhancement of the City Centre whilst at the same time allowing for development and change. Careful consideration will need to be given to future planning policy in the Oak Street area and a detailed planning study of this area is in preparation.

16 As stated in Circular 8/87 'Historic Buildings and Conservation Areas – Policy and Procedures', Conservation Area designation is not a matter of 'preserving in aspic', the emphasis is to control rather than prevent development to allow the area to remain alive and prosperous, but at the same time ensure that any new development accords with its special architectural and visual qualities.

17 Encouragement will be given to promoting good quality modern design which will complement the existing character of the City Centre, be compatible with Local Plan policies, take account of existing built and natural features and respect the historic context of its setting. Circular 8/87 expands on this to state that in determining applications in Conservation Areas account should be taken of any formally adopted local policies for the area and special regard should be taken to matters such as bulk, height, materials, colour, vertical or horizontal emphasis and design.

ENVIRONMENTAL IMPLICATIONS

18 The designation of the whole of the historic medieval City Centre of Norwich would ensure additional environmental protection and enhancement of the City's historic fabric, both built and natural.

CONCLUSIONS

19 The boundaries of the existing central Conservation Areas have been drawn previously to focus mainly on the existing Listed Buildings and other historic features and structures within the historic core. Other areas within the City walls which form an essential part of this 'core' were not included. In re-examining the central conservation areas now we suggest that the City Centre be considered as a whole in terms of its historic growth and development. This area is identified as being within the archaeological sensitive area defined by the Norfolk Archaeological Unit. Extending the existing central Conservation Areas would ensure more comprehensive protection of the historic City centre whilst allowing for appropriate development and change consistent with Local Plan policies.

EXAMPLE OF SUB-AREA CHARACTER ASSESSMENT

A North-Western Area – St Martins/Anglia

Existing Character: From Barn Road/River Wensum in the west up to and including Anglia Square in the east to include the retail, commercial and service trades located along Oak Street and Pitt Street, the retail, office and commercial uses within and adjacent to Anglia Square, and the 19th Century terraced housing areas of Esdelle Street, Leonard Street and Magpie Road.

Historic Features: Oak Street and Pitt Street are historic routes into the centre. St Martins Gate, one of the main historic gateways into the City from the north, was located at Oak Street/St Martins Road. The river formed an important boundary to the west alongside which industries located originally making use of the river for transport. Warehouses, factories and other associated works became established within this general vicinity. The area also contains 10 statutory Listed Buildings and the course of the City walls.

Key Issues

- Major riverside redevelopment along Oak Street.
- Major redevelopment in and around Anglia Square and between two existing Conservation Areas.
- Impact of redevelopment on the historic core.

APPENDIX 2

COMMENTS ON THE CROWNGATE SHOPPING CENTRE SCHEME FROM WORCESTER CITY DEPARTMENT OF TECHNICAL SERVICES

Since the bulk of this research was completed, a number of significant changes have occurred to the Blackfriars Square and Huntingdon Arcades schemes proposed by Centrovincial Estates PLC. Centrovincial was taken over by Ford Sellar-Morris, who attracted the Crown Estate into the scheme. The development has been changed in detail, and is now known as Crowngate. The following comments were supplied in correspondence with the author by Worcester City Department of Technical Services.

BACKGROUND

Negotiations on the central area shopping scheme commenced in 1979 and throughout the process from concept drawings to the final details, the City Council had the primary objective of minimising the disturbance to the existing historic fabric and, where possible, to enhance the character and appearance of the central conservation area.

The scheme is in two separate parts. The Friary Walk side is simply a refurbishment and partial rebuilding of the late 1960s development of Blackfriars Square. No listed buildings have been directly affected and it could be said that the setting of others has been enhanced; it has been a catalyst in this respect for Broad Street and Angel Place. The Chapel Walk area (previously referred to as Huntingdon Arcades) is a new build scheme on a backland area to the west of the High Street. This has been achieved at a cost to listed buildings. The entrance to the High Street at Bull Entry involved the gutting of the ground floor of 85 High Street, a listed building, and the entrance to Broad Street also affected the ground floor of Unicorn Chambers, 55 Broad Street and 56 Broad Street (previously the Long Stop public house). No listed buildings have been demolished and, again, a number have been refurbished or are likely to be refurbished as a consequence of the development.

The City Council consider that the future of the existing historic buildings have been taken fully into account in minimising impact, retaining servicing, and encouraging repair and refurbishment by providing an economic stimulus. The development package has also allowed further traffic management measures to be introduced and has paid for the repaving of Broad Street, Angel Place, Bank Street and the remaining section of the High Street. These measures enhance both the setting of the buildings and their economic viability.

During the final design and construction stages of the development, a number of changes have occurred which have had an adverse impact on the listed buildings of 55–56 Broad Street. Owing to fire regulations, the mall width has been increased to 6m, which has compromised the original objectives of retaining the original coaching inn entrance. Cellars were also revealed, which required a damage limitation exercise. The momentum and scale of the construction project has not been sensitive to the detailed repair and refurbishment of the individual historic buildings, but to put this in perspective the buildings are being repaired and works are still in progress, and it is therefore unreasonable to judge the finer details at this stage.

The Countess of Huntingdon's Hall, a Grade II* listed building, is at the centre of the scheme. This has been used as a focal point to this section of the development, and the new buildings and spaces have been designed to respect and enhance this building. Once proposed for demolition, it has since been repaired and restored and recently acquired by the Crown; thus securing its future as a concert hall and music school.

THE DEVELOPERS

Centrovincial Estates was taken over by a company known as Ford Sellar-Morris (FSM). It reviewed its investment in Worcester and decided to invite a Fund to become involved in the scheme. It attracted the interest of the Crown Estate. At this stage the situation changed, with FSM taking the role of the developer, and the Crown Estate becoming the owner, or owner upon completion, of the development. In 1991, FSM had financial difficulties, and the Crown Estate took over the completion of the scheme and its ownership. FSM are no longer involved in the scheme. It is difficult to know whether or not Centrovincial would have retained a long-term financial interest in the scheme if it were not taken over or if the Crown Estate had not become involved.

CHANGES TO THE SCHEME

It is difficult to identify the changes to the scheme that have been a direct result of the change in developer. The name is obvious, together with changes to the signs, logos and image. There were a whole series of minor design changes to the scheme at the time of negotiations with the Crown Estate, but many of these resulted from the Fire Officer's requirements, comments by structural engineers and tenants, and other design refinements.

Two aspects of the scheme that can be attributed to the Crown Estate are first the quality of the materials, and second the concern with peripheral areas of the scheme. In terms of materials, the original scheme negotiated with the City Council did include high-quality finishes, including the natural materials of brick, oak, clay tiles, and slate within the conservation area. In the final months of FSM's involvement, both materials and design features were threatened. For example, attempts were made to substitute concrete tiles on one of the major shop units, to delete roof details on the car-park and to reduce the enclosed area of the bus station concourse. The Crown Estate has safeguarded the quality of the scheme. It also appears to be placing an emphasis on the surroundings and entrances to the development. An example of this is the recent expansion of the scheme into 86/87 High Street, with the resulting widening of the mall entrance on to High Street.

APPENDIX 3

APPROACHES TO MANAGING HISTORICAL URBAN LANDSCAPES

INTRODUCTION

This appendix presents two significant documents that describe approaches to the management of historical urban landscapes. The first document is a general charter, applicable worldwide, compiled for the General Assembly of the International Commission on Monuments and Sites by the Eger International Committee on Historic Towns, and adopted by ICOMOS in 1987. This document is reproduced with the consent of ICOMOS UK. The second document is by English Heritage, otherwise known as the Royal Commission on Historic Monuments (England). It relates primarily to historic towns in England, and the particular problem of retail development. It was published in 1988 (English Heritage 1988b), a copy being sent to every English local planning authority, and is reproduced here with the permission of English Heritage. Both documents are reproduced as published, without corrections or interpolations.

DOCUMENT 1

INTERNATIONAL CHARTER FOR THE PROTECTION OF HISTORIC TOWNS (INTERNATIONAL COMMISSION ON MONUMENTS AND SITES)

I. Definitions

1. 'Historic towns' and 'historic districts' will be defined as all groups of buildings and space that comprise human settlements and whose unity and integration into the landscape endows them with historic, artistic, architectural, urbanistic or scientific value. Such values exist irrespective of the period and the culture that gave birth to them and do not depend on the manner of their construction, which may have been planned or spontaneous. A historic town may comprise one or more historic districts. As living entities, and subject to cultural, economic and social evolution, historic towns and districts must inevitably change, as they have done in the past.
2. 'Protection' will be defined as all actions necessary for preserving a

historic town or district and promoting its harmonious evolution. This action includes identification, conservation, restoration, rehabilitation, maintenance and revitalization.

II. General principles and objectives

3. The principles set forth in the Venice Charter apply to historic towns and districts as long as it is understood that the priority objective of protection is rehabilitation.
4. The protection of historic towns and districts should be part of a coherent policy of economic and social development and of town planning as well. Protection must recognize the diversities in settings, cultures and economic development in the urban area, the region and the country concerned.
5. The protection of historic towns and districts must satisfy the needs and aspirations of residents. It must not only meet the demands of contemporary life, but also assure the preservation of cultural and architectural values.
6. The success of a protection plan depends on the participation of the residents, which must begin as soon as preliminary studies are undertaken and continue throughout the protection process.
7. Districts must not be cut off from one another, and a historic district must be linked to other districts by visible integration and through a clear definition of its role.
8. Wherever possible, local life styles should be preserved and encouraged. New uses of space and new activities should be compatible with those already existing, and the creation of 'museum' towns and districts destined only for tourists must be avoided. The rights and aspirations of the population must be respected, as its social and economic activities often depend on the organization of the setting.
9. In a historic town or district the following physical features are to be preserved:
   a) form and shape, including the distribution of buildings, their height, mass and overall appearance, as well as the general character of the streets and squares and their layout, the rhythm of space and the distribution of land parcels;
   b) urban fabric, meaning the connections between different districts and the road network;
   c) the overall aspect of the town depending on the angle from which it is viewed, the relationship of masses, perspectives from within the town including breakaway views;
   d) the harmonious relationship between the buildings and the natural setting which, together, form a single landscape;
   e) the specific quality of the historic town or district as reflected in the contributions made by various cultures to the architectural heritage of the ensemble;

f) works of symbolic value such as town halls, towers, archaeological monuments, etc.;
g) fortifications (walls, ramparts, towers, bastions, gates, etc.);
h) connections with historic monuments that are located outside a historic town or district;
i) the other elements that lend the townscape its specific character, such as construction materials, colours, roofs and inner courts and all decorative elements (statues, grilles, the pavement, street furniture, etc.);
j) parks, gardens and open spaces, and bodies of water or streams;
k) traditional institutions and centres that contribute to cultural and social life, including universities, places of worship, markets, shopping districts, public promenades;
l) traditional crafts and business activities that are basic to the community's cultural identity and daily life.

III. Actions and methods

10. Planning for the protection of historic towns and districts must be a multidisciplinary effort involving a wide range of professionals and specialists, including archaeologists, art historians, architects, town planners, restorers, photogrammeters, civil, structural, traffic and soil mechanics engineers, jurists, sociologists, economists, etc.
11. When a historic town or district is located within a larger urban or metropolitan area, the plan for the historic area should be integrated into town, metropolitan and regional plans by determining the principal function and role of the heritage and values to be protected.
12. The protection plan must clearly set forth the principal line of action, but it must be flexible enough to allow for changes in life styles and provision should be made for periodic review of the plan.
13. Rehabilitation must aim at improving housing, sanitation systems and necessary public utilities. It should aim at increasing employment opportunities and promoting new economic activities, as well as encouraging those traditional activities compatible with the role and function of the area and its values as determined beforehand.
14. Wherever possible, demolition in historic towns or districts must be avoided. If new buildings are necessary their architecture should be harmonious with the historic town or district's existing scale, character, buildings and construction materials. They must also be compatible with the town or quarter as originally conceived and they should add to the enhancement of the area. Concern for harmony must determine the choice of supports, cables, antennas and signs, as well as street furniture and pavement.
15. If integration of new districts within historic towns or districts is difficult to achieve, and if there is a risk that new buildings are not compatible, consideration should be given to the creation of transition

zones, possibly composed of green belts.

16. The road network must be located outside the area while providing access to it. New roads must be compatible with the townscape. Solutions must be imagined to resolve the contradictory demands of traffic and the values to be preserved. Pedestrian zones and public transport must be favoured and parking facilities should be planned outside the district or even the agglomeration.
17. Historic towns and districts must be protected against the pollution, noise, shocks and vibrations caused especially by traffic. Preventative measures must be taken to protect the ensembles against the consequences of natural disasters such as earthquakes and floods.
18. Legislative or administrative measures must be enacted in order to provide the protection plan with the legal force necessary for rapid and efficient action. An administrative mechanism must be set up to assure financial support for the protection plan.

IV. Social participation

19. The rehabilitation of a historic town or district must satisfy present needs and aspirations and meet those of the future, especially social demands. Social and economic measures must be taken to encourage the residents to remain in the town or district concerned.
20. The residents must be informed and their interest in the protection process awakened so that their participation will stimulate the efforts of public authorities. Technical and financial assistance must be available to encourage action on the part of the residents and reduce the inconveniences of the protection process.

DOCUMENT 2
SHOPPING IN HISTORIC TOWNS: A POLICY STATEMENT
(ENGLISH HERITAGE (1988b))

Shopping has become a powerful influence on the character of urban redevelopment. The high degree of investment concentrated in the retail sector has produced a sudden increase in the size and number of new shopping centres. The historic areas of English towns and cities retain their appeal to shoppers, and the pressure to accommodate the new giant units in the centres of those towns is intense. Towns have come to compete with each other to attract the 'major multiples'. Many have sacrificed planning and environmental controls for fear of losing out to rival centres, either out-of-town or in adjacent towns.

English Heritage feels that these developments are posing a threat to the visual and architectural integrity of English towns, as great as that posed by the comprehensive housing and office developments of the 1960s. The threat is particularly severe in the smaller market towns which largely escaped earlier waves of redevelopment, many of them still predominantly of a Georgian or early Victorian character.

We accept that shopping is the essence of a market town. It attracts people and prosperity, and the town grows and changes in response to new demands. Many towns are now worried at the growth of out-of-town shopping, based on intensive car use and motorway access, and feel that they need to be able to offer 'the multiples' central sites on which to expand. They need also to offer their customers a place in which to park their cars.

Against this must be set the public's concern to maintain the economic variety of historic cities and towns, as well as their visual and architectural character, in the longer term. If, however, that character is sacrificed to large-scale development undertaken in response to what could be a short-term upswing in the retail investment cycle, it cannot be recovered. Once lost, historic character is irretrievable, and the town itself loses the ability to adapt in the future to a new cycle of demand, which might well include small-scale, tourist-based specialist shopping, or town-centre housing. Many historic town centres could be left as wastelands of disused retail warehouses. Many American towns are bitterly regretting the unplanned destruction of their historic centres, now that inner city investment is seeking character and variety in the urban landscape.

Our task is thus to accept the need for historic towns to attract retail investment and provide for new retail development, while seeking to ensure that the process does not do irreversible damage to their architectural and visual integrity. This statement is an attempt to lay down general principles to that end.

English Heritage takes the view that very large-scale retail developments are inappropriate to the town centres, at least, of smaller market towns. Not only are they vast in themselves, usually requiring the demolition of whole neighbourhoods and the elimination of variegated economic activity, but they can blight much of the rest of the town with their traffic and by drawing custom from other shops. It is virtually impossible to accommodate modern large-scale development in historic streets without losing their essential scale and character. Since evidence of overcapacity of this sort is starting to emerge, planning authorities have a duty to look to the longer term in assessing applications whose apparent purpose is short term speculative gain. They should look most carefully at the wider and longer-term social costs of such investment.

A large number of retail stores are now coming forward with planning approval. Their scale should be kept to a minimum and their siting and access be made as respectful as possible of existing buildings and streetscape. We would draw attention to the following considerations:

- New shopping areas should be designed to respect existing listed buildings and conservation areas, including those in towns and cities which have suffered as the result of previous phases of urban renewal. Such respect should extend beyond individual buildings and groups. The character of English historic towns derives as much from the continuity of plot sizes, the survival of back (or burgage) plots, the pattern of lanes and alleyways, and the general historic topography, which together make up the 'grain' of the town, as from the architectural styles of the buildings, the shop fronts, and the street furniture which provide the townscape. While it may not be appropriate to preserve all such features intact in every redevelopment, respect for the scale and variety they have produced is vital.

- The external form of the structures should seek to minimise the scale and bulk of their internal volumes, while being designed to be convertible to other uses, should market circumstances change. The external detailing and materials of the structures should respect the existing character of their surroundings.
- We recommend planning authorities to encourage competitive designs for public consultation. As a first step, the deliberation involved in such competition helps to avoid mistakes being made and provides architects and developers with ideas and options for improving their schemes. Financial bids should not form part of this initial process.
- Part of the character of a town lies not just in the façades of buildings to the streets, but in their integrity as historic structures. Buildings in conservation areas, therefore, should be preserved intact wherever possible. Whilst façade preservation is preferable to wholesale demolition (and reproduction to total oblivion), there should be a presumption against 'façadism' in conservation areas.
- Vehicular access has proved even more damaging to many smaller towns than retail development itself. The demand on the part of retailers for access, delivery bays, and turning circles for large European lorries and the need to supply the shops with car parking can double or treble the site area required by one shop alone. Since towns are, by their nature, generally congested places, such a destruction of townscape in the interest of extra vehicular access can swiftly become self-defeating (as many 1960s shopping centres found). It is crucial that vehicular access should be kept unobtrusive and that priority be given to pedestrian circulation and the quality of the public spaces.
- Schemes which allow the dual use of streets – for servicing or for pedestrians – separated by time should be considered. Conscious efforts should be made to improve the appearance, attraction, and use of historic buildings, streets, and areas with preference being given to people rather than to motor vehicles.

REFERENCES

Although this is a full list of references used in this book, it has also been designed as a conservation bibliography for use as a free-standing resource.

Abercrombie, N., Hill, S. and Turner, B. (1980) *The Dominant Ideology Thesis*, London: Allen and Unwin.

Aeroplane Monthly (1995) 'Vintage news', *Aeroplane Monthly* August: 2–9.

AG 1824 (1995) *Invitation to Young European Architects: the Reconstruction of a Historic Brussels Street 1989–1995*, Exhibition catalogue, Brussels: AG 1824.

Alberge, D. (1995) 'Christie's ordered to repay £700,000 for sale of forgery', *The Times* 12 January.

Aldous, T. (1975) *Goodbye, Britain?*, London: Sidgwick and Jackson.

Aldous, T. (1978) 'Defending vernacular', *Building* 15 December: 24–5.

Alfrey, J. and Putnam, T. (1991) *The Industrial Heritage: Managing Resources and Uses*, London: Routledge.

Ali, R.H. (1990) 'Urban conservation in Pakistan: a case study of the walled city of Lahore', in The Aga Khan Trust for Culture (ed.) *Architectural and Urban Conservation in the Islamic World*, Geneva: The Aga Khan Trust for Culture.

Alió, M.A. (1987) 'Els expedients d'obres particulars com a eina d'anàlisi del procés urbà. Vilafranca del Pendès, 1845–1945', in Morell, R. and Vilagrasa, J. (eds) *Les Ciutats Petites i Mitjanes a Catalunya: Evolució Recent i Problemàtica Actuel*, Barcelona: Institut Cartogràfic de Catalunya.

Allaby, M. (1971) *The Eco-Activists – Youth Fights for the Environment*, London: Knight.

Alliance Building Society (1978) *Buying a House for the First Time*, London: Alliance Building Society.

Amery, C. and Cruickshank, D. (1975) *The Rape of Britain*, London: Elek.

Ancient Monuments Society (1989) 'The Society's casework 1987–8: Dee House, Little St John Street, Chester', *Transactions of the Ancient Monuments Society* 33: 207.

Andrews, G.E. (ed.) (1980) *Tax Incentives for Historic Preservation*, Washington, DC: Preservation Press.

Archer, J.H.G. (1982) 'A civic achievement: the building of Manchester Town Hall', *Transactions of the Lancashire and Cheshire Antiquarian Society* 81: 3–41.

Architectural Design (1980a) 'St. Eugène and the Gothic debate', in *Architectural Design, Viollet-le-Duc*, London: Academy Editions.

Architectural Design (1980b) 'Le Château de Pierrefonds', in *Architectural Design, Viollet-le-Duc*, London: Academy Editions.

Arnison, C.J. (1980) 'The Speculative Development of Leamington Spa: 1800–1830', unpublished MPhil thesis, Leicester: Department of Economic and Social History, University of Leicester.

Ashworth, G.J. (1984) 'The management of change: conservation policy in Groningen, The Netherlands', *Cities* 1, 4: 605–16.

Ashworth, G.J. (1992) 'Heritage and tourism: an argument, two problems and three solutions', in Fleischer van Rooijen, C.A.M. (ed.) *Spatial Implications of Tourism*, Groningen: Geo Pers.

Ashworth, G.J. (1993) Review of Whitehand, J.W.R. and Larkham, P.J. (eds) (1992) *Urban Landscapes: International Perspectives* in *Tijdschrift voor Economische en Sociale Geografie* 84, 3: 235–6.

Ashworth, G.J. (1994) 'From history to heritage: from heritage to identity: in search of concepts

and models', in Ashworth, G.J. and Larkham, P.J. (eds) *Building a New Heritage: Tourism, Culture and Identity in the New Europe*, London: Routledge.
Ashworth, G.J. and Tunbridge, J.E. (1990) *The Tourist-Historic City*, London: Belhaven.
Ashworth, G.J. and Voogd, H. (1990) *Selling the City*, London: Belhaven.
Aslet, C. (1982) *The Last Country Houses*, New Haven: Yale University Press.
Baker, N.J. (1988) *The Archaeology of the Charles Darwin Centre*, Birmingham: Field Archaeology Unit, University of Birmingham.
Baker, N.J. and Slater, T.R. (1992) 'Morphological regions in English mediaeval towns', in Whitehand, J.W.R. and Larkham, P.J. (eds) *Urban Landscapes: International Perspectives*, London: Routledge.
Baldwin, N. (1995) 'Commercially speaking: other people's thoughts', *The Automobile* 13, 7: 62.
Ball, M. (1983) *Housing Policy and Economic Power*, London: Methuen.
Bandini, M. (1986) contribution to Choay, F. and Merlin, P. (eds) *A Propos de la Morphologie Urbaine*, Paris: Université de Paris VIII, Institut d'Urbanisme de L'Académie de Paris (full English version available from author).
Bandini, M. (1992) 'Some architectural approaches to urban form', in Whitehand, J.W.R. and Larkham, P.J. (eds) *Urban Landscapes: International Perspectives*, London: Routledge.
Bar-Hillel, M. (1988a) 'Only 18 post-war buildings are listed', *Chartered Surveyor Weekly* 14 April: 28.
Bar-Hillel, M. (1988b) 'Legislating now to preserve history', *Chartered Surveyor Weekly* 7 January: 19.
Bar-Hillel, M. (1991) 'Conservationists' nightmare or developers' dream?', *Chartered Surveyor Weekly* 14 March: 22.
Barker, A. (1976) *The Local Amenity Movement*, London: Civic Trust.
Barker, A. and Farmer, D. (1974) 'The British amenity society movement', paper given at the Urban Politics Workshop, second annual joint session, European Consortium for Political Research, Strasbourg.
Barrett, H. and Larkham, P.J. (1994) *Disguising Development: Façadism in City Centres*, Research Paper 11, Birmingham: Faculty of the Built Environment, University of Central England.
Barrett, H. and Phillips, J. (1987) *Suburban Style: the British Home, 1840–1960*, London: Macdonald Orbis.
Bearman, R. (1982) Member of the Stratford Society, quoted in Beard, P., 'All the world's a stage', *Sunday Times* 28 March: 14.
Bentley, I. (1983) *Bureaucratic Patronage and Local Urban Form*, Research Note 15, Oxford: Joint Centre for Urban Design, Oxford Polytechnic.
Bentley, I. (1984) *User Choice and Urban Form: the Impact of Commercial Redevelopment*, Research Note 18, Oxford: Joint Centre for Urban Design, Oxford Polytechnic.
Bentley, I., Alcock, A., Murrain, P., McGlynn, S. and Smith, G. (1992) *Responsive Environments: a Manual for Designers*, London: Architectural Press (second edition).
Biddle, M. (ed.) (1980) *Old and New Architecture: Design Relationship*, Washington DC: Preservation Press.
Binney, M. (1978) 'Assassination of the High Street', *Design Magazine* 353: 50–3.
Binney, M. (1981) 'Opposition to obsession', in Binney, M. and Lowenthal, D. (eds) *Our Past Before Us: Why do we Save It?*, London: Temple Smith.
Bishop, J. and Davison, I. (1989) *Development Densities: a Discussion Paper*, Amersham: Housing Research Foundation.
Bishop, R. (1982) 'The perception and importance of time in architecture', unpublished PhD thesis, Guildford: University of Surrey.
Blowers, A. (1980) *The Limits of Power: the Politics of Local Planning Policy*, Oxford: Pergamon.
Blowers, A. (1986) 'Town planning: paradoxes and prospects', *The Planner* 72 (April): 11–18.
Booth, E. (1993) 'Enhancement in conservation areas', *The Planner* 79, 4: 22–3.
Bourguignon, F. (1971) 'Les secteurs sauvegardés: premier bilan des réalisations, premiers enseignements', *Administration* 72: 52–3.
Bourne, L.S. (1967) *Private Redevelopment of the Central City*, Research Paper 112, Chicago: Department of Geography, University of Chicago.
Boyer, C. (1987) 'The return of the aesthetic to city planning', paper presented to the Conference on Planning Theory in the 1990s, Center for Urban Policy Research, Rutgers University.
Bradshaw, M. (1989) 'The loneliness of the long-distance conservationist', *Context* 23: 6–8.

Briggs, A. (1952) *History of Birmingham vol. II: Borough and City, 1865–1938*, Oxford: Clarendon Press.
Briggs, A. (1975) 'The philosophy of conservation', *Royal Society of Arts Journal* 123: 685–95.
Broady, M. (1968) *Planning for People*, London: Bedford Square Press.
Brolin, B. (1980) *Architecture in Context*, New York: Van Nostrand Reinhold.
Brookes, S.N., Jordan, A.G., Kimber, R.H. and Richardson, J.J. (1976) 'The growth of the environment as a political issue in Britain', *British Journal of Political Science* NS 6: 245–55.
Brown, G.B. (1905) *The Care of Ancient Monuments*, Cambridge: Cambridge University Press.
Brown, H.J.J. (1967) 'Civic Amenities Act 1967', *Journal of Planning and Property Law*: 692–8.
Bruton, M.J. and Nicholson, D.J. (1984) 'The use of non-statutory local planning instruments in development control and Section 36 Appeals: Part 1', *Journal of Planning and Environment Law* August: 552–65.
Bruton, M.J. and Nicholson, D.J. (1987) *Local Planning in Practice*, Cheltenham: Thornes.
Buchanan, C. and Partners (1968) *Bath: a Study in Conservation*, London: HMSO.
Buissink, J.D. and de Widt, D.J. (1967) 'Some aspects of the development of the shopping centre of the city of Utrecht (The Netherlands)', in Heinemeyer, W.F., van Hulten, M. and de Vries Reilingh (eds) *Urban Core and Inner City: Proceedings of the International Study Week, Amsterdam, 1966*, Leiden: Brill.
Buller, H. and Lowe, P. (1980) 'The campaign for the preservation of rural England', *Vole* 3: 31–4.
Burns, A. (1990) 'Reinstatement identical or just the same?', *Chartered Surveyor Weekly* 2 August: 51.
Buswell, R.J. (1984a) *Changing Approaches to Urban Conservation: a Case Study of Newcastle upon Tyne*, Occasional Series in Geography 8, Newcastle upon Tyne: School of Geography and Environmental Studies, Newcastle upon Tyne Polytechnic.
Buswell, R.J. (1984b) 'Reconciling the past with the present: conservation policy in Newcastle upon Tyne', *Cities* 1, 4: 500–14.
Butler, R. (1980) 'The concept of a tourism area cycle of evolution', *Canadian Geographer* 24: 5–12.
Callis, S.E. (1986) 'Redevelopment in Epsom commercial centre, 1898–1984: a spatial analysis of agents and architectural styles', unpublished BSc dissertation, Birmingham: Department of Geography, University of Birmingham.
Cannadine, D. (1980) *Lords and Landlords: the Aristocracy and the Towns*, Leicester: Leicester University Press.
Cantell, T.C. (1975) 'Why conserve?', *The Planner* 61, 1: 6–10.
Cantell, T.C. (1981) 'Comment on the Local Government Planning and Land (no. 2) Bill', *Journal of the Royal Institute of British Architects* 98, 4: 532.
Carlson, A.W. (1978) 'Designating historical rural areas', *Landscape* 22.
Carter, H. (1970) 'A decision-making approach to town plan analysis: a case study of Llandudno', in Carter, H. and Davies, W.K.D. (eds) *Urban Essays: Studies in the Geography of Wales*, London: Longman.
Carter, H. (1984) review of Whitehand, J.W.R. (ed.) (1981) *The Urban Landscape: Historical Development and Management: Papers by M.R.G. Conzen*, Institute of British Geographers, Special Publication 13, London: Academic Press, in *Progress in Human Geography* 8, 1: 145–7.
Chapman, D.W. and Larkham, P.J. (1992) *Discovering the Art of Relationship: Urban Design, Aesthetic Control and Design Guidance*, Research Paper 9, Birmingham: Faculty of the Built Environment, Birmingham Polytechnic.
Chapman, D.W., Larkham, P.J. and Street, A. (1995) *The Use of Article 4 Directions in Planning Control, with Particular Reference to Conservation Areas*, Research Paper 15, Birmingham: Faculty of the Built Environment, University of Central England.
Chapman, H. (1975) 'The machinery of conservation: finance and planning problems', *Town Planning Review* 46, 4: 365–82.
Chartered Surveyor Weekly (1988) 'Fund seeks Pavilions buyer', *Chartered Surveyor Weekly* 21 July: 6.
Chartered Surveyor Weekly (1991) 'Special report on Listed buildings', *Chartered Surveyor Weekly* 1 August: 18–19.
Cherry, G.E. (1974) *The Evolution of British Town Planning*, London: Hill.
Cherry, G.E. (1975) 'The conservation movement', *The Planner* 61: 3–5.
Cherry, G.E. (1981) 'An historical approach to urbanisation', *The Planner* 67: 146–7.
Cherry, G.E. (1982) *The Politics of Town Planning*, London: Longman.
Chester Archaeological Society (1984) 'Galleries which they call the Rows', special volume,

Journal of the Chester Archaeological Society 67.
Civic Trust (1980a) 'Legislating for enhancement: the Trust's greatest Act', *Civic Trust News* 78: 4.
Civic Trust (1980b) 'Local Government Planning and Land Bill', *Civic Trust News* 77.
Civic Trust (1981) 'Conservation Officers: the gilt on the gingerbread?', *Heritage Outlook* 1: 36–7.
Civic Trust (1988) 'The Lord Duncan-Sandys, CH', *Planning Outlook* 8, 1: 2–3.
Civic Trust (1989) 'Conservation areas – the important "Steinberg test"', *Heritage Outlook* 9, 1: 1.
Civic Trust Bulletin (1962) Issue 10.
Clark, Lord K. (1969) *Civilisation: a Personal View*, London: BBC.
Cloke, P.J. and Park, C.C. (1985) *Rural Resource Management*, London: Croom Helm.
Coing, H. (1966) *Rénovation Urbaine et Changement Social*, Paris: Editions Ouvrières.
Collins, G.R. and Collins, C.C. (1965) *Camillo Sitte and the Birth of Modern City Planning*, New York: Rizzoli.
Collins, R.C. (1980) 'Changing views on historical conservation in cities', *Annals of the American Academy of Political and Social Science* 451: 86–97.
Conzen, M.P. (1978) 'Analytical approaches to the urban landscape', in Butzer, K.W. (ed.) *Dimensions of Human Geography: Essays in some Familiar and Neglected Themes*, Research Paper 186, Chicago: Department of Geography, University of Chicago.
Conzen, M.R.G. (1958) 'The growth and character of Whitby', in Daysh, G.H.J. (ed.) *A Survey of Whitby and the Surrounding Area*, Eton: Shakespeare Head Press.
Conzen, M.R.G. (1960) *Alnwick, Northumberland: a Study in Town-Plan Analysis*, Publication 27, London: Institute of British Geographers.
Conzen, M.R.G. (1962) 'The plan analysis of an English city centre', in Norborg, K. (ed.) 'Proceedings of the IGU Symposium on urban geography, Lund, 1960', *Lund Studies in Geography* 24B.
Conzen, M.R.G. (1966) 'Historical townscapes in Britain: a problem in applied geography', in House, J.W. (ed.) *Northern Geographical Essays in Honour of G.H.J. Daysh*, Newcastle upon Tyne: Oriel Press.
Conzen, M.R.G. (1975) 'Geography and townscape conservation', in Uhlig, H. and Lienau, C. (eds) 'Anglo-German Symposium in Applied Geography, Giessen-Würzburg-München, 1973', *Giessener Geographische Schriften* 1975: 95–102.
Conzen, M.R.G. (1988) 'Morphogenesis, morphological regions and secular human agency in the historic townscape, as exemplified by Ludlow', in Denecke, D. and Shaw, G. (eds) *Urban Historical Geography: Recent Progress in Britain and Germany*, Cambridge: Cambridge University Press.
Coventry City Council (1976) *Kenilworth Road Control Plan*, Coventry: City Council.
Coventry City Council (1979) *Chapelfields Conservation Area*, Coventry: Department of Architecture and Planning, Coventry City Council.
Coventry City Council (1981) *Kenilworth Road Development Control Plan* (revised version), Coventry: City Council.
Cowan, P. (1963) 'Studies in the growth, change and ageing of buildings', *Transactions of the Bartlett Society* 1: 55–84.
Crewe, L.J. and Hall-Taylor, M. (1991) 'The restructuring of the Nottingham Lace Market: industrial relic or new model?', *East Midland Geographer* 14: 14–30.
Crewe, S. (ed.) (1986) *Visionary Spires*, London: Waterstone.
Crook, J.M. (1989) *The Dilemma of Style: Architectural Ideas from the Picturesque to the Post-Modern*, London: Murray.
Crosland, A. (1971) *A Social Democratic Britain*, Fabian Tract 404, London: the Fabian Society.
Crossman, R. (1975) *The Diaries of a Cabinet Minister: volume 1: Minister of Housing, 1964–1966*, London: Hamilton and Cape.
Cullen, G. (1961) *Townscape*, London: Architectural Press.
Cullen, G. (1971) *The Concise Townscape*, London: Architectural Press.
Cullingworth, J.B. (1992) 'Historic preservation in the US: from landmarks to planning perspectives', *Planning Perspectives* 7: 65–79.
Cullingworth, J.B. and Nadin, V. (1994) *Town and Country Planning in Britain*, London: Routledge (11th edition).
Curl, J.S. (1977) *The Erosion of Oxford*, Oxford: Oxford Illustrated Press.
Curr, G.G. (1984) 'Who saved York? The roles of William Etty and the Corporation of York', *York Historian* 5: 25–38.

Darley, G. (1975) *Villages of Vision*, London: Paladin.

Darlington, G. (1993) 'Non-settlement based conservation areas within a National Park', in Airs, M. (ed.) *Conservation Areas: the First 25 Years*, Proceedings of a conference of the Association of Conservation Officers, Oxford: Department of Continuing Education, University of Oxford.

Davies, H.W.E. *et al.* (1984) *The Relationship between Development Plans and Development Control. Vol 1: Summary and Conclusions*, Draft Final Report to the Department of the Environment, Reading: Joint Centre for Land Development Studies and College of Estate Management, University of Reading.

Davies, H.W.E., Edwards, D. and Rowley, A.R. (1989) *The Approval of Reserved Matters Following Outline Planning Permission*, London: HMSO.

Davies, W.K.D. (1968) 'The morphology of central places: a case study', *Annals, Association of American Geographers* 58, 2: 91–110.

Debenham, Tewson and Chinnocks (1987) *Shop Rent and Rates Report*, London: Debenham, Tewson and Chinnocks.

Dellheim, C. (1982) *The Face of the Past: the Preservation of the Medieval Inheritance in Victorian England*, Cambridge: Cambridge University Press.

Denton, J.H. (1970) *English Royal Free Chapels, 1100–1300*, Manchester: Manchester University Press.

Department of National Heritage (1992) Circular 1/92 *Responsibility for Conservation Policy and Casework*, London: HMSO.

Department of the Environment (1972a) Circular 52/72 *Town and Country Planning Act 1971*, London: HMSO.

Department of the Environment (1972b) Circular 86/72 *Town and Country Planning (Amendment) Act 1972*, London: HMSO.

Department of the Environment (1973) Circular 12/73 *Town and Country Planning General Development Order 1973*, London: HMSO.

Department of the Environment (1974) Circular 147/74 *Town and Country Amenities Act 1974*, London: HMSO.

Department of the Environment (1977a) Circular 23/77 *Historic Buildings and Conservation Areas – Policy and Procedure*, London: HMSO.

Department of the Environment (1977b) *Town and Country Planning (Listed Buildings and Buildings in Conservation Areas) Regulations 1977*, London: HMSO.

Department of the Environment (1978) *The Under-Use of Upper Floors in Historic Town Centres*, London: HMSO.

Department of the Environment (1980) Circular 22/80 *Development Control: Policy and Practice*, London: HMSO.

Department of the Environment (1981a) Circular 23/81 *Local Government Planning and Land Act: Town and Country Planning: Development Plans*, London: HMSO.

Department of the Environment (1981b) *Town and Country Planning (National Parks, Areas of Outstanding Natural Beauty and Conservation Areas) Special Development Order*, London: HMSO.

Department of the Environment (1981c) Decision letter on Appeal T/APP/5105/A/81/5802/G6, Bristol: Department of the Environment Planning Inspectorate.

Department of the Environment (1985a) *Development Control Statistics: England 1979/80–1982/83*, London: Department of the Environment SPPG Division.

Department of the Environment (1985b) Circular 31/85 *Aesthetic Control*, London: HMSO.

Department of the Environment (1987a) Circular 8/87 *Historic Buildings and Conservation Areas – Policy and Procedures*, London: HMSO.

Department of the Environment (1987b) *Development Control Statistics: England 1983/84–1985/86*, London: Department of the Environment Land and General Statistics Division.

Department of the Environment (1988a) Report of the Local Inquiry held to determine applications for planning permissions and listed building consents by Centrovincial Estates PLC. Unpublished report, Bristol: DoE Planning Inspectorate (copy in Worcester City planning files).

Department of the Environment (1988b) Decision letter on Appeals T/APP/A/87/74274/P3, 87/75275/P3, 88/086114/P3 and 88/086115/P3, Bristol: Department of the Environment Planning Inspectorate.

Department of the Environment (1991) *Planning Policy Guidance: General Policy and Principles*, Consultation paper on draft revision of PPG 1, London: Department of the Environment.

Department of the Environment (1992) Planning Policy Guidance Note 1 *General Policy and Principles*, London: HMSO (second edition).
Department of the Environment and Department of National Heritage (1994) Planning Policy Guidance Note 15 *Planning and the Historic Environment*, London: HMSO.
Diefendorf, J.M. (ed.) (1990) *Rebuilding Europe's Bombed Cities*, London: Macmillan.
Dix, G. (1985) 'Design and conservation in the city', *Town Planning Review* 56, 2: 127–34.
Dobby, A. (1978) *Conservation and Planning*, London: Hutchinson.
von der Dollen, B. (1983) 'City planning, conservation, and urban historical geography in Germany', *Planning History Bulletin* 5: 39–43.
Downs, A. (1972) 'Up and down with ecology: the issue-attention cycle', *Public Interest* 28: 38–50.
Dron, T. (1991) 'Risen ghosts', *Thoroughbred & Classic Cars* September: 26–31.
Dupont, J. (1966) 'Viollet-le-Duc and restoration in France', in *Historic Preservation Today*, Charlottesville, Va.: US National Trust for Historic Preservation.
Edwards, A.M. (1981) *The Design of Suburbia*, London: Pembridge Press.
Edwards, R. (1995) 'Conservation area designation and the funding of enhancement through schemes of partnership', unpublished Diploma in Town Planning project, Birmingham: School of Planning, University of Central England.
Elkin, T., McLaren, D. and Hillman, M. (1991) *Reviving the City: Towards Sustainable Urban Development*, London: Friends of the Earth.
Empson, W. (1955) *Seven Types of Ambiguity*, New York: Meridian Books.
English Heritage (1988a) Unpublished report to the Worcester Centrovincial Development Inquiry (copy in Worcester City planning files).
English Heritage (1988b) 'Shopping in historic towns: a policy statement', *Conservation Bulletin* 5: 1–2.
English Heritage (1991) 'Framing opinions', supplement to *Conservation Bulletin* 14.
English Heritage (1993) *Conservation Area Practice* Guidance note, London: English Heritage.
English Heritage Monitor (annual) London: English Tourist Board.
English Historic Towns Forum (1992) *Townscape in Trouble*, Bath: EHTF.
Esher, L. (1981) *A Broken Wave: the Rebuilding of England, 1940–1980*, London: Allen Lane.
Eversley, D. (1974) 'Conservation for the minority?', *Built Environment* 3: 14–15.
Falk, N. (1993) 'Regeneration and sustainable development', in Berry, J.N., McGreal, W.S. and Deddis, W.G. (eds) *Urban Regeneration: Property Investment and Development*, London: Spon.
Faludi, A. (1973) *Planning Theory*, Oxford: Pergamon.
Faulkner, P.A. (1978) 'Definition and evaluation of the historic heritage', 'Is preservation possible?', 'Preservation within a philosophy', Bossom Lectures, *Royal Society of Arts Journal* CXXVI: 452–80.
Fennell, R.I. (1982) 'Theory into practice: the effect of planning on the urban form of Chichester', paper presented to the Institute of British Geographers conference 'Urban morphology: research and practice in geography and urban design', Birmingham.
Ferguson, C.M. (1990) 'Steinberg reconsidered', *Journal of Planning and Environment Law* January: 8–10.
Fergusson, A. (1973) *The Sack of Bath*, Salisbury: Compton Russel.
Fitch, J.M. (1980) 'A funny thing happened on our way to the Eighties', *Journal of the American Institute of Architects* 4: 66–8.
Fitch, J.M. (1982) *Historic Preservation: Curatorial Management of the Built World*, New York: McGraw-Hill.
Fleming, S.C. and Short, J.R. (1984) 'Committee rules OK? An examination of planning committee action on officer recommendations', *Environment and Planning A*, 16: 965–73.
Foote, K. (1985) 'Velocities of change in a built environment, 1880–1980: evidence from the photoarchives of Austin, Texas', *Urban Geography* 6, 4: 220–45.
Fowler, P. (1989) 'Heritage: a post-modern perspective', in Uzzell, D.L. (ed.) *Heritage Interpretation: the Natural and Built Environment*, London: Belhaven.
Frampton, K. (1980) *Modern Architecture: a Critical History*, London: Thames and Hudson.
Frederick Gibberd Coombes & Partners (1988) 'Blackfriars Square and Huntingdon Arcades, Worcester: Statement of the history of the proposals' (revised edition). Unpublished: copy in Worcester City planning files.
Freeman, J. (1982) 'Mortgage myopia', *Chartered Surveyor* March: 291–8.
Freeman, M. (1986) 'The nature and agents of central-area change: a case study of Aylesbury and Wembley town centres, 1935–1983', unpublished PhD thesis, Birmingham: Department of Geography, University of Birmingham.

Freyer, H. (1934) *Theorie des objektiven Geistes: eine Einleitung in die Kulturphilosophie*, Leipzig (reprinted 1966, Stuttgart: Teubner).
Gad, G. and Holdsworth, D. (1988) 'Streetscape society: the changing built environment of King Street, Toronto', in Hall, R., Westfall, W. and MacDowell, K.S. (eds) *Patterns of the Past: Interpreting Ontario's History*, Toronto: Dundurn Press.
Gale, D.E. (1991) 'The impacts of historic district designation: planning and policy implications', *Journal of the American Planning Association* 57, 3: 325–40.
Gamston, D. (1975) *The Designation of Conservation Areas: a Survey of the Yorkshire Region*, Research Paper 9, York: Institute of Advanced Architectural Studies, University of York.
Gardiner, S. (1985) 'Sainsbury's other sites', *The Observer* 26 May.
Gibbon, E. (1782–8) *The History of the Decline and Fall of the Roman Empire*, London.
Glancey, J. (1989) *New British Architecture*, London: Thames and Hudson.
Gold, J.R. and Ward, S.V. (1994) *Place Promotion*, Chichester: Wiley.
Goldthorpe, J.H., Lockwood, D., Bechofer, D. and Platt, J. (1969) *The Affluent Worker in the Class Structure*, Cambridge: Cambridge University Press.
Gosling, D. and Maitland, B. (1984) *Concepts of Urban Design*, London: Academy Editions.
Gould, J. (1977) *Modern Houses in Britain*, Architectural History Monograph 1, London: Society of Architectural Historians of Great Britain.
Granelli, R. (1973) 'Conservation?', *Architecture West Midlands* 15: 14–16.
Graves, P. and Ross, S. (1991) 'Conservation areas: a presumption to conserve', *Estates Gazette* 9137: 108–10.
Great Britain (1993) *Aspects of Britain: Conservation*, London: HMSO.
Greenslade, M.W., Tringham, N.J. and Johnson, D.A. (1984) 'Tettenhall', in *Victoria County History of Staffordshire*, XX, Oxford: Oxford University Press.
Griffiths, A. (1985) 'Planning blighters', *Sunday Times Magazine* 16 March.
Groat, L. (1988) 'Contextual compatibility in architecture', in Canter, D., Krampen, M. and Stea, D. (eds) *Ethnoscapes*, Aldershot: Gower.
A Guide to Delineating the Edges of Historic Districts (1976) Washington DC: Preservation Press.
Haines, G.H. (1974) 'Conservation in Europe', *Housing and Planning Review* 36: 2–5.
Harrison, P. (1984) 'The code and the new clients', *Royal Institute of British Architects Journal* 91: 92.
Harvey, J.H. (1993) 'The origin of listed buildings', *Transactions of the Ancient Monuments Society* 37: 1–20.
Hawtin, D. (1995) '1924 11hp Aston Martin', *The Automobile* 13, 6: 34–40.
Hayward, R. and McGlynn, S. (eds) (1993) *Making Better Places: Urban Design Now*, London: Butterworth.
Heap, D. (1986) 'Planning participation – is it just a bit of a fraud?', letter to the Editor, the *Daily Telegraph* September.
Heighway, C.M. (1972) *The Erosion of History*, London: Council for British Archaeology.
Hendry, J. (1993) 'Conservation areas in Northern Ireland: an alternative approach', *Town Planning Review* 64, 4: 415–34.
Herbert, D.T. (ed.) (1995) *Heritage, Tourism and Society*, London: Mansell.
Hewison, R. (1987) *The Heritage Industry: Britain in a State of Decline*, London: Methuen.
Hillman, J. (1990) *Planning for Beauty*, London: HMSO.
Hipple, W.J. (1957) *The Beautiful, the Sublime and the Picturesque in Eighteenth Century British Aesthetic Theory*, Carbondale, Ill.: Southern Illinois University Press.
HRH The Prince of Wales (1987) Speech given at the Mansion House, London, 1 December.
HRH The Prince of Wales (1989) *A Vision of Britain*, London: Doubleday.
Holdsworth, D. (ed.) (1985) *Reviving Main Street*, Toronto: University of Toronto Press.
Holdsworth, D. (1986) 'Cottages and castles for Vancouver home-seekers', *BC Studies* 69/70: 11–32.
Holmes, R.J. (1982) 'Conservation easement: at last, conservation pays', *Urban Land* 41: 3–9.
Holzner, L. (1970) 'The role of history and tradition in the urban geography of West Germany', *Annals of the Association of American Geographers* 60: 315–39.
Horne, M. (1993) 'The listing process in Scotland and the statutory protection of vernacular building types', *Town Planning Review* 64, 4: 375–93.
Horsey, M. (1985) 'Speculative housebuilding in London in the 1930s: official control and popular taste', paper presented to the Construction History Group Conference, London.
House of Lords (1992) *Judgement: South Lakeland District Council v. Secretary of State for the Environment and Others*, London: HMSO.

Hubbard, P. (1993) 'The value of conservation: a critical review of behavioural research', *Town Planning Review* 64, 4: 359–73.

Hughes, D.J. (1995) 'Planning and conservation areas – where do we stand following PPG 15, and whatever happened to Steinberg?', *Journal of Planning and Environment Law* August: 679–91.

Hunt, J.D. and Willis, P. (eds) (1988) *The Genius of the Place: the English Landscape Garden 1620–1820*, Cambridge, Mass.: MIT Press.

Huntingford, G. (1991) Letter to the Editor, *The Planner* 77, 24: 2.

Hutchinson, M. (1989) *The Prince of Wales: Right or Wrong? An Architect Replies*, London: Faber.

International Congress of Architects and Technicians of Historic Monuments (1964) *Statement of Principles* (the 'Venice Charter'), Venice: Second International Congress.

Jakle, J.A. (1987) *The Visual Elements of Landscape*, Amherst, Mass.: University of Massachusetts Press.

Jarman, D. (1992) 'Drawing the line on conservation areas', *Planning* 965 (24 April): 16–17.

Jencks, C.A. (1985) *Modern Movements in Architecture*, Harmondsworth: Pelican (second edition).

Jencks, C.A. (1988) *The Prince, the Architects and the New Wave Monarchy*, London: Academy Editions.

Jencks, C.A. (1991) *The Language of Post-Modern Architecture*, London: Academy Editions (sixth edition).

Jencks, C.A. and Chaitkin, W. (1982) *Current Architecture*, London: Academy Editions.

Jennings-Smith, D. (1977) 'Guide or rule book? Rethinking design control applications', *Built Environment Quarterly* 3, 3: 245–7.

Johnston, R.J. (1969) 'Towards an analytical study of the townscape: the residential building fabric', *Geografiska Annaler* 51B: 20–32.

Johnston, R.J. (1984) 'The world is our oyster', *Transactions of the Institute of British Geographers* NS 9, 4: 443–59.

Jones, A.N. (1991) 'The management of residential townscapes', unpublished PhD thesis, Birmingham: School of Geography, Universty of Birmingham.

Jones, A.N. and Larkham, P.J. (1993) *The Character of Conservation Areas*, Report commissioned from Plan Local for the Conservation and Built Environment Panel, London: Royal Town Planning Institute.

Jones, B. (1974) *Follies and Grottoes*, London: Constable.

Jones, M. (ed.) (1990) *Fake? The Art of Deception*, London: British Museum Publications.

Journal of Planning and Environment Law (1992) 'Notes of cases: R. v. Canterbury City Council, ex parte Halford', *Journal of Planning and Environment Law* (September): 851–5.

Journal of Planning and Environment Law (1993) 'Notes of cases: Chorley and James et al. v. Secretary of State for the Environment and Basingstoke and Deane Borough Council', *Journal of Planning and Environment Law* October: 927–32.

Journal of Planning and Environment Law Bulletin (1991) 'Conservation areas – the duty under Section 277(8) of the Town and Country Planning Act 1971', *JPEL Bulletin* March: 3–4.

Kain, R.J.P. (1981) 'Conservation and planning in France: policy and practice in the Marais, Paris', in Kain, R.J.P. (ed.) *Planning for Conservation: an International Perspective*, London: Mansell.

Kain, R.J.P. (1982) 'Europe's model and exemplar still? The French approach to urban conservation', *Town Planning Review* 54, 4: 403–22.

Kain, R.J.P. (1986) 'Gothic revival, restoration and preservation of historic buildings in nineteenth-century England', paper presented to the Eastern Historical Geography Association, Savannah, Georgia.

Kearns, G. and Philo, C. (eds) (1993) *Selling Places: the City as Cultural Capital, Past and Present*, Oxford: Pergamon.

Kennett, W. (1972) *Preservation*, London: Temple Smith.

Knevitt, C. (1985) *Space on Earth*, London: Thames Methuen.

Knevitt, C. (1986) *Perspectives: an Anthology of 1001 Architectural Quotations*, London: Lund Humphries.

Knox, P.L. (1991) 'The restless urban landscape: economic and sociocultural change and the transformation of Metropolitan Washington, DC', *Annals of the Association of American Geographers* 81, 2: 181–209.

Knox, P.L. (1992) 'The packaged landscapes of post-suburban America', in Whitehand, J.W.R. and Larkham, P.J. (eds) *Urban Landscapes: International Perspectives*, London: Routledge.

Kong, L. and Yeoh, B.S.A. (1994) 'Urban conservation in Singapore: a survey of state policies and popular attitudes', *Urban Studies* 31, 2: 247–65.
Kostof, S. (1991) *The City Shaped*, London: Thames and Hudson.
Kropf, K.S. (1993) 'The definition of built form in urban morphology', unpublished PhD thesis, Birmingham: School of Geography, University of Birmingham.
Lambert, D. (1993) 'Historic gardens', in Airs, M. (ed.) *Conservation Areas: the First 25 Years*, Proceedings of a conference of the Association of Conservation Officers, Oxford: Department of Continuing Education, University of Oxford.
Larkham, P.J. (1985) *Voluntary Amenity Societies and Conservation Planning*, Working Paper 30, Birmingham: Department of Geography, University of Birmingham.
Larkham, P.J. (1986) 'Conservation, planning and morphology in West Midlands conservation areas, 1968–84', unpublished PhD thesis, Birmingham: Department of Geography, University of Birmingham.
Larkham, P.J. (1988a) 'Agents and types of change in the conserved townscape', *Transactions of the Institute of British Geographers* NS 13, 2: 148–64.
Larkham, P.J. (1988b) 'Changing conservation areas in the English midlands: evidence from local planning records', *Urban Geography* 9, 5: 445–65.
Larkham, P.J. (1988c) 'The style of superstores: the response of J. Sainsbury PLC to a planning problem', *International Journal of Retailing* 3, 1: 44–59.
Larkham, P.J. (1989) 'Planning and development in historical town centres', paper presented to the conference 'The hidden history of the High Street', The Ironbridge Institute.
Larkham, P.J. (1990a) 'Conservation and the management of historical townscapes', in Slater, T.R. (ed.) *The Built Form of Western Cities*, Leicester: Leicester University Press.
Larkham, P.J. (1990b) 'The concept of delay in development control', *Planning Outlook* 33, 2: 101–7.
Larkham, P.J. (1990c) 'Development control information and planning research', *Local Government Studies* 16, 2: 1–7.
Larkham, P.J. (1992) 'Conservation and the changing urban landscape', *Progress in Planning* 37, 2: 83–181.
Larkham, P.J. (1995a) 'Heritage as planned and conserved', in Herbert, D.T. (ed.) *Heritage, Tourism and Society*, London: Mansell.
Larkham, P.J. (1995b) 'Constraints of urban history and form upon redevelopment', *Geography* 80, 2: 111–24.
Larkham, P.J. and Freeman, M. (1988) 'Twentieth-century British commercial architecture', *Journal of Cultural Geography* 9, 1: 1–16.
Larkham, P.J. and Jones, A.N. (1993) 'Strategies for increasing residential density', *Housing Studies* 8, 2: 83–97.
Larkham, P.J. and Pompa, N.D. (1988) 'Research in urban morphology', *Planning History* 9, 3: 12–15.
Larkham, P.J. and Pompa, N.D. (1989) 'Planning problems of large retail centres: the West Midlands County, 1987', *Cities* 7, 4: 309–16.
Lassar, T.J. (1989) *Carrots and Sticks: New Zoning Downtown*, Washington, DC: Urban Land Institute.
Layfield, F.H.B. (1971) 'Powers for conservation', *Journal of the Town Planning Institute* 57, 3: 142–51.
Lee, F. (1954) 'A new theory of the origins and early growth of Northampton', *Archaeological Journal* 110: 164–74.
Lewis, J.P. (1965) *Building Cycles and Britain's Growth*, London: Macmillan.
Lewis, P.F. (1975) 'The future of the past: our clouded vision of historic preservation', *Pioneer America* 7: 1–20.
Lichfield, N. (1988) *Economics in Urban Conservation*, Cambridge: Cambridge University Press.
Listokin, D. (1985) *Living Cities*, New York: Priority Press Publications.
Llewelyn-Davies (1993) *The Gun Quarter: Planning and Urban Design Framework*, London: Llewelyn-Davies.
Loftman, P. and Nevin, B. (1992) *Urban Regeneration and Social Equity – a Case Study of Birmingham 1986–1992*, Research Paper 8, Birmingham: Faculty of the Built Environment, University of Central England.
Long, A.R. (1975) 'Participation and the community', *Progress in Planning* 5, 2: 61–134.
Lorentz, S. (1966) 'Reconstruction of the old town centres of Poland', in *Historic Preservation Today*, Charlottesville, Va.: US National Trust for Historic Preservation.
Loudon, J.C. (1850) *The Villa Gardener*, London.

Low, N. (1991) *Planning, Politics and the State: Political Foundations of Planning Thought*, London: Unwin Hyman.
Lowe, P.D. and Goyder, J. (1983) *Environmental Groups in Politics*, London: Allen and Unwin.
Lowenthal, D. (1966) 'The American way of history', *Columbia University Forum* 9: 27–32.
Lowenthal, D. (1975) 'Past time, present place: landscape and memory', *Geographical Review* 65: 1–36.
Lowenthal, D. (1977) 'The bicentennial landscape: a mirror held up to the past', *Geographical Review* 67: 253–67.
Lowenthal, D. (1979) 'Environmental perception: preserving the past', *Progress in Human Geography* 3, 4: 549–59.
Lowenthal, D. (1985) *The Past is a Foreign Country*, Cambridge: Cambridge University Press.
Lowenthal, D. (1990) 'Forging the past', in Jones, M. (ed.) *Fake? The Art of Deception*, London: British Museum Publications.
Lozano, E.E. (1974) 'Visual needs in the built environment', *Town Planning Review* 45, 3: 351–74.
Lumley, R. (ed.) (1988) *The Museum Time-Machine*, London: Routledge.
Lyall, S. (1980) *The State of British Architecture*, London: Architectural Press.
Lynch, K. (1960) *The Image of the City*, Cambridge, Mass.: MIT Press.
MacCannell, D. (1976) *The Tourist: a New Theory of the Leisure Class*, New York: Schocken.
MacKail, J. (1899) *The Life of William Morris* (2 vols), London.
McNamara, P. (1985) 'The control of office development in central Edinburgh, 1959–1979', unpublished PhD thesis, Edinburgh: Department of Geography, University of Edinburgh.
Maguire, P. (1980) 'Rehabilitation in the next decade', *Journal of the Royal Institute of British Architects* 97: 25–7.
Maltese Planning Authority (1993) *Design Guidelines for Development Control within Urban Conservation Areas*, Valletta: Planning Authority.
Mander, G.P. (1933) *Wolverhampton Antiquary* 3.
Manser, M. (1980) 'An excuse for lousy buildings', *Journal of the Royal Institute of British Architects* 87, 2: 49.
Marriott, O. (1967) *The Property Boom*, London: Hamish Hamilton.
Martin, G.H. (1968) 'The town as a palimpsest', in Dyos, H.J. (ed.) *The Study of Urban History*, London: Edward Arnold.
Martin, S. and Buckler, J. (1988) 'Major shopping developments and their impact on small towns', *Heritage Outlook* 8, 1: 4–7.
Marx, K. and Engels, F. (1864; translation 1965) *The German Ideology*, London: Lawrence and Wishart.
Matson, S. (1982) Letter to the Editor, *Architectural Preservation* 1: 12.
Metropolitan Borough of Solihull (1985) *Ashleigh Road Conservation Area*, Solihull: Metropolitan Borough Council.
Millichap, D. (1989a) 'Conservation areas – Steinberg and after', *Journal of Planning and Environment Law* April: 233–40.
Millichap, D. (1989b) 'Conservation areas and Steinberg – the Inspectorate's response', *Journal of Planning and Environment Law* July: 499–504.
Millichap, D. (1992) 'Neglected area of conservation law', *Planning* 965 (24 April): 17.
Millichap, D. (1993) 'Exceptions to the rule', *Planning* 1002 (22 January): 17.
Ministry of Housing and Local Government (1962) Planning Bulletin 1: *Town Centres: Approach to Renewal*, London: HMSO.
Ministry of Housing and Local Government (1963) Planning Bulletin 4: *Town Centres: Current Practices*, London: HMSO.
Ministry of Housing and Local Government (1967a) Circular 53/67 *Civic Amenities Act 1967: Parts I and II*, London: HMSO.
Ministry of Housing and Local Government (1967b) *Historic Towns: Preservation and Change*, London: HMSO.
Ministry of Housing and Local Government (1968) Circular 61/68 *Town and Country Planning Act 1968: Part V: Historic Buildings and Conservation*, London: HMSO.
Ministry of Housing and Local Government (1969) Development Control Policy Note 7 *Preservation of Historic Buildings and Areas*, London: HMSO.
Minoprio, A. and Spencely, H. (1946) *Worcester Plan: an Outline Development Plan for Worcester*, Worcester: City Council.
Montagu of Beaulieu, Lord (1967) *The Gilt and the Gingerbread: or How to Live in a Stately Home and Make Money*, London: Michael Joseph.
Montgomery, J. (1992) 'Dressed to kill off urban culture', *Planning* 989: 6–7.

Moodie, G.C. and Studdert-Kennedy, G. (1970) *Opinions, Publics and Pressure Groups*, London: Allen and Unwin.
Morris, C.J. (1978) 'Townscape images: studying meaning and classification', unpublished PhD thesis, Department of Geography, University of Exeter.
Morris, C.J. (1981) 'Townscape images', in Kain, R.J.P. (ed.) *Planning for Conservation*, London: Mansell.
Morris, M. (ed.) (1910–15) *Collected Works of William Morris* (10 vols), London: Longmans, Green.
Morris, W. (1877) 'Restoration', reprinted by the Society for the Protection of Ancient Buildings as their *Manifesto*, London: SPAB.
Morton, D.M. (1991) 'Conservation areas: has saturation point been reached?', *The Planner* 77, 17: 5–8.
Morton, D.M. (1994) 'The designation debate', in Larkham, P.J. (ed.) *Conservation Areas: Issues and Management*, Birmingham: Faculty of the Built Environment, University of Central England.
Morton, H.V. (1928) *The Call of England*, London.
Muthesius, S. (1972) *The High Victorian Movement in Architecture, 1850–1870*, London: Routledge and Kegan Paul.
Muthesius, S. (1981) 'The origins of the German conservation movement', in Kain, R.J.P. (ed.) *Planning for Conservation: an International Perspective*, London: Mansell.
Mynors, C. (1984) 'Conservation areas: protecting the familiar and cherished local scene', *Journal of Planning and Environment Law* March: 144–57; April: 235–47.
Nelson, S. (1991) 'Value added tax: a disincentive to sensitive repairs', Framing Opinions, supplement to *Conservation Bulletin* 14.
Newby, P.T. (1994) 'Tourism: support or threat to heritage?', in Ashworth, G.J. and Larkham, P.J. (eds) *Building a New Heritage*, London: Routledge.
Newcomb, R.M. (1983) *A Business and a Charity: Conservation in Transition*, Geographical Paper 83, Reading: Department of Geography, University of Reading.
Newsom, M.D. (1971) 'Blacks and historic preservation', *Law and Contemporary Problems* 36: 423–6.
New Statesman (1973) 'Networks: the Sons of the Earth', *New Statesman* 3 August.
Norberg-Schulz, C. (1980) *Genius Loci: Towards a Phenomenology of Architecture*, New York: Rizzoli.
North Norfolk District Council (1989) *Planning and Design Guide*, Cromer: Planning Department, North Norfolk District Council.
Nuffield Foundation (1986) *Town and Country Planning*, London: Nuffield Foundation.
Oakeshott, W.F. (ed.) (1975) *Oxford Stone Restored*, Oxford: Oxford Historic Buildings Fund.
Oliver, K.A. (1982) 'Places, conservation and the care of streets in Hartlepool', in Gold, J.R. and Burgess, J. (eds) *Valued Environments*, London: Allen and Unwin.
Olsen, D.J. (1976) *The Growth of Victorian London*, London: Batsford.
Olsen, D.J. (1986) *The City as a Work of Art: London, Paris, Vienna*, New Haven, Conn.: Yale University Press.
O'Rourke, T. (1987) 'Conservation, preservation and planning', *The Planner* 73, 2: 13–18.
Palliser, D.M. (1974) 'Preserving our heritage: the city of York', in Kimber, R. and Richardson, J.J. (eds) *Campaigning for the Environment*, London: Routledge and Kegan Paul.
Palmer, F.R. and Crowquill, A. (*c.* 1846) *Wanderings of a Pen and Pencil*, no publisher given.
Palser, G. (1984a) 'Shops: getting into shape', *Property Business* April: 18–19.
Palser, G. (1984b) 'Burtons shops tops under Halpern's lead', *Property Business* April: 20–2.
Pearce, G., Hems, L. and Hennessy, B. (1990) *The Conservation Areas of England*, London: English Heritage.
Pepper, M. and Swenarton, M. (1980) 'Neo-Georgian maison-type', *Architectural Review* CLXVIII: 87–92.
Petherick, A. (1990) *Living Over the Shop*, London: National Housing and Town Planning Council.
Pevsner, N. (1968) 'The picturesque in architecture', in *Studies in Art, Architecture and Design*, London: Thames and Hudson.
Pevsner, N. (1972) *Some Architectural Writers of the Nineteenth Century*, Oxford: Clarendon Press.
Pevsner, N. (1976) 'Scrape and anti-scrape', in Fawcett, J. (ed.) *The Future of the Past: Attitudes to Conservation 1174–1974*, London: Thames and Hudson.
Pevsner, N. (1980) 'Ruskin and Viollet-le-Duc: Englishness and Frenchness in the appreciation

of Gothic architecture', in *Architectural Design, Viollet-le-Duc*, London: Academy Editions.

Planning (1994) 'Conservation area partners watch the cash start to flow', *Planning* 1061: 25.

Planning Advisory Group (1965) *The Future of Development Plans*, London: HMSO.

Plant, H. (1993) 'The use and abuse of conservation area designation powers by Local Planning Authorities in the West Midlands region', unpublished MA thesis, Birmingham: School of Planning, University of Central England.

Powell, K. and Fieldhouse, J. (1982) *Manchester: the Disappearing Cathedral Conservation Area*, London: SAVE Britain's Heritage.

Prentice, R.C. (1993) *Tourism and Heritage Attractions*, London: Routledge.

Preservation Policy Group (1970) *Report to the Minister of Housing and Local Government*, London: HMSO.

Price, C. (1981) 'The built environment: the case against conservation', *The Environmentalist* 1: 39–41.

Price, U. (1794) *Essay on the Picturesque*, London.

Pugin, A.W.N. (1836) *Contrasts*, London.

Pugin, A.W.N. (1841) *The True Principles of Pointed or Christian Architecture*, London.

Punter, J.V. (1985) *Office Development in the Borough of Reading, 1954–1984: a Case Study in the Role of Aesthetic Control within the Planning Process*, Working Papers in Land Management and Development Environment Policy 6, Reading: Department of Land Management and Development, University of Reading.

Punter, J.V. (1986a) 'The contradictions of aesthetic control under the Conservatives', *Planning Practice and Research* 1: 8–13.

Punter, J.V. (1986b) 'A history of aesthetic control, Part I: 1909–1953', *Town Planning Review* 57, 4: 351–81.

Punter, J.V. (1986c) 'Aesthetic control within the development process: a case study', *Land Development Studies* 3: 197–212.

Punter, J.V. (1989) 'Development control – case studies', in Hebbert, M. (ed.) *Development Control Data: a Research Guide*, Swindon: Economic and Social Research Council.

Punter, J.V. (1990a) 'The Ten Commandments of architecture and urban design', *The Planner* 76, 39: 10–14.

Punter, J.V. (1990b) *Design Control in Bristol, 1940–1990*, Bristol: Redcliffe Press.

Punter, J.V., Carmona, M. and Platts, A. (1994) 'The design content of development plans', *Planning Practice and Research* 9, 3: 199–220.

Racine, E. and Creutz, Y. (1975) 'Planning and housing: France, a developer's view', *The Planner* 61: 83–5.

Rasmussen, S.E. (1951) *Towns and Buildings*, Liverpool: University Press of Liverpool.

Rawlinson, C. (1989) 'Design and development control: an analysis of refusals', *The Planner* 75: 17–20.

Ray, P. (1978) *Conservation Area Advisory Committees*, London: Civic Trust.

Reade, E. (1991) 'The little world of Upper Bangor: part 1: how many conservation areas are slums?', *Town and Country Planning* 60, 11/12: 340–3.

Reade, E. (1992a) 'The little world of Upper Bangor: part 2: professionally prestigious projects or routine public administration?', *Town and Country Planning* 61, 1: 25–7.

Reade, E. (1992b) 'The little world of Upper Bangor: part 3: what is planning for anyway?', *Town and Country Planning* 61, 2: 44–7.

Redmayne, R. (ed.) (1950) *Ideals in Industry*, Leeds, no publisher given.

Reed, T.J. (1969) 'Land use controls in historic areas', *Notre Dame Lawyer* 44, 3.

Relph, E. (1982) 'The landscapes of the consumer society', in Sadler, B. and Carlson, A. (eds) *Environmental Aesthetics: Essays in Interpretation*, Western Geographical Series 20, Victoria, NSW: Department of Geography, University of Victoria.

Robertson, M. (ed.) (1993) 'Listed buildings: the national resurvey of England', *Transactions of the Ancient Monuments Society* 37: 21–94.

Robinson, A. (1982) 'The evaluation of conservation areas', in Grant, E. and Robinson, A. (eds) *Landscape and Industry: Essays in Memory of Geoffrey Gullett*, Enfield: Middlesex Polytechnic.

Robinson, J.M. (1984) *The Latest Country Houses*, London: Bodley Head.

Rock, D. (1974) 'Conservation: a confusion of ideas', *Built Environment* 3: 363–6.

Rose, E.A. (1974) 'The search for environment', in Cherry, G.E. (ed.) *Urban Planning Problems*, London: Prentice-Hall.

Ross, M. (1991) *Planning and Heritage: Policy and Practice*, London: Spon.

Rowntree, L.B. and Conkey, M.W. (1980) 'Symbolism and the cultural landscape', *Annals of the*

Association of American Geographers 70: 459–74.
Royal Institution of Chartered Surveyors and English Heritage (1993) *The Investment Performance of Listed Buildings*, London: RICS and English Heritage.
Ruskin, J. (1849) *The Seven Lamps of Architecture*, London.
Russell, J. (supposed author) (1750) *Letters from a Young Painter Abroad to his Friends in England*, London.
Sabelberg, E. (1983) 'The persistence of palazzi and intra-urban structures in Tuscany and Sicily', *Journal of Historical Geography* 9, 3: 247–64.
Samuel, I. (1990) 'The philosophy of brick', *New Formations* 11: 45–55.
Samuels, I. (1985) 'Theorie des mutations urbains en pays développés', unpublished report for the French Ministry of Urbanism, Housing and Transport, Oxford: Joint Centre for Urban Design, Oxford Polytechnic.
Samuels, I. (1990) 'Architectural practice and urban morphology', in Slater, T.R. (ed.) *The Built Form of Western Cities*, Leicester: Leicester University Press.
Samuels, I. (1993) 'The Plan d'Occupation des Sols for Asnières sur Oise: a morphological design guide', in Hayward, R. and McGlynn, S. (eds) *Making Better Places: Urban Design Now*, London: Butterworth.
Sauder, R.A. and Wilkinson, T. (1989) 'Preservation planning and geographic change in New Orleans' Vieux Carré', *Urban Geography* 10, 1: 41–61.
SAVE Britain's Heritage (1978) 'Report on shopfronts', *Building Design Magazine* June (unpaginated).
SAVE Britain's Heritage (1980) *Drowning in VAT*, London: SAVE.
Scanlon, K., Edge, A., Wilmott, T. *et al.* (1994) *The Economics of Listed Buildings*, Discussion Paper 43, Cambridge: Department of Land Economy, University of Cambridge.
Schlüter, O. (1899) 'Bemerkungen zur Siedlungsgeographie', *Geographische Zeitschrift* 5: 65–84.
Schouten, F.F.J. (1995) 'Heritage as historical reality', in Herbert, D.T. (ed.) *Heritage, Tourism and Society*, London: Mansell.
Schwind, M. (1951) 'Kulturlandschaft als objektivierter Geist', *Deutsche geographische Bletter* 46: 5–28.
Sharman, F.A. (1981) 'The new law on ancient monuments', *Journal of Planning and Environment Law* November: 785–91.
Sharp, T. (1946) *Exeter Phoenix*, London: Architectural Press.
Sharp, T. (1948) *Oxford Replanned*, London: Architectural Press.
Sharp, T. (1969) *Town and Townscape*, London: Murray.
Short, J.R., Fleming, S.C. and Witt, S.J.C. (1986) *Housebuilding, Planning and Community Action: the Production and Regulation of the Built Environment*, London: Routledge and Kegan Paul.
Sim, D.F. (1982) *Change in the City*, Aldershot: Gower.
Simmie, J. (1981) *Power, Property and Corporatism: the Political Ideology of Planning*, London: Macmillan.
Simms, A.G. (1978) *The Effect of Consultation on Delays in Development Control*, Note N9/78, Watford: Building Research Establishment.
Simpson, M.A. and Lloyd, T.H. (eds) (1977) *Middle Class Housing in Britain*, Newton Abbot: David and Charles.
Skea, R. (1988) 'Recent aspects of Dutch urban conservation', *The Planner* 74, 6: 17–21.
Skeffington (Chairman) (1969) *Report of the Committee on Public Participation in Planning*, London: HMSO.
Skovgaard, J.A. (1978) 'Conservation planning in Denmark', *Town Planning Review* 49, 4: 519–39.
Slater, T.R. (1977) 'Landscape parks and the form of small towns in Great Britain', *Transactions of the Institute of British Geographers* NS 2, 3: 314–31.
Slater, T.R. (1978) 'Family, society and the ornamental villa on the fringes of the English country town', *Journal of Historical Geography* 4, 2: 129–44.
Slater, T.R. (1985) 'A medieval new town and port: a plan analysis of Hedon, East Yorkshire', *Yorkshire Archaeological Journal* 57.
Slater, T.R. (1987) 'Ideal and reality in English episcopal medieval town planning', *Transactions of the Institute of British Geographers* NS 12, 2: 191–203.
Slater, T.R. (1990a) 'English medieval new towns with composite plans: evidence from the Midlands', in Slater, T.R. (ed.) (1990) *The Built Form of Western Cities*, Leicester: Leicester University Press.
Slater, T.R. (ed.) (1990b) *The Built Form of Western Cities*, Leicester: Leicester University Press.
Slater, T.R. (1994) 'Practice examined: the Jewellery Quarter, Birmingham', in Larkham, P.J. (ed.)

Conservation Areas: Issues and Management, Faculty of the Built Environment, University of Central England.
Smith, D. (1969a) 'Outlook for conservation', *Town and Country Planning* 37, 7: 302–7.
Smith, D. (1969b) 'The Civic Amenities Act', *Town Planning Review* 40, 2: 149–62.
Smith, D.L. (1974) *Amenity and Urban Planning*, London: Crosby Lockwood Staples.
Smith, P.B. (1979) 'Conserving Charleston's architectural heritage', *Town Planning Review* 50, 4: 459–76.
Smith, P.F. (1974a) 'Familiarity breeds contentment', *The Planner* 60: 901–4.
Smith, P.F. (1974b) *The Dynamics of Urbanism*, London: Hutchinson.
Smith, P.F. (1975) 'Façadism used to be a dirty word', *Built Environment Quarterly* 1, 1: 77–80.
Smith, P.F. (1977) *The Syntax of Cities*, London: Hutchinson.
Smith, P.F. (1979) *Architecture and the Human Dimension*, London: Goodwin.
Smith, P.F. (1981) 'A question of scale', *Journal of the Royal Institute of British Architects* 88, 6: 45–9.
Soane, J. (1994) 'The renaissance of cultural vernacularism in Germany', in Ashworth, G.J. and Larkham, P.J. (eds) *Building a New Heritage*, London: Routledge.
Society for Medieval Archaeology (1993) 'Barley Hall, York', *Newsletter, Society for Medieval Archaeology* 1993.
Society for the Protection of Ancient Buildings (1903) *Notes on the Repair of Ancient Buildings*, London: SPAB.
Society for the Protection of Ancient Buildings (1992) News item on Barley Hall, York, *SPAB News* 13, 3.
Sorlin, F. (1968) 'The French system for conservation and revitalisation in historic centres', in Ward, P. (ed.) *Conservation and Development in Historic Towns and Cities*, Newcastle upon Tyne: Oriel Press.
Soucy, C. (1974) 'Restauration immobilière et changement sociale', *Les Monuments Historiques de la France* 20, 4.
Spiers, M. (1976) *Victoria Park, Manchester: a Nineteenth-Century Suburb in its Social and Administrative Context*, Chetham Society (third series) XXIII, Manchester: Manchester University Press.
Spranger, E. (1936) *Probleme der Kulturmorphologie*, Berlin: Sanderausgabe aus dem Sitzungsberichte der Preussischen Akademie der Wissenschaften, Philosophisch-Historische Klasse.
Stacey, M. (1960) *Tradition and Change: a Study of Banbury*, Oxford: Oxford University Press.
Stacey, M. (1991) 'The protection of historic parks and gardens in the planning process', unpublished BTP thesis, Bristol Polytechnic.
Stacey, M. (1992) *The Protection of Historic Parks and Gardens in the Planning System*, Working Paper 29, Bristol: Faculty of the Built Environment, University of the West of England.
Stanley, N. (1991) 'The Bath Society case – a Pandora's box?', *Journal of Planning and Environment Law* November: 1014–15.
Stanton, P.B. (1971) *Pugin*, London: Thames and Hudson.
Steegman, J. (1970) *Victorian Taste*, London: Nelson.
Strong, R. (1990) *Lost Treasures of Britain: Five Centuries of Creation and Destruction*, London: Viking.
Stubbs, M. and Lavers, A. (1991) 'Steinberg and after: decision making and development control in conservation areas', *Journal of Environment and Planning Law* January: 9–14.
Stungo, A. (1972) 'The Malraux Act 1962–72', *Journal of the Royal Town Planning Institute* 59: 357–62.
Suddards, R.W. (1982) 'Listed buildings – keeping the past present', *Valuer* 51, 9: 190–1.
Suddards, R.W. (1988) *Listed Buildings*, London: Sweet and Maxwell (second edition).
Suddards, R.W. and Morton, D.M. (1991) 'The character of conservation areas', *Journal of Planning and Environment Law* November: 1011–13.
Summerson, J. (1945) *Georgian London*, London: Pleiades.
Summerson, J. (1980) 'Viollet-le-Duc and the rational point of view', in *Architectural Design, Viollet-le-Duc*, London: Academy Editions.
Sunday Times (1993) 'The best buildings of the year', *Sunday Times* 20 June: 24–5.
Sutcliffe, A.R. (1981a) 'Why planning history?', *Built Environment* 7: 65–7.
Sutcliffe, A.R. (1981b) *Towards the Planned City*, Oxford: Blackwell.
Tarn, J.N. (1985) 'Urban regeneration: the conservation dimension', *Town Planning Review* 56, 2: 245–68.
Taunton Deane Borough Council (1984) *Wellington Local Plan: Written Statement and*

Proposals Maps, Taunton: Taunton Deane Borough Council.
Taylor, S.M. and Konrad, V.A. (1980) 'Scaling dispositions towards the past', *Environment and Behaviour* 12, 3: 283–307.
Telling, A.E. (1967) 'Civic Amenities Act 1967', *Town and Country Planning* 35, 11: 567–9.
Thomas, A.D. (1983) 'Planning in residential conservation areas', *Progress in Planning* 20: 178–256.
Thompson, M.W. (1981) *Ruins: their Preservation and Display*, London: Colonnade.
Thompson, P. (1967) *The Work of William Morris*, London: Heinemann.
Tibbalds, F. (1988) 'Urban design: Tibbalds offers the Prince his ten commandments', *The Planner* 74, 12: mid-month supplement, 1.
Tibbalds Colbourne Karski Williams (1990) *Birmingham Urban Design Studies: Stage 1: City Centre Design Strategy*, Birmingham: Birmingham City Council.
Tiesdell, S. (1995) 'Tensions between revitalization and conservation: Nottingham's Lace Market', *Cities* 12, 4: 231–41.
Toffler, A. (1970) *Future Shock*, New York: Bantam.
Trafford-Owen, D. (1991) 'The conservation area criterion', *Conservation Bulletin* 14: 16.
Tuan, Y.-F. (1977) *Space and Place: the Perspective of Experience*, London: Edward Arnold.
Tugnutt, A. (1991) 'Design – the wider aspects of townscapes', *The Planner* 13 December: 19–22.
Tugnutt, A. and Robertson, M. (1987) *Making Townscape*, London: Batsford.
Tunbridge, J.E. (1984) 'Whose heritage to conserve? Cross-cultural reflections upon political dominance and urban heritage conservation', *Canadian Geographer* 28: 171–80.
Tunbridge, J.E. (1986) 'Clarence Street, Ottawa: contemporary change in an inner-city "zone of discard"', *Urban History Review* 14, 3: 247–57.
Tunbridge, J.E. (1994) 'Whose heritage? Global problem, European nightmare', in Ashworth, G.J. and Larkham, P.J. (eds) *Building a New Heritage*, London: Routledge.
Tyack, G. (1982) 'The Freemans of Fawey and their buildings', *Records of Buckinghamshire* 24: 130–43.
Roger Tym and Partners (1993) *Merry Hill Impact Study: Final Report*, London: HMSO.
United States Department of Treasury (1947) *Bulletin F*, Washington, DC: US Department of Treasury.
United States Department of Treasury (1962) *New Depreciation Rules*, Washington, DC: US Department of Treasury.
Urban Redevelopment Authority (1975) *Annual Report* 1974/75, Singapore: URA.
Urban Redevelopment Authority (1986) *Conservation Master Plan*, Singapore: URA.
Urban Renewal Department (Singapore) (1986) *Annual Report*, Singapore: URA.
Vallis, R. (1994) 'Character: meaning and measurement', in Larkham, P.J. (ed.) *Conservation Areas: Issues and Management*, Birmingham: Faculty of the Built Environment, University of Central England.
Venturi, R. (1966) *Complexity and Contradiction in Architecture*, New York: Museum of Modern Art (illustrations clearer in second edition of 1977).
Vernez Moudon, A. (1994) 'Getting to know the built landscape: typomorphology', in Franck, K.A. and Schneekloth, L. (eds) *Type and the Ordering of Space*, New York: Van Nostrand Reinhold.
Vilagrasa, J. (1984) 'Creixement urbà i producció de l'espai a Lleida (1940–1980)', *Revista Catalana de Geografia* II, 5: 33–50.
Vilagrasa, J. (1990) *Centre Històric i Activitat Commercial: Worcester 1947–1988*, Lleida: Estudi General de Lleida.
Vilagrasa, J. (1992) 'Recent change in two historical city centres: an Anglo-Spanish comparison', in Whitehand, J.W.R. and Larkham, P.J. (eds) *Urban Landscapes: International Perspectives*, London: Routledge.
Vilagrasa, J. and Larkham, P.J. (1995) 'Post-war redevelopment and conservation in Britain: ideal and reality in the historic core of Worcester', *Planning Perspectives* 10, 2: 149–72.
Viollet-le-Duc, E.E. (1854–68) *Dictionnaire Raisonné de l'Architecture Française du XIe au XVIe Siècle* (10 vols), Paris.
Viollet-le-Duc, E.E. (1863) *Entretiens sur l'Architecture*, Paris.
Viollet-le-Duc, E.E. (1872) *Entretiens sur l'Architecture*, Paris.
van Voorden, F.W. (1981) 'The preservation of monuments and historic townscapes in the Netherlands', *Town Planning Review* 52, 4: 433–54.
Waley, M. (1983) 'A very ordinary village house, or, see how they grow', *Wiltshire Archaeology and Natural History Magazine* 77: 93–101.
Wallis, I. (1991) 'Folly in combining judge and jury', *Town and Country Planning* 60, 4: 106–7.

Walter, E.V. (1988) *Placeways: a Theory of the Human Environment*, Chapel Hill, NC: University of North Carolina Press.
Watkin, D. (1977) *Morality and Architecture*, Oxford: Clarendon Press.
Watson, I. (1991) 'Demolition upheld in No 1 Poultry decision', *Chartered Surveyor Weekly* 28 March: 73.
Watts, D.G. (1967) 'Stivichall', in *Victoria County History of Warwick*, VIII, Oxford: Oxford University Press.
Whitehand, J.W.R. (1984) *Rebuilding Town Centres: Developers, Architects and Styles*, Occasional Publication 19, Birmingham: Department of Geography, University of Birmingham.
Whitehand, J.W.R. (1986) 'Taking stock of urban geography', *Area* 18, 2: 147–51.
Whitehand, J.W.R. (1987a) *The Changing Face of Cities*, Institute of British Geographers Special Publication 21, Oxford: Blackwell.
Whitehand, J.W.R. (1987b) 'Urban morphology', in Pacione, M. (ed.) *Historical Geography: Progress and Prospect*, London: Croom Helm.
Whitehand, J.W.R. (1987c) 'M.R.G. Conzen and the intellectual parentage of urban morphology', *Planning History Bulletin* 9, 2: 35–41.
Whitehand, J.W.R. (1988) 'The changing urban landscape: the case of London's high-class residential fringe', *Geographical Journal* 154, 3: 351–66.
Whitehand, J.W.R. (1989a) *Residential Development Under Restraint: a Case Study in London's Rural-Urban Fringe*, Occasional Publication 28, Birmingham: School of Geography, University of Birmingham.
Whitehand, J.W.R. (1989b) 'Development pressure, development control and suburban townscape change: case studies in south-east England', *Town Planning Review* 60, 4: 403–21.
Whitehand, J.W.R. (1990a) 'Townscape management: ideal and reality', in Slater, T.R. (ed.) *The Built Form of Western Cities*, Leicester: Leicester University Press.
Whitehand, J.W.R. (1990b) 'Makers of the residential landscape: conflict and change in outer London', *Transactions of the Institute of British Geographers* NS 15, 1: 87–101.
Whitehand, J.W.R. (1992a) *The Making of the Urban Landscape*, Oxford: Blackwell.
Whitehand, J.W.R. (1992b) 'The makers of British towns: architects, builders and property owners, *c.* 1850–1939', *Journal of Historical Geography* 18, 4: 417–38.
Whitehand, J.W.R. and Larkham, P.J. (1991a) 'Housebuilding in the back garden: reshaping suburban townscapes in the Midlands and South-East England', *Area* 23, 1: 57–65.
Whitehand, J.W.R. and Larkham, P.J. (1991b) 'Suburban cramming and development control', *Journal of Property Research* 8, 2: 147–59.
Whitehand, J.W.R. and Larkham, P.J. (1992)'The urban landscape: issues and perspectives', in Whitehand, J.W.R. and Larkham, P.J. (eds) *Urban Landscapes: International Perspectives*, London: Routledge.
Whitehand, J.W.R., Larkham, P.J. and Jones, A.N. (1992) 'The changing suburban landscape in post-war England', in Whitehand, J.W.R. and Larkham, P.J. (eds) *Urban Landscapes: International Perspectives*, London: Routledge.
Whitehand, J.W.R. and Whitehand, S.M. (1983) 'The study of physical change in town centres: research procedures and types of change', *Transactions of the Institute of British Geographers* NS 8, 4: 483–507.
Whitehand, J.W.R. and Whitehand, S.M. (1984) 'The physical fabric of town centres: the agents of change', *Transactions of the Institute of British Geographers* NS 9, 2: 231–47.
Whitfield, C.J. (1995) 'Conservation and planning in Chester and Shrewsbury', unpublished PhD thesis, Birmingham: School of Geography, University of Birmingham.
Wilcock, R. (1990) 'A house fit for a Queen', *Royal Institute of British Architects Journal* 97, 6: 71–4.
Williams, P.R. (1978) 'Building societies in the inner city', *Transactions of the Institute of British Geographers* NS 3, 1: 23–34.
Wise, D. (1990) 'Conservation and the creative dilemma', unpublished typescript of lecture to the AGM of the Historic Houses Association, London, November.
Witt, S.J.C. and Fleming, S.C. (1984) *Planning Councillors in an Area of Growth: Little Power but All the Blame?*, Geographical Paper 85, Reading: Department of Geography, University of Reading.
Wolverhampton Chronicle (1812) Advertisements for houses in Tettenhall, *Wolverhampton Chronicle* 23 December.
Wools, R. (1978) 'Conservation in the counties', *Building Design* 1 December (unpaginated).
Worcester City Council (1985) *City Centre Draft District Plan*, Worcester: City Council.

Worcester Evening News (1991) 'No change over bus station plans', *Worcester Evening News* 10 January: 10.

Worskett, R. (1969) *The Character of Towns*, London: Architectural Press.

Worskett, R. (1975) 'Great Britain: progress in conservation', *Architectural Review* CLVII, 935: 9–18.

Worskett, R. (1982) 'New buildings in historic areas 1: Conservation: the missing ethic', *Monumentum* 25: 151–61.

Worthington-Williams, M. (1995) 'Finds and discoveries: Edwardian Austin', *The Automobile* 13, 7: 68.

Wright, J.B. (1994) 'Designing and applying conservation easements', *Journal of the American Planning Association* 60, 3: 380–8.

Yeomans, D. (1994) 'Rehabilitation and historic preservation: a comparison of American and British approaches', *Town Planning Review* 65, 2: 159–78.

Yeomans, G. (1974) 'The subtleties of Welsh conservation', *Built Environment* 3: 22–5.

Young, A. (1932 edition) *Tours in England and Wales by Arthur Young*, reprints of Scarce Tracts in Economics and Political Science 14, London: London School of Economics.

Young, G. (1977) *Conservation Scene*, Harmondsworth: Penguin.

INDEX

Note: references in italics refer to figures and tables.